AF443297

Topics in Current Chemistry Collections

Aims and Scope

The series *Topics in Current Chemistry Collections* presents critical reviews from the journal *Topics in Current Chemistry* organized in topical volumes. The scope of coverage is all areas of chemical science including the interfaces with related disciplines such as biology, medicine and materials science.

The goal of each thematic volume is to give the non-specialist reader, whether in academia or industry, a comprehensive insight into an area where new research is emerging which is of interest to a larger scientific audience.

Each review within the volume critically surveys one aspect of that topic and places it within the context of the volume as a whole. The most significant developments of the last 5 to 10 years are presented using selected examples to illustrate the principles discussed. The coverage is not intended to be an exhaustive summary of the field or include large quantities of data, but should rather be conceptual, concentrating on the methodological thinking that will allow the non-specialist reader to understand the information presented.

Contributions also offer an outlook on potential future developments in the field.

More information about this series at http://www.springer.com/series/14181

Yoshito Tobe • Takashi Kubo
Editors

Physical Organic Chemistry of Quinodimethanes

With contributions from

Juan Casado • Chunyan Chi • Justin C. Johnson • Akihito Konishi
Takashi Kubo • Josef Michl • Masayoshi Nakano • Xueliang Shi
Yoshito Tobe

 Springer

Editors
Yoshito Tobe
The Institute of Scientific
and Industrial Research
Osaka University
Ibaraki, Osaka, Japan

Takashi Kubo
Department of Chemistry
Graduate School of Science
Osaka University
Toyonaka, Osaka, Japan

Partly previously published in Top Curr Chem (Z) Volume 375 (2017); Top Curr Chem (Z) Volume 376 (2018).

ISSN 2367-4067
Topics in Current Chemistry Collections
ISBN 978-3-319-93301-6

Library of Congress Control Number: 2018944626

Printed on acid-free paper

This Springer imprint is published by the registered company Springer International Publishing AG part of Springer Nature.
The registered company address is: Gewerbestrasse 11, 6330 Cham, Switzerland

Contents

Preface

Quinodimethanes (QDMs) belong to a class of reactive intermediates constructed by connecting two methylene ($:CH_2$) groups onto phenylene (C_6H_4) via three possible manners, 1,2 (ortho), 1,3 (meta), and 1,4 (para). The electronic configurations and the relevant properties are critically dependent on the connectivity; whereas p-QDM **1** and o-QDM **2** adopt closed-shell singlet ground states, m-QDM **3** has an open-shell triplet configuration as illustrated by its canonical structures, in which no Kekulé structure can be drawn (non- Kekulé hydrocarbon).

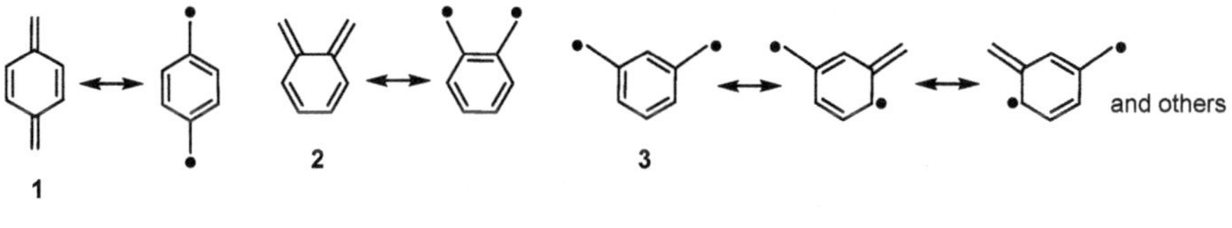

Although stabilized derivatives of p-QDM **1**, Thiele's hydrocarbon **4**[1] and Chichibabin's hydrocarbon **5**[2], were synthesized at the dawn of physical organic chemistry in connection with triphenylmethyl radical, chemists had to wait for many decades to understand their electronic structures until modern physical chemistry was established. On the other hand, **1** and its derivatives have been utilized as monomer units of various polymers and synthetic intermediates for [2.2]paracyclophanes taking advantage of their high reactivity.[3] Similarly, m-QDM **2** serves as a versatile building block for a naphthalene backbone by making use of facile [4+2] cycloaddition with various dienophiles.[4] Physical organic chemistry of **1–3** has advanced significantly during the last few decades of the last century as represented by spectroscopic characterization of all of the parent QDMs[5] and the sensible use of triplet m-QDM **3** as a building block of high-spin molecules by ferromagnetic interactions.[6,7]

Ph—⟨═⟩—Ph   Ph—⟨═⟩—⟨═⟩—Ph

4   **5**

While these are glorious scientific achievements in this research field, the recent advances in open-shell polycyclic aromatic compounds, which contain (a) QDM unit(s) as a key component, have opened up a new window to QDMs as open-shell singlet diradicaloids. As a result of intense research activity and underlying prospects as new organic materials, a number of review articles have been published for open-shell *singlet* diradicaloids including those written by two main contributors in this field, M. Haley and J. Wu, in Top. Curr. Chem.[8,9] Therefore, this collection focuses on physical organic aspects of QDMs including theoretical backgrounds of open-shell character and its relevance to physical properties, structural, physical and spectroscopic properties specific to various kind of QDMs, and singlet fission relevant to open-shell character of QDMs.

M. Nakano gives theoretical backgrounds of open-shell character and its relevance to physical properties, more specifically non-linear optical responses which singlet diradicaloids typically exhibit. A. Konishi and T. Kubo focus on the electronic structure of benzenoid quinodimethanes and also show the importance of aromatic sextet formation to the expression of open-shell character in diradicaloids and multiradicaloids. Y. Tobe provides an overview of structures and physical properties of QDMs incorporated into non-benzenoid aromatic frameworks which display different characteristics from those of benzenoid counterparts basically due to irregular molecular orbital levels and distributions. X. Xi and C. Chi cover new diradicaloids containing heterocyclic subunits, revealing the role of heteroatoms in the conjugation and their effect on the diradical characters. J. Casado discusses structural aspects, especially from bond length alternation (BLA), of diradicaloids on the basis of Raman spectroscopic measurements and their application as semi-conductors and optical materials. Finally, J. C. Johnson and J. Michl discuss physical aspects of singlet fission, a phenomenon generating two triplet excited states from one singlet excited state, of QDMs mostly taking isobenzofuran as an example on both theoretical and experimental basis to exemplify the role of open-shell character of QDMs.

As describe in this collection, significant advances have been achieved in the chemistry of QDMs in relevance to their open-shell character during the last two decades, though there remain many unsolved or unexplored issues. These include, rational design for materials exhibiting efficient singlet fission, molecular designs for spin-state control and multiradicaloid species, and supramolecular chemistry of open-shell singlet molecules to take just a few examples. Besides these challenges, there are many research opportunities in open-shell molecules particularly at the interface of disciplines. We hope this collection will give some insight to those people not only who are already involved in this field to confirm the current status of the research but also who wish to start new study related to quinodimethanes.

We sincerely thank the contributors who have participated to complete this collection of focused articles. Our thanks are also due to reviewers and editorial staff of Topics in Current Chemistry who contributed to the improvement of this collection. Finally, we are grateful to the editorial board for encouraging us to publish a collection on this timely topic.

1. Thiele, J.; Balhorn, H. *Ber. Dtsch. Chem. Ges.* **1904**, 37, 1463-1470.
2. Tschitschibabin. A. E. *Ber. Dtsch. Chem. Ges.* **1907**, 40, 1810-1819.
3. Iwatsuki, S. *Adv. Polym. Sci.* **1984**, 58, 93-120.
4. Segura J. L.; Martín, N. *Chem. Rev.* **1999**, 99, 3199-4126.
5. Platz, M. in *Diradicals*, Borden, W. T. Ed, 1982, Wiley, New York, pp 195-258.
6. Iwamura, H. *Adv. Phys. Org. Chem.* **1990**, 26, 179-253.
7. Rajca, A. *Adv. Phys. Org. Chem.* **2005**, 40, 153-199.
8. Fix, A. G.; Chase, D. T.; Haley, M. M. *Top. Curr. Chem.* **2014**, 349, 159-196.
9. Sun Z.; Wu, J. *Top. Curr. Chem.* **2014**, 349, 159-196

Top Curr Chem (Z) (2017) 375:47
DOI 10.1007/s41061-017-0134-7

Electronic Structure of Open-Shell Singlet Molecules: Diradical Character Viewpoint

Masayoshi Nakano[1,2]

Received: 23 December 2016 / Accepted: 20 March 2017 / Published online: 4 April 2017
© Springer International Publishing Switzerland 2017

Abstract This chapter theoretically explains the electronic structures of open-shell singlet systems with a wide range of open-shell (diradical) characters. The definition of diradical character and its correlation to the excitation energies, transition properties, and dipole moment differences are described based on the valence configuration interaction scheme using a two-site model with two electrons in two active orbitals. The linear and nonlinear optical properties for various polycyclic aromatic hydrocarbons with open-shell character are also discussed as a function of diradical character.

Keywords Diradical character · Open-shell singlet · Excitation energy and property · Valence configuration interaction · Nonlinear optical property

1 Introduction

Recently, polycyclic aromatic hydrocarbons (PAHs) have attracted great attention from various science and engineering fields due to their unique electronic structures and fascinating physicochemical functionalities, e.g., low-energy gap between the singlet and triplet ground states [1, 2], geometrical dependences of open-shell character such as unpaired electron density distributions on the zigzag edges of

---

Chapter 1 was originally published as Nakano, M. Top Curr Chem (Z) (2017) 375: 47. DOI 10.1007/ s41061-017-0134-7.

✉ Masayoshi Nakano
  mnaka@cheng.es.osaka-u.ac.jp

[1] Department of Materials Engineering Science, Graduate School of Engineering Science, Osaka University, Toyonaka, Osaka 560-8531, Japan

[2] Center for Spintronics Research Network (CSRN), Graduate School of Engineering Science, Osaka University, Toyonaka, Osaka 560-8531, Japan

acenes, which leads to the high reactivity on those region [2], significant near-infrared absorption [3], enhancement of nonlinear optical (NLO) properties including two-photon absorption [4–9], and small stacking distance (less than van der Waals radius) and high electronic conductivity in π–π stack open-shell aggregates [10]. These features are known to originate in the open-shell character in the ground electronic states of those open-shell singlet systems [11–18]. The open-shell nature of PAHs is qualitatively understood by resonance structures. For example, benzenoid and quinoid forms of the resonance structures of zethrene species and diphenalenyl compounds correspond to the closed-shell and open-shell (diradcial) states, respectively (Fig. 1a, b). Also, for acenes, considering Clar's aromatic π-sextet rule [19], which states that the resonance forms with the largest number of disjoint aromatic π-sextets (benzenoid forms) contribute most to the electronic ground states of PAHs, it is found that the acenes tend to have radical distributions on the zigzag edges as increasing the size (Fig. 1c). Indeed, recent highly accurate quantum chemical calculations including density matrix renormalization group (DMRG) method clarify that the electronic ground states of long acenes and several graphene nanoflakes (GNFs) are open-shell singlet multiradical states [20–24]. Also, the local aromaticity of such compounds is turned out to be well correlated to the benzenoid moieties in the resonance structures [25, 26].

Although the resonance structures with Clar's sextet rule and aromaticity are useful for qualitatively estimating the open-shell character of the ground-state PAHs, we need a quantitative estimation scheme of the open-shell character and chemical design guidelines for tuning the open-shell character, which contribute to deepening the understanding of the electronic structures of these systems and also to realizing applications of open-shell based unique functionalities. In this chapter, we first provide a quantum-chemically well-defined open-shell character, i.e., diradical character [16, 18, 27–31], and clarify the physical and chemical meaning of this factor. Next, the relationships between the excitation energies/properties and diradical character are revealed based on the analysis of a simple two-site molecular model with two electrons in two active orbitals using the valence configuration interaction (VCI) method [7]. On the basis of this result, linear and

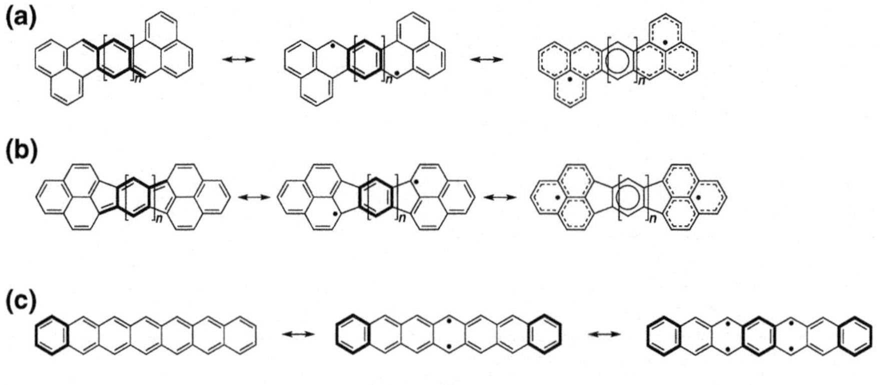

(a)

(b)

(c)

Fig. 1 Resonance structures of zethrene series (**a**), diphenalenyl compounds (**b**) and heptacene (**c**). *Bold lines* indicate the benzenoid and quinoid structures in (**a**) and (**b**), and Clar's sextets in (**c**)

nonlinear optical properties are investigated from the viewpoint of diradical character. Such analysis is also extended to asymmetric open-shell systems. Several realistic open-shell singlet molecular systems are also investigated from the viewpoint of the relationship between the diradical character and resonance structures.

2 Electronic Structures of Open-Shell Singlet Systems

2.1 Classification of Electronic States Based on Diradical Character

The simplest understanding of the open-shell character can be achieved by the single bond dissociation of a homodinuclear molecule (see Fig. 2), which is described by the highest occupied molecular orbital (HOMO) and the lowest unoccupied MO (LUMO) in the symmetry-adapted approach like restricted Hartree–Fock (RHF) method. Namely, the bond dissociation process is described by the decrease in the HOMO–LUMO gap, i.e., the correct wavefunction is described by the mixing between the HOMO (bonding) and LUMO (antibonding), and the wavefunction at the dissociation limit is composed of the equally weighted mixing of the HOMO and LUMO, which creates localized spatial distribution on each atom site and thus no distribution between the atoms. More precisely, as increasing the bond distance, the double excitation configuration from the HOMO to

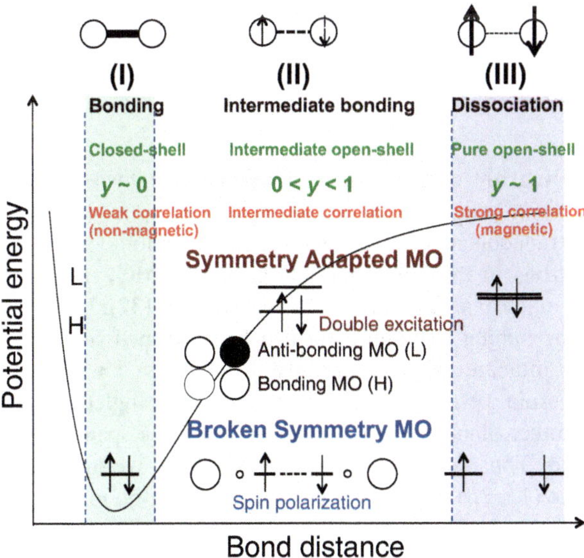

Fig. 2 Bond dissociation process of a homodinuclear molecule, where the variations of the HOMO and LUMO levels in the symmetry-adapted approach as well as of the magnetic orbitals for the α and β spins in the broken-symmetry approach are also shown as a function of bond distance. The physical and chemical meanings of diradical character (y) are also shown in the three regimes (I)–(III) of the electronic states in the bond dissociation process

the LUMO becomes mixed into the doubly occupied configuration in the HOMO. On the other hand, in the spin-unrestricted (broken-symmetry) approach, the MO could have different spatial distribution for the α and β spins, e.g., α spin distributes mainly on the left-hand side, while the β spin mainly on the right-hand side as increasing the bond distance. This picture (approximation) seems to be more intuitive than the symmetry-adapted approach, but this suffers from the intrinsic deficiency, i.e., spin contamination [16, 29], where high spin states such as triplet states are mixed in the singlet wavefunction. The bond dissociation process is qualitatively categorized into three regimes, i.e., stable bond regime (I), intermediate bond regime (II) and bond dissociation (weak bond) regime (III). As shown in later, these regimes are characterized by "diracial character" y, which takes a value between 0 and 1: small y (~ 0) for (I), intermediate y for (II) and large y (~ 1) for (III) (see Fig. 2). In other words, $1-y$ indicates an "effective bond order" [29]. This description is employed in chemistry, while in physics, these three regimes are characterized by the degree of "electron correlation": weak correlation regime (I), intermediate correlation regime (II) and strong correlation regime (III) (see Fig. 2). This physical picture is also described by the variation in the degree of delocalization of two electrons on two atomic sites: strong delocalization (weak localization) (I), intermediate delocalization (intermediate localization) (II) and weak delocalization (strong localization) (III). Namely, the effective repulsion interaction between two electrons means the electron correlation, so that the delocalization decreases (the localization increases) when the correlation increases. Namely, in physics, the bond dissociation limit is considered to be caused by the strong correlation limit (strong localization limit). Thus, the "diradical character" is a fundamental factor for describing the electronic states and could be a key factor bridging between chemical and physical concepts on the electronic structures [16, 18].

2.2 Schematic Diagram of Electronic Structure of a Two-Site Model

In this section, let us consider a one-dimensional (1D) homodinuclear molecule A–B with two electrons in two orbitals (HOMO and LUMO) in order to understand schematically its electronic structure, i.e., wavefunction [32]. In this case, the spatial distribution of the singlet wavefunction can be described on the (1α, 2β) plane, where 1α and 2β indicate the real coordinate of electron 1 with α spin and that of electron 2 with β spin, respectively. More exactly, the singlet wavefunction is also distributed on another plane (1β, 2α), but this is the same spatial distribution as that on (1α, 2β) plane. Thus, we can discuss the singlet wavefunction using only the distribution on the (1α, 2β) plane without loss of generality. Figure 3a shows the 1D two-electron system A–B and the 2D plane (1α, 2β), on which the spatial distribution of the singlet wavefunction is plotted. On the (1α, 2β) plane, the dotted lines represent the positions of nuclei A and B, and the diagonal dashed line indicates the Coulomb wall. The two electrons undergo large Coulomb repulsion near the Coulomb wall, while those receive attractive forces from nuclei A and B near the dotted lines. The covalent (or diradical) configuration (where mutually antiparallel spins are distributed on A and B, respectively) is described by the black dots symmetrically distributed with respect to the diagonal dashed line, while the zwitterionic configuration (where a pair of α and β spins is distributed on A or B) is done by the black dots on the diagonal dashed line.

We can here consider the spatial distribution of the singlet wavefunctions composed of the HOMO and LUMO. As shown in Fig. 3b, the HOMO and LUMO are represented by two while circles and a pair of white and black circles, respectively, where white and black indicate positive and negative phase of the MO. Using various electron configurations in the HOMO (ϕ_H) and LUMO (ϕ_L), we can describe the symmetry-adapted wavefunctions. For example, the double-occupied configuration in the HOMO gives the HF singlet ground state ψ_G, which is represented by the Slater determinant:

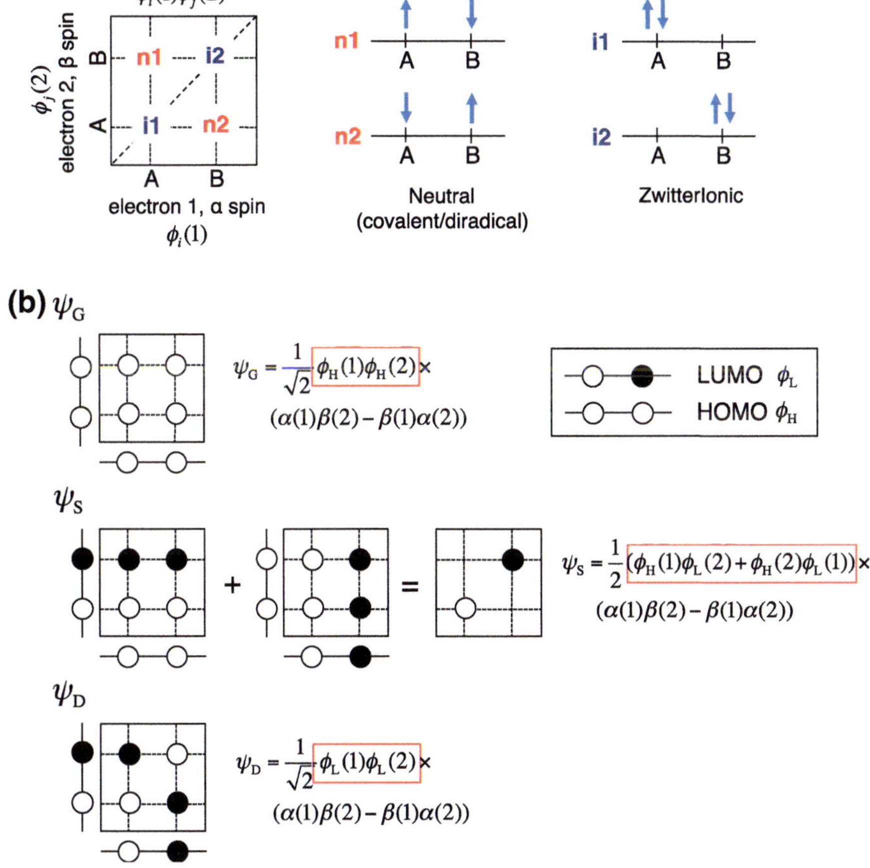

Fig. 3 Schematic diagram of 1D two-electron system A–B and the 2D (1α, 2β) plane (**a**) and the singlet spatial wavefunctions, ψ_G (HF ground state determinant), ψ_S (singly excited determinant), and ψ_D (doubly excited determinant) on the (1α, 2β) plane with the HOMO and LUMO distributions (**b**)

$$\psi_G = \frac{1}{\sqrt{2}} \begin{vmatrix} \phi_H(1)\alpha(1) & \phi_H(1)\beta(1) \\ \phi_H(2)\alpha(2) & \phi_H(2)\beta(2) \end{vmatrix} \equiv \psi(\phi_H \bar{\phi}_H)$$
$$= \frac{1}{\sqrt{2}} \phi_H(1)\phi_H(2)(\alpha(1)\beta(2) - \beta(1)\alpha(2)). \tag{1}$$

As mentioned before, the $(1\alpha, 2\beta)$ plane corresponds to $\phi_H(1)\phi_H(2)\alpha(1)\beta(2)$, so that spatial part $\phi_H(1)\phi_H(2)$ is a product of the HOMO(1) and HOMO(2) as shown in Fig. 3b. Apparently, the distribution of each black dot (intersection points of dotted lines) is found to be equal in the amplitude and phase. This implies that covalent (neutral) and ionic configurations are equally mixed in the HF singlet ground state wavefunction, which is a well-known feature of mean field approximation, i.e., no electron correlation. Next, we consider singly excited configuration from the HOMO to LUMO. The singly excited singlet Slater determinant is represented by

$$\psi_S = \frac{1}{\sqrt{2}} \left\{ \frac{1}{\sqrt{2}} \begin{vmatrix} \phi_H(1)\alpha(1) & \phi_L(1)\beta(1) \\ \phi_H(2)\alpha(2) & \phi_L(2)\beta(2) \end{vmatrix} + \frac{1}{\sqrt{2}} \begin{vmatrix} \phi_L(1)\alpha(1) & \phi_H(1)\beta(1) \\ \phi_L(2)\alpha(2) & \phi_H(2)\beta(2) \end{vmatrix} \right\}$$
$$\equiv \frac{1}{\sqrt{2}} [\psi(\phi_H \bar{\phi}_L) + \psi(\phi_L \bar{\phi}_H)] \tag{2}$$
$$= \frac{1}{2} (\phi_H(1)\phi_L(2) + \phi_H(2)\phi_L(1))(\alpha(1)\beta(2) - \beta(1)\alpha(2)).$$

The spatial part is composed of two components, $\phi_H(1)\phi_L(2)$ and $\phi_H(2)\phi_L(1)$, which are needed to satisfy the symmetry for exchange between the real coordinates for electron 1 and 2. As shown in Fig. 3b, the spatial distribution of ψ_S is pure ionic, i.e., only diagonal distribution, and has a node line along the anti-diagonal line. The doubly excited Slater determinant is expressed as

$$\psi_D = \frac{1}{\sqrt{2}} \begin{vmatrix} \phi_L(1)\alpha(1) & \phi_L(1)\beta(1) \\ \phi_L(2)\alpha(2) & \phi_L(2)\beta(2) \end{vmatrix} \equiv \psi(\phi_L \bar{\phi}_L)$$
$$= \frac{1}{\sqrt{2}} \phi_L(1)\phi_L(2)(\alpha(1)\beta(2) - \beta(1)\alpha(2)) \tag{3}$$

In this case, the spatial distribution is described by $\phi_L(1)\phi_L(2)$, which indicates the doubly occupied in the LUMO. The spatial distribution is the same as that of ψ_G except for the phase, where $\phi_L(1)\phi_L(2)$ has two node lines, i.e., neutral (covalent or diradical) and ionic distributions possess mutually opposite phase. Note here that the spatial parts of these wavefunctions are easily constructed from the HOMO and/or LUMO and that the symmetry of the spatial part for exchanging electron 1 and 2 is straightforward. For these Slater determinants, apparently, neutral and ionic distribution amplitudes in ψ_G and ψ_D are equal to each other, which indicates that the electron correlation is not considered. Next, we consider the effect of electron correlation on the spatial distribution of these wavefunctions.

In the ground state, the ionic distribution should be smaller than the neutral (covalent) distributions in order to more stabilize the ground state by avoiding the strong Coulomb repulsion on the ionic distribution. By mixing the spatial wavefunctions of ψ_G and ψ_D, we can construct such wavefunction distribution as shown in Fig. 4. From symmetry, the HF ground state wavefunction ψ_G and doubly excited wavefunction ψ_D are correlated (mixed) with mutually opposite phase in the ground state and with the same phase in the second excited state, which leads to the

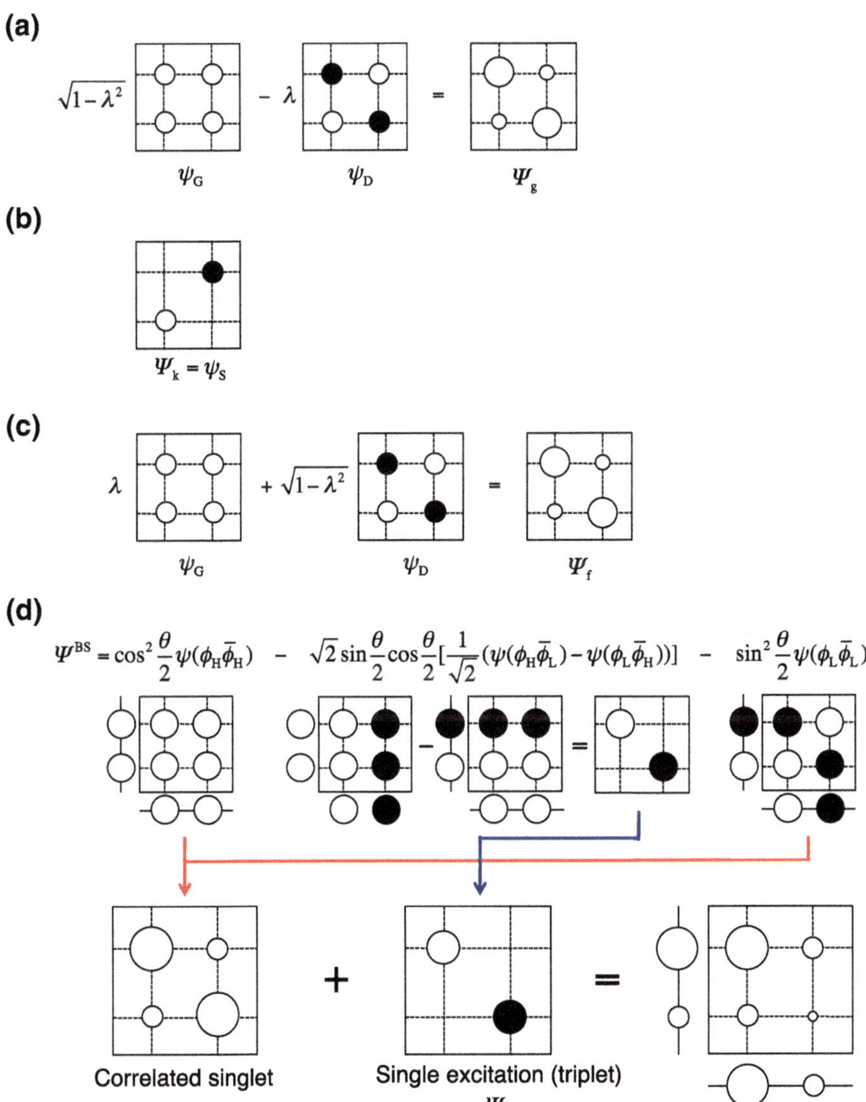

Fig. 4 Schematic diagram of electron correlated wavefunctions for 1D two-electron system A–B on the $(1\alpha, 2\beta)$ plane: the ground state Ψ_g (**a**), the first excited state Ψ_k (**b**) and the second excited state Ψ_f (**c**). The BS wavefunction described by the symmetry-adapted determinants, ψ_G, ψ_T, and ψ_D, is also shown (**d**) (see Eq. 9)

increase (decrease) in the neutral component and decrease (increase) in the ionic component in the ground state g (the second excited state f). Note here that the first excited state k is not mixed with other wavefunctions and is a pure ionic state. Thus, the correct wavefunctions for states {g, k, f} are described by

$$\Psi_{\mathrm{g}} = \sqrt{1 - \lambda^2}\psi_{\mathrm{G}} - \lambda\psi_{\mathrm{D}} \tag{4}$$

$$\Psi_{\mathrm{k}} = \psi_{\mathrm{S}} \tag{5}$$

$$\Psi_{\mathrm{f}} = \lambda\psi_{\mathrm{G}} + \sqrt{1 - \lambda^2}\psi_{\mathrm{D}} \tag{6}$$

The λ^2, which is a weight of the doubly excited configuration in the ground state, is found to be able to change from 0 to 1/2, which indicates the change from the mean field wavefunction ψ_{G} (MO limit) to the pure neutral (diradical) component (atomic orbital (AO) limit in the bond dissociation system). Accordingly, the second excited state Ψ_{f} changes from the mean field wavefunction ψ_{D} to the pure ionic component. As a result, considering the bond dissociation model, the change of $2\lambda^2$ from 0 to 1 corresponds to the change from the stable bond region to the bond dissociation limit. Namely, the $2\lambda^2$ is regarded as the "diradical character", which is indeed the original definition of the diradical character [27–29].

2.3 Broken-Symmetry Approach with Spin-Projection Scheme for Evaluation of Diradical Character

We consider the spin-unrestricted [broken-symmetry (BS)] ground state wavefunction using the symmetry-adapted wavefunctions. Using the BS HOMOs χ and η, the ground state BS wavefunction is expressed as

$$\Psi^{\mathrm{BS}}(\chi\bar{\eta}) = \frac{1}{\sqrt{2}}\begin{vmatrix} \chi(1)\alpha(1) & \chi(2)\alpha(2) \\ \eta(1)\beta(1) & \eta(2)\beta(2) \end{vmatrix}$$
$$= \frac{1}{\sqrt{2}}(\chi(1)\eta(2)\alpha(1)\beta(2) - \eta(1)\chi(2)\beta(1)\alpha(2)). \tag{7}$$

Here, the BS orbitals χ and η are represented by symmetry-adapted MOs ϕ_{H} and ϕ_{L} as [28, 29]

$$\chi = \cos\frac{\theta}{2}\phi_{\mathrm{H}} + \sin\frac{\theta}{2}\phi_{\mathrm{L}}, \text{ and } \eta = \cos\frac{\theta}{2}\phi_{\mathrm{H}} - \sin\frac{\theta}{2}\phi_{\mathrm{L}} \tag{8}$$

where θ is a mixing parameter ranging from 0 to $\pi/2$. For $\theta = 0$, $\chi = \eta = \phi_{\mathrm{H}}$, while $\theta = \pi/2$, $\chi = \frac{1}{\sqrt{2}}(\phi_{\mathrm{H}} + \phi_{\mathrm{L}}) \equiv a$ and $\eta = \frac{1}{\sqrt{2}}(\phi_{\mathrm{H}} - \phi_{\mathrm{L}}) \equiv b$, where a and b are referred to as magnetic orbitals (localized natural orbitals (LNOs)) and are nearly equal to AO φ_{A} and φ_{B}, respectively. Namely, the BS orbitals can represent the variation from the MO limit to the AO limit by changing θ from 0 to $\pi/2$. Using Eq. 8, the ground state BS wavefunction $\Psi^{\mathrm{BS}}(\chi\bar{\eta})$ is expressed as [28, 29]

$$\Psi^{BS}(\chi\bar{\eta}) = \cos^2\frac{\theta}{2}\psi(\phi_H\bar{\phi}_H) - \sqrt{2}\sin\frac{\theta}{2}\cos\frac{\theta}{2}\left[\frac{1}{\sqrt{2}}(\psi(\phi_H\bar{\phi}_L) - \psi(\phi_L\bar{\phi}_H))\right]$$
$$- \sin^2\frac{\theta}{2}\psi(\phi_L\bar{\phi}_L) \tag{9}$$

where the first, second, and third terms involve the singlet ground state determinant $\psi_G(=\psi(\phi_H\bar{\phi}_H))$ (Eq. 1), the triplet determinant $\psi_T = \frac{1}{\sqrt{2}}(\psi(\phi_H\bar{\phi}_L) - \psi(\phi_L\bar{\phi}_H))$, and singlet double excited determinant $\psi_D = \psi(\phi_L\bar{\phi}_L)$ (Eq. 3), respectively. As seen from Fig. 4d, the qualitatively correct spatial distribution of the singlet ground state wavefunction is built from superposition of ψ_G and ψ_D, while the incorrect spin component (triplet) ψ_T is also mixed into the wavefunction. This triplet component, which is anti-symmetric with respect to the exchange of the real coordinate between electron 1 and 2, is schematically shown to asymmetrize the neutral components as shown in Fig. 4d. This is the reason why this wavefunction is called "broken symmetry" (neither symmetric nor anti-symmetric with respect to the exchange between electron 1 and 2), and is found to be made of broken-symmetry HOMOs χ and η. Although the BS wavefunction suffers from a spin contamination, which is known to sometimes give improper relative energies for different spin states and erroneous physicochemical properties [29, 33, 34], the BS approach has an advantage of being able to include partial electron-correlation, qualitatively correct singlet spatial distribution in the present case, by just using a simple single determinant calculation scheme instead of high-cost multi-reference calculation schemes. Indeed, Yamaguchi applied the perfect-pairing type spin-projection scheme to the BS solution and developed an easy evaluation method of diradical character y [28, 29]. Using the overlap between χ and η, i.e., $T \equiv \langle \chi \mid \eta \rangle = \cos^2\frac{\theta}{2} - \sin^2\frac{\theta}{2} = \cos\theta$, we rewrite Eq. 9 as

$$\Psi^{BS} = \frac{1+T}{2}\psi_G - \sqrt{\frac{1-T^2}{2}}\psi_T - \frac{1-T}{2}\psi_D \tag{10}$$

The perfect-pairing type spin-projection implies the removal of the second term from the BS wavefunction with keeping the weight ratio of the first and third terms, $[(1+T)/(1-T)]^2$. Thus, the spin-projected wavefunction is expressed by

$$\Psi^{PU} = \frac{1+T}{\sqrt{2(1+T^2)}}\psi_G - \frac{1-T}{\sqrt{2(1+T^2)}}\psi_D \tag{11}$$

From the definition of the diradical character y, i.e., twice the weight of the doubly excitation configuration, we obtain the expression of diradical character in the PUHF formalism [28, 29]:

$$y^{PU} = 1 - \frac{2T}{1+T^2} \tag{12}$$

Here, let us consider the one-electron reduced density using the BS wavefunction Eq. 7,

$$\rho(\mathbf{r}) = |\chi(\mathbf{r})|^2 + |\eta(\mathbf{r})|^2 = 2\cos^2\frac{\theta}{2}|\phi_{\mathrm{H}}(\mathbf{r})|^2 + 2\sin^2\frac{\theta}{2}|\phi_{\mathrm{L}}(\mathbf{r})|^2$$
$$= (1+T)|\phi_{\mathrm{H}}(\mathbf{r})|^2 + (1-T)|\phi_{\mathrm{L}}(\mathbf{r})|^2. \tag{13}$$

This equation indicates that $\phi_{\mathrm{H}}(\mathbf{r})$ and $\phi_{\mathrm{L}}(\mathbf{r})$ are the HONO (the highest occupied natural orbital) and LUNO (the lowest occupied natural orbital) of the BS solution with the occupation numbers of $1 + T(\equiv n_{\mathrm{HONO}})$ and $1 - T(\equiv n_{\mathrm{LUNO}})$, respectively. On the other hand, the occupation numbers of the HONO and LUNO of the spin-projected wavefunction Eq. 11 are expressed by

$$n_{\mathrm{HONO}}^{\mathrm{PU}} = \frac{(1+T)^2}{1+T^2} = \frac{n_{\mathrm{HONO}}^2}{1+T^2} = 2 - y^{\mathrm{PU}} \tag{14}$$

and

$$n_{\mathrm{LUNO}}^{\mathrm{PU}} = \frac{(1-T)^2}{1+T^2} = \frac{n_{\mathrm{LUNO}}^2}{1+T^2} = y^{\mathrm{PU}} \tag{15}$$

where $n_{\mathrm{HONO}} = 1 + T$ and $n_{\mathrm{LUNO}} = 1 - T$ are employed (see Eq. 13). This expression can be extended to a $2n$-radical system, the perfect-pairing type (i.e., considering a doubly excitation from $\mathrm{HONO} - i$ to $\mathrm{LUNO} + i$) spin-projected diradical characters and occupation numbers are defined as [28, 29]

$$y_i^{\mathrm{PU}} = 1 - \frac{2T_i}{1+T_i^2} \tag{16}$$

and

$$n_{\mathrm{HONO}-i}^{\mathrm{PU}} = 2 - y_i^{\mathrm{PU}}, \text{ and } n_{\mathrm{LUNO}+i}^{\mathrm{PU}} = y_i^{\mathrm{PU}} \tag{17}$$

where T_i is the overlap between the corresponding orbitals χ_i and η_i, and the occupation number of $\mathrm{LUNO} + i$ $(n_{\mathrm{LUNO}+i})$ is given by $1 - T_i$.

3 Electronic States of Two-Site Model by the Valence Configuration Interaction Method

3.1 Ground and Excited Electronic States and Diradical Character

For a symmetric two-site diradical system with two electrons in two orbitals (LNOs), a and b, with the z-component of spin angular momentum $M_{\mathrm{s}} = 0$ (singlet and triplet), we can consider two neutral

$$\psi(a\bar{b}) = \frac{1}{\sqrt{2}}(a(1)b(2)\alpha(1)\beta(2) - b(1)a(2)\beta(1)\alpha(2))$$

$$\psi(b\bar{a}) = \frac{1}{\sqrt{2}}(b(1)a(2)\alpha(1)\beta(2) - a(1)b(2)\beta(1)\alpha(2)) \tag{18}$$

and two ionic determinants:

$$\psi(a\bar{a}) = \frac{1}{\sqrt{2}}(a(1)a(2)\alpha(1)\beta(2) - a(1)a(2)\beta(1)\alpha(2))$$

$$\psi(b\bar{b}) = \frac{1}{\sqrt{2}}(b(1)b(2)\alpha(1)\beta(2) - b(1)b(2)\beta(1)\alpha(2)) \tag{19}$$

The spatial distributions of these wavefunctions on the $(1\alpha, 2\beta)$ plane are described by $a(1)b(2)$, $b(1)a(2)$, $a(1)a(2)$, and $b(1)b(2)$, respectively (see $n1$, $n2$, $i1$, and $i2$, respectively, shown in Fig. 3a). The valence configuration interaction (VCI) matrix of the electronic Hamiltonian H is represented by using the LNO basis [7, 35]:

$$\begin{pmatrix} \langle a\bar{b}|H|a\bar{b}\rangle & \langle a\bar{b}|H|b\bar{a}\rangle & \langle a\bar{b}|H|a\bar{a}\rangle & \langle a\bar{b}|H|b\bar{b}\rangle \\ \langle b\bar{a}|H|a\bar{b}\rangle & \langle b\bar{a}|H|b\bar{a}\rangle & \langle b\bar{a}|H|a\bar{a}\rangle & \langle b\bar{a}|H|b\bar{b}\rangle \\ \langle a\bar{a}|H|a\bar{b}\rangle & \langle a\bar{a}|H|b\bar{a}\rangle & \langle a\bar{a}|H|a\bar{a}\rangle & \langle a\bar{a}|H|b\bar{b}\rangle \\ \langle b\bar{b}|H|a\bar{b}\rangle & \langle b\bar{b}|H|b\bar{a}\rangle & \langle b\bar{b}|H|a\bar{a}\rangle & \langle b\bar{b}|H|b\bar{b}\rangle \end{pmatrix}$$

$$= \begin{pmatrix} 0 & K_{ab} & t_{ab} & t_{ab} \\ K_{ab} & 0 & t_{ab} & t_{ab} \\ t_{ab} & t_{ab} & U & K_{ab} \\ t_{ab} & t_{ab} & K_{ab} & U \end{pmatrix} \tag{20}$$

Here, $\langle a\bar{b}|H|b\bar{b}\rangle \equiv \int \psi^*(a\bar{b})H\psi(b\bar{b})d\tau_1 d\tau_2$ and so on. The energy of the neutral determinant, $\langle a\bar{b}|H|a\bar{b}\rangle = \langle b\bar{a}|H|b\bar{a}\rangle$, is taken as the energy origin (0). U represents the difference between on- and neighbor-site Coulomb repulsions, referred to as effective Coulomb repulsion:

$$U \equiv U_{aa} - U_{bb}$$

$$= \int a^*(1)a(1)r_{12}^{-1}a^*(2)a(2)d\mathbf{r}_1 d\mathbf{r}_2 - \int b^*(1)b(1)r_{12}^{-1}b^*(2)b(2)d\mathbf{r}_1 d\mathbf{r}_2 \tag{21}$$

$$= (aa|aa) - (bb|bb).$$

K_{ab} is a direct exchange integral [$K_{ab} = (ab|ba) \geq 0$], and t_{ab} is a transfer integral [$t_{ab} = \langle a\bar{b}|H|b\bar{b}\rangle = \langle a|f|b\rangle \leq 0$, where f is the Fock operator in the LNO representation] [16, 36].

We obtain the following four solutions by diagonalizing the CI matrix of Eq. 20 [4, 5, 7, 16, 18].

(A) Neutral triplet state (with u symmetry)

$$\Psi_{T_{1u}} = \frac{1}{\sqrt{2}}(\psi(a\bar{b}) - \psi(b\bar{a})) \text{ with energy } {}^3E_{1u} = -K_{ab} \tag{22}$$

(B) Ionic singlet state (with u symmetry)

$$\Psi_{S_{1u}} = \frac{1}{\sqrt{2}}(\psi(a\bar{a}) - \psi(b\bar{b})) \text{ with } {}^1E_{1u} = U - K_{ab} \tag{23}$$

(C) Lower singlet state (with g symmetry)

$$\Psi_{S_{1g}} = \kappa(\psi(a\bar{b}) + \psi(b\bar{a})) + \eta(\psi(a\bar{a}) + \psi(b\bar{b})), \tag{24a}$$

where $2(\kappa^2 + \eta^2) = 1$ and $\kappa > \eta > 0$. Thus, state S_{1g} has a larger weight of neutral determinant (the first term) than that of ionic one (the second term). The energy is

$$^1E_{1g} = K_{ab} + \frac{U - \sqrt{U^2 + 16t_{ab}^2}}{2} \tag{24b}$$

(D) Higher singlet state (with g symmetry)

$$\Psi_{S_{2g}} = -\eta(\psi(a\bar{b}) + \psi(b\bar{a})) + \kappa(\psi(a\bar{a}) + \psi(b\bar{b})), \tag{25a}$$

where $2(\kappa^2 + \eta^2) = 1$ and $\kappa > \eta > 0$. In contrast to S_{1g}, state S_{2g} has a larger weight of ionic determinant (the second term) than that of neutral one (the first term). The energy is

$$^1E_{2g} = K_{ab} + \frac{U + \sqrt{U^2 + 16t_{ab}^2}}{2}. \tag{25b}$$

Here, κ and η are functions of $|t_{ab}/U|$ [4, 5, 7, 16, 18], which indicates the ease of the electron transfer, i.e., the degree of delocalization, between atoms A and B. As seen from Fig. 5, as decreasing r_t, the κ (the coefficient of the neutral determinant) increases toward $1/\sqrt{2}$ at $r_t = 0$, while the η (the coefficient of the ionic determinant) decreases toward 0 at $r_t = 0$. From this behavior, the mobility of electrons, i.e., the delocalization nature, between sites A and B is found to determine the relative neutral (covalent) and ionic natures of the state, i.e., the diradical nature.

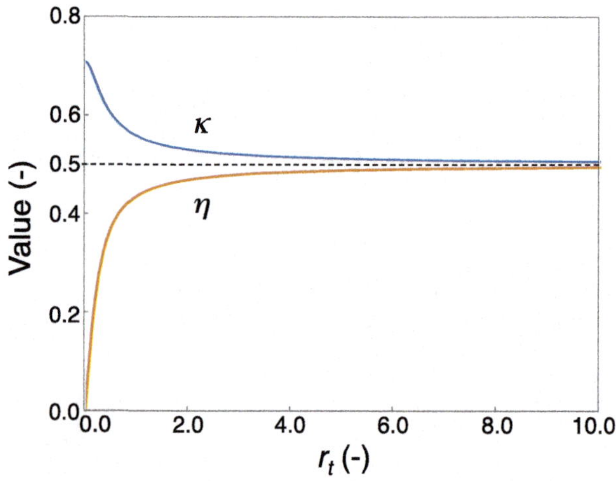

Fig. 5 Variations of κ and η as a function of r_t

Using the relationship between BS orbitals $\{a, b\}$ and symmetry-adapted MOs $\{\phi_H, \phi_L\}$, i..e, $a = \frac{1}{\sqrt{2}}(\phi_H + \phi_L)$ and $b = \frac{1}{\sqrt{2}}(\phi_H - \phi_L)$, the lower singlet state Eq. 24a is also expressed by

$$\Psi_{S_{1g}} = (\kappa + \eta)\psi(g\bar{g}) + (\kappa - \eta)\psi(u\bar{u}). \tag{26}$$

Thus, the diradical character y, which is defined as twice the weight of the doubly excitation configuration, $2\zeta^2 = 2(\kappa - \eta)^2 = 1 - 4\kappa\eta$, is represented by

$$y = 1 - \frac{1}{\sqrt{1 + \left(\frac{U}{4t_{ab}}\right)^2}} = 1 - \frac{1}{\sqrt{1 + \left(\frac{1}{4r_t}\right)^2}} \tag{27}$$

The variation of y as a function of $|U/t_{ab}|(\equiv 1/r_t)$ is shown in Fig. 6. As increasing $|U/t_{ab}|$, y value is shown to increase from 0 to 1, which correspond to $|U/t_{ab}| \leq \sim 1$ $(r_t \geq \sim 1)$ and $|U/t_{ab}| \to \infty$ $(r_t \to 0)$, respectively. From the physical meaning of the transfer integral t_{ab} and the effective Coulomb repulsion U, $y \to 1$ at $|U/t_{ab}| \to \infty$ $(r_t \to 0)$ implies the localization of electrons on each site, i.e., a pure diradical state, while $y \to 0$ at $|U/t_{ab}| \leq \sim 1$ $(r_t \geq \sim 1)$ implies the delocalization of electrons over two sites, i.e., a closed-shell stable bond state. Namely, this represents that the diradical character y indicates the degree of electron correlation $|U/t_{ab}|$ in the physical sense. On the other hand, this variation in delocalization over two sites according to the variation in y substantiates the variation of diradical character during the bond dissociation of a homodinuclear system discussed in Sect. 2.1. Indeed, from Eq. 17, we obtain

$$1 - y_i = \frac{n_{\text{HONO}-i} - n_{\text{LUNO}-i}}{2}, \tag{28}$$

which represents that $1 - y_i$ indicates the effective bond order concerned with bonding (HONO $- i$) and antibonding (LUNO $+ i$) orbitals [29]. This is demonstrated in Fig. 6 by the variation of 1–y from 1 (stable bond region) to 0 (bond

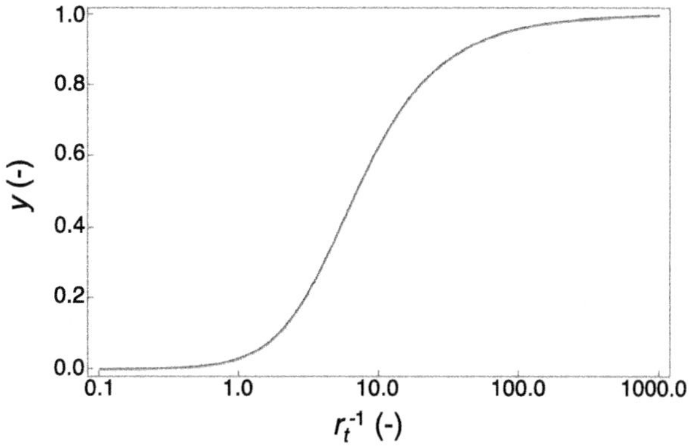

Fig. 6 Variation of y as a function of $|U/t_{ab}|(\equiv 1/r_t)$

breaking region) with increasing the electron correlation $|U/t_{ab}|$. Namely, y indicates the bond weakness in the chemical sense. In summary, the diradical character y is a fundamental factor for describing electronic states and can bridge the two pictures for electronic states between physics, i.e., electron correlation, and chemistry, i.e., effective chemical bond.

3.2 Diradical Character Dependence of Excitation Energies and Properties

From Eqs. 22–25b and 27, we obtain excitation energies ($E_{S_{1u},S_{1g}}$, $E_{S_{2g},S_{1g}}$) and transition moments squared ($(\mu_{S_{1g},S_{1u}})^2$, $(\mu_{S_{1u},S_{2g}})^2$) (see Fig. 7a, b):

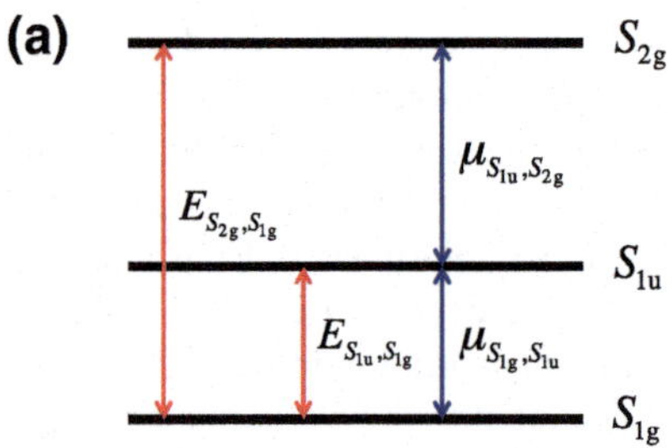

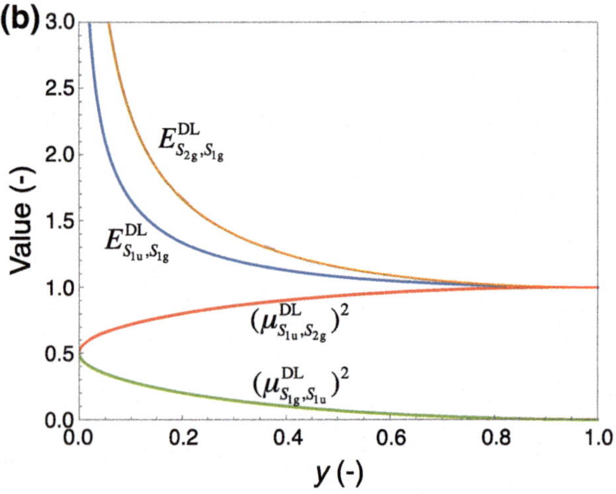

Fig. 7 **a** Electronic states of a two-site diradical model: three singlet states (S_{1g}, S_{1u}, S_{2g}) and a triplet state (T_{1u}). The excitation energies ($E_{S_{1u},S_{1g}}$, $E_{S_{2g},S_{1g}}$) and transition moments ($\mu_{S_{1g},S_{1u}}$, $\mu_{S_{1u},S_{2g}}$) are also shown. Note here that the transition between S_{1g} and S_{2g} is optically forbidden. **b** Diradical character dependences of dimensionless excitation energies ($E_{S_{1u},S_{1g}}^{DL} \equiv E_{S_{1u},S_{1g}}/U$, $E_{S_{2g},S_{1g}}^{DL} \equiv E_{S_{2g},S_{1g}}/U$) and dimensionless transition moments squared ($(\mu_{S_{1g},S_{1u}}^{DL})^2 \equiv (\mu_{S_{1g},S_{1u}})^2/R_{BA}^2$, $(\mu_{S_{1u},S_{2g}}^{DL})^2 \equiv (\mu_{S_{1u},S_{2g}})^2/R_{BA}^2$) for $r_K = 0$

$$E_{S_{1u},S_{1g}} \equiv {}^1E_{1u} - {}^1E_{1g} = \frac{U}{2}\left\{ 1 - 2r_K + \frac{1}{\sqrt{1-(1-y)^2}} \right\} \qquad (29)$$

$$E_{S_{2g},S_{1g}} \equiv {}^1E_{2g} - {}^1E_{1g} = \frac{U}{\sqrt{1-(1-y)^2}} \qquad (30)$$

$$(\mu_{S_{1g},S_{1u}})^2 = \frac{R_{BA}^2}{2}\left\{ 1 - \sqrt{1-(1-y)^2} \right\} \qquad (31)$$

and

$$(\mu_{S_{1u},S_{2g}})^2 = \frac{R_{BA}^2}{2}\left\{ 1 + \sqrt{1-(1-y)^2} \right\}. \qquad (32)$$

Here, $R_{BA} \equiv R_{bb} - R_{aa} = (b|r|b) - (a|r|a)$ is an effective distance between the two radicals. In these formulae, U and R_{BA} play roles for their units, energy and length, respectively. Except for Eq. 29, which includes the dimensionless direct exchange $r_K (\equiv 2K_{ab}/U)$, these quantities are as functions of y. These dimensionless excitation energies ($E_{S_{1u},S_{1g}}^{DL} \equiv E_{S_{1u},S_{1g}}/U$, $E_{S_{2g},S_{1g}}^{DL} \equiv E_{S_{2g},S_{1g}}/U$) and dimensionless transition moments squared (($\mu_{S_{1g},S_{1u}}^{DL})^2 \equiv (\mu_{S_{1g},S_{1u}})^2/R_{BA}^2$, $(\mu_{S_{1u},S_{2g}}^{DL})^2 \equiv (\mu_{S_{1u},S_{2g}})^2/R_{BA}^2$) are plotted as functions of y (see Fig. 7b for $r_K = 0$), where $E_{S_{1u},S_{1g}}^{DL}$ approaches $1 - r_K$ at $y \to 1$.

Both the dimensionless transition moments squared, $(\mu_{S_{1g},S_{1u}}^{DL})^2$ and $(\mu_{S_{1u},S_{2g}}^{DL})^2$, monotonically increase toward 1 and decrease toward 0, respectively, from 0.5 at $y = 0$, as increasing y from 0 to 1. This is understood by the fact that the ground (S_{1g}) and the second (S_{2g}) excited states are correlated as described in Sect. 3.1 and become primary-diradical (neutral) and primary-ionic states as increasing the ground state diradical character y, while the first optically allowed excited state (S_{1u}) remains in a pure ionic state. Namely, as increasing y, the overlap between the ground (S_{1g}) and the first (S_{1u}) excited states, transition density corresponding to $(\mu_{S_{1g},S_{1u}}^{DL})^2$ decreases, while that, transition density corresponding to $(\mu_{S_{1g},S_{1u}}^{DL})^2$, between the first ($S_{1u}$) and second ($S_{2g}$) excited states increases. On the other hand, for $r_K = 0$, with increasing y, both the dimensionless first and second excitation energies, $E_{S_{1u},S_{1g}}^{DL}$ and $E_{S_{2g},S_{1g}}^{DL}$, rapidly decrease in the small y region, and they gradually decrease toward 1 and then achieve a stationary value (1) from the intermediate to large y region. The reduction rate in the small y region is significant in $E_{S_{1u},S_{1g}}^{DL}$ as compared with $E_{S_{2g},S_{1g}}^{DL}$. It is also found that as increasing r_K, the converged value of $E_{S_{1u},S_{1g}}^{DL}$ is decreased, i.e., $E_{S_{1u},S_{1g}}^{DL} \to 1 - r_K$ at $y \to 1$ (see Eq. 29). Here, we consider the relationship between the first optically allowed excitation energy $E_{S_{1u},S_{1g}}$ and diradical character y. From Eq. 27, y tends to increase when

U becomes large. Considering the y dependence of $E^{DL}_{S_{2g},S_{1g}}$ (Fig. 7) and $E_{S_{2g},S_{1g}} = UE^{DL}_{S_{2g},S_{1g}}$, it is predicted that the excitation energy $E_{S_{1u},S_{1g}}$ decreases, reaches a stationary value, and for very large U, it increases again with increasing y values [16, 37]. Usually, the extension of π-conjugation length causes the decrease of the HOMO–LUMO gap ($-2t_{ab}$) and the increase of U, so that the extension of the size of molecules with non-negligible diradical character y tends to decrease the first excitation energy in the relatively small y region, while tends to increase again in the intermediate/large y region. This behavior is contrast to the well-known feature that a closed-shell π-conjugated system exhibits a decrease of the excitation energy with increasing the π-conjugation length.

4 Asymmetric Open-Shell Singlet Systems

4.1 Ground/Excited Electronic States and Diradical Character Using the Valence Configuration Interaction Method

As explained in Sect. 3, the neutral (diradical) and ionic components in a wavefunction play a complementary role, so that the asymmetric charge distribution, referred to as, asymmetricity, tends to reduce the diradical character. This feature seems to be qualitatively correct, but "asymmetricity" and primary "ionic" contribution is not necessarily the same concept. In this section, we show the feature of the wavefunctions of the ground and excited states based on an asymmetric two-site model A˙–B˙ with two electrons in two orbitals in order to clarify the effects of an asymmetric electronic distribution on the excitation energies and properties of open-shell molecular systems [38].

The asymmetric two-site model A˙–B˙ is placed along the bond axis (x-axis). Using the AOs for A and B, i.e., χ_A and χ_B, with overlap S_{AB}, bonding and antibonding MOs, g and u can be defined as in the symmetric system:

$$g = \frac{1}{\sqrt{2(1 + S_{AB})}}(\chi_A + \chi_B), \text{ and } u = \frac{1}{\sqrt{2(1 - S_{AB})}}(\chi_A - \chi_B) \qquad (33)$$

Note here that these are not the canonical MOs of the asymmetric systems when A $\neq$ B. Using these MOs, we can define the localized natural orbitals (LNOs), a and b,

$$a = \frac{1}{\sqrt{2}}(g + u), \text{ and } b = \frac{1}{\sqrt{2}}(g - u) \qquad (34)$$

which become the corresponding AOs, χ_A and χ_B, at the dissociation limit. Using the LNOs, the VCI matrix for zero z-component of spin angular momentum ($M_s = 0$, singlet and triplet) is expressed by [38],

$$
\begin{pmatrix}
\langle a\bar{b}|\hat{H}|a\bar{b}\rangle & \langle a\bar{b}|\hat{H}|b\bar{a}\rangle & \langle a\bar{b}|\hat{H}|a\bar{a}\rangle & \langle a\bar{b}|\hat{H}|b\bar{b}\rangle \\
\langle b\bar{a}|\hat{H}|a\bar{b}\rangle & \langle b\bar{a}|\hat{H}|b\bar{a}\rangle & \langle b\bar{a}|\hat{H}|a\bar{a}\rangle & \langle b\bar{a}|\hat{H}|b\bar{b}\rangle \\
\langle a\bar{a}|\hat{H}|a\bar{b}\rangle & \langle a\bar{a}|\hat{H}|b\bar{a}\rangle & \langle a\bar{a}|\hat{H}|a\bar{a}\rangle & \langle a\bar{a}|\hat{H}|b\bar{b}\rangle \\
\langle b\bar{b}|\hat{H}|a\bar{b}\rangle & \langle b\bar{b}|\hat{H}|b\bar{a}\rangle & \langle b\bar{b}|\hat{H}|a\bar{a}\rangle & \langle b\bar{b}|\hat{H}|b\bar{b}\rangle
\end{pmatrix}
$$
$$
=
\begin{pmatrix}
0 & K_{ab} & t_{ab(aa)} & t_{ab(bb)} \\
K_{ab} & 0 & t_{ab(aa)} & t_{ab(bb)} \\
t_{ab(aa)} & t_{ab(aa)} & -h + U_a & K_{ab} \\
t_{ab(bb)} & t_{ab(bb)} & K_{ab} & h + U_b
\end{pmatrix}.
\tag{35}
$$

Here, the matrix elements are similar to those of symmetric case Eq. 20. On the other hand, some additional and modified physical parameters are introduced to describe the asymmetric two-site system. For example, h represents the one-electron core Hamiltonian difference, $h \equiv h_{bb} - h_{aa}$, where $h_{pp} \equiv \langle p|h(1)|p\rangle = \langle \bar{p}|h(1)|\bar{p}\rangle \leq 0$ and $h \geq 0$ ($h_{aa} \leq h_{bb}$). Since the transfer integrals include the two-electron integral between the neutral and ionic determinants, there are two types of transfer integrals, e.g., $t_{ab(aa)} \equiv \langle a\bar{b}|\hat{H}|a\bar{a}\rangle$ and $t_{ab(bb)} \equiv \langle a\bar{b}|\hat{H}|b\bar{b}\rangle$, which are different since $(ab|aa) \neq (ab|bb)$. Thus, the average transfer integral, $t_{ab} \equiv (t_{ab(aa)} + t_{ab(bb)})/2$, is introduced. In the present case, U_a and U_b represent effective Coulomb repulsions, $U_a \equiv U_{aa} - U_{ab}$ and $U_b \equiv U_{bb} - U_{ab}$, and we define the average effective Coulomb repulsion U [$\equiv (U_a + U_b)/2$] [38].

Similar to the symmetric diradical system in Sect. 3, we introduce dimensionless quantities [38],

$$
\frac{|t_{ab}|}{U} \equiv r_t(\geq 0), \quad \frac{2K_{ab}}{U} \equiv r_K(\geq 0), \quad \frac{h}{U} \equiv r_h(\geq 0), \quad \frac{U_a}{U_b} \equiv r_U(\geq 0), \text{ and } \left|\frac{t_{ab(aa)}}{t_{ab(bb)}}\right|
$$
$$
\equiv r_{tab}(\geq 0)
\tag{36}
$$

We here introduce a parameter y_S,

$$
y_S = 1 - \frac{4r_t}{\sqrt{1 + 16r_t^2}}
\tag{37}
$$

which indicates the diradical character before introducing the asymmetricity, i.e., $(r_h, r_U, r_{tab}) = (0, 1, 1)$ [38]. Note here that this is not the diradical character for the asymmetric two-site model (referred to as y_A) and is referred to as "pseudo-diradical character". The diradical character of the asymmetric two-site model is represented by y_A, which is a function of $(r_t, r_K, r_h, r_U, r_{tab})$. For simplicity, we consider the case that the asymmetricity is caused by changing r_h between 0 and 2 with keeping $(r_U, r_{tab}) = (1, 1)$, which means that the asymmetricity is governed by the difference of ionization potentials of the constitutive atoms A and B. The dimensionless Hamiltonian matrix, $\mathbf{H}_{DL}(\equiv \mathbf{H}/U)$, in the case of $(r_U, r_{tab}) = (1, 1)$ is expressed by [38]

$$\boldsymbol{H}_{\mathrm{DL}} = \begin{pmatrix} 0 & \dfrac{r_K}{2} & -r_t & -r_t \\ \dfrac{r_K}{2} & 0 & -r_t & -r_t \\ -r_t & -r_t & 1 - r_h & \dfrac{r_K}{2} \\ -r_t & -r_t & \dfrac{r_K}{2} & 1 + r_h \end{pmatrix} \tag{38}$$

From this expression, the eigenvalues and eigenvectors of $\boldsymbol{H}_{\mathrm{DL}}$ are found to depend on the dimensionless quantities $(r_t,\ r_K,\ r_h,\ r_U\ r_{tab})$, i.e., $(y_S,\ r_K,\ r_h,\ r_U,\ r_{tab})$. The eigenvectors for state $\{j\} = \{\mathrm{T, g, k, f}\}$ (T: triplet state, and g, k, f: singlet states) are represented by

$$\Psi_j = C_{a\bar{b},j}\psi(a\bar{b}) + C_{b\bar{a},j}\psi(b\bar{a}) + C_{a\bar{a},j}\psi(a\bar{a}) + C_{b\bar{b},j}\psi(b\bar{b}). \tag{39}$$

It is found that $C_{a\bar{b},j} = C_{b\bar{a},j}$ and $\left|C_{a\bar{a},j}\right| \neq \left|C_{b\bar{b},j}\right|$ for the asymmetric singlet states, while $C_{a\bar{b},\mathrm{T}} = -C_{b\bar{a},\mathrm{T}} = 1/\sqrt{2}$ and $C_{a\bar{a},\mathrm{T}} = C_{b\bar{b},\mathrm{T}} = 0$ for the triplet state. Using the MOs (g and u) in Sect. 3.1, we can construct an alternative basis set $\{\psi_{\mathrm{G}},\ \psi_{\mathrm{S}},\ \psi_{\mathrm{D}}\} = \{\psi(g\bar{g}),\ (\psi(g\bar{u}) + \psi(u\bar{g}))/\sqrt{2},\ \psi(u\bar{u})\}$ for the singlet states. By using this basis set, the singlet ground state is expressed by

$$\Psi_{\mathrm{g}} = \xi\psi_{\mathrm{G}} + \eta\psi_{\mathrm{S}} - \zeta\psi_{\mathrm{D}} \tag{40}$$

where the normalization condition, $\xi^2 + \eta^2 + \zeta^2 = 1$, is satisfied. By comparing Eq. 39 with 40, we obtain the relationships:

$$\xi = C_{a\bar{b},\mathrm{g}} + \frac{1}{2}\left(C_{a\bar{a},\mathrm{g}} + C_{b\bar{b},\mathrm{g}}\right),\ \eta = \frac{1}{\sqrt{2}}\left(C_{a\bar{a},\mathrm{g}} - C_{b\bar{b},\mathrm{g}}\right),\ \text{and}\ \zeta$$
$$= C_{a\bar{b},\mathrm{g}} - \frac{1}{2}\left(C_{a\bar{a},\mathrm{g}} + C_{b\bar{b},\mathrm{g}}\right) \tag{41}$$

The diradical character y_A of the two-site asymmetric model is defined, in the same way as in the symmetric model in Sect. 2.3, as the occupation number (n_{LUNO}) of the LUNO of the singlet ground state Ψ_{g}. The diradical character y_A is expressed by

$$y_A \equiv n_{\mathrm{LUNO}} = 1 - |\xi - \zeta|\sqrt{2 - (\xi - \zeta)^2}$$
$$= 1 - \left|C_{a\bar{a},\mathrm{g}} + C_{b\bar{b},\mathrm{g}}\right|\sqrt{2 - \left(C_{a\bar{a},\mathrm{g}} + C_{b\bar{b},\mathrm{g}}\right)^2} \tag{42}$$

which reduces to the usual definition of y for symmetric systems:

$$y_A = y_S = 2\zeta^2 \tag{43}$$

Figure 8 shows the y_S and y_A relationship, which reveals that y_A is smaller than y_S, in particular for $y_S \sim 0.5$ as increasing the asymmetricity r_h. As seen from Fig. 8a, if $y_S = 1$ then $y_A = 1$ for $r_h < 1$ but $y_A = 0$ for $r_h > 1$, while y_A is close to ~ 0.134 for $r_h = 1$ $(r_K = 0)$. This behavior corresponds to the exchange of the dominant

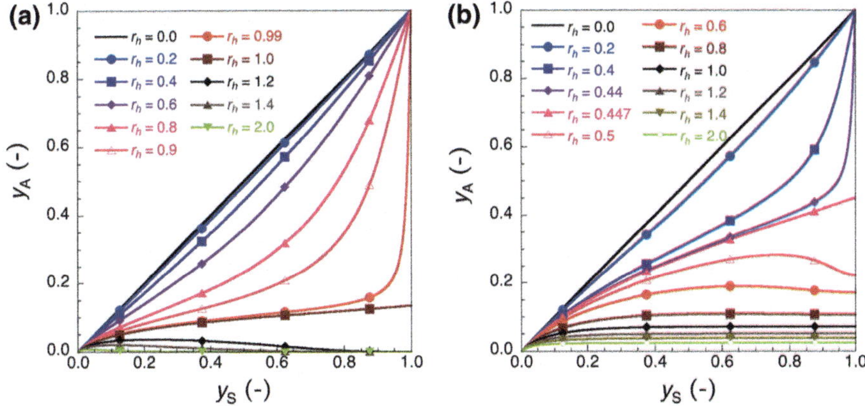

Fig. 8 y_S versus y_A plots with $r_h = 0.0$–2.0 for $r_K = 0.0$ (**a**) and 0.8 (**b**)

configurations (neutral/ionic) in state g, i.e., $P_N = \left|C_{a\bar{b},g}\right|^2 + \left|C_{b\bar{a},g}\right|^2$ and $P_I = \left|C_{a\bar{a},g}\right|^2 + \left|C_{b\bar{b},g}\right|^2$, between $r_h < 1$ and $r_h > 1$ for $y_S > 0$ and $r_K = 0$ (see Fig. 9).

4.2 Asymmetricity (r_h) and Direct Exchange (r_K) Dependence of Energies, Wavefunctions, and Diradical Character

As shown in Fig. 9, for $r_K = 0$, the asymmetricity r_h causes the exchange between the dominant configurations (neutral/ionic) in each state and the variation in the diradical character y_A. We here focus on the effects of r_K on these variations. In order to capture the feature of r_K effect, we consider the analytical expressions of energies and wavefunctions of each state {g, k, f} in the case of $(y_S, r_U, r_{tab}) = (1, 1, 1)$ $(r_h > 0)$. The solutions are classified in the following three regions based on the amplitude relationship between r_h^2 and $1 - r_K$ [39].

For $r_h^2 < 1 - r_K$,

$$E_g = \frac{r_K}{2}, \; \left|\Psi_g\right\rangle = \frac{1}{\sqrt{2}}\left|G\right\rangle - \frac{1}{\sqrt{2}}\left|D\right\rangle = \frac{1}{\sqrt{2}}\left|a\bar{b}\right\rangle + \frac{1}{\sqrt{2}}\left|b\bar{a}\right\rangle, y_A = 1, \tag{44a}$$

$$E_k = 1 - \sqrt{r_h^2 + \left(\frac{r_K}{2}\right)^2}, \; \left|\Psi_k\right\rangle = \frac{1}{\sqrt{2(1+A^2)}}\left(\left|G\right\rangle + \sqrt{2}A\left|S\right\rangle + \left|D\right\rangle\right)$$

$$= \frac{1}{\sqrt{2(1+A^2)}}\left\{(1+A)\left|a\bar{a}\right\rangle + (1-A)\left|b\bar{b}\right\rangle\right\}. \tag{44b}$$

$$E_f = 1 + \sqrt{r_h^2 + \left(\frac{r_K}{2}\right)^2}, \; \left|\Psi_f\right\rangle = \frac{1}{\sqrt{2(1+B^2)}}\left(\left|G\right\rangle + \sqrt{2}B\left|S\right\rangle + \left|D\right\rangle\right)$$

$$= \frac{1}{\sqrt{2(1+B^2)}}\left\{(1+B)\left|a\bar{a}\right\rangle + (1-B)\left|b\bar{b}\right\rangle\right\}. \tag{44c}$$

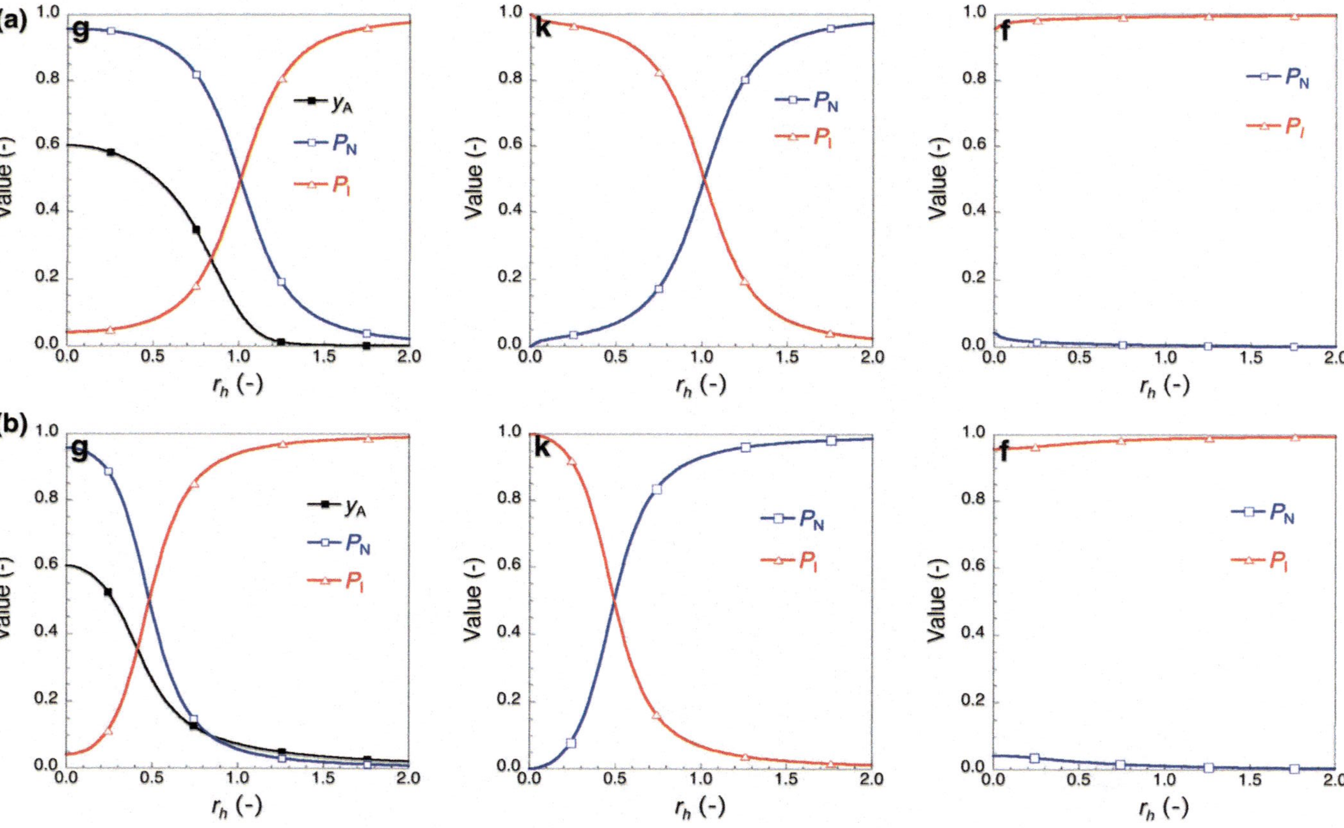

Fig. 9 r_h dependences of P_N and P_I for states g, k and f at $y_S = 0.6$ for $r_K = 0.0$ (**a**), and 0.8 (**b**). The variations in y_A for state g are also shown

For $r_h^2 = 1 - r_K$,

$$E_g = E_k = \frac{r_K}{2}, \; E_f = 2 - \frac{r_K}{2} \tag{44d}$$

For $r_h^2 > 1 - r_K$,

$$E_g = 1 - \sqrt{r_h^2 + \left(\frac{r_K}{2}\right)^2},$$

$$|\Psi_g\rangle = \frac{1}{\sqrt{2(1+A^2)}}\left(|G\rangle + \sqrt{2}A|S\rangle + |D\rangle\right)$$

$$= \frac{1}{\sqrt{2(1+A^2)}}\left\{(1+A)|a\bar{a}\rangle + (1-A)|b\bar{b}\rangle\right\}, \tag{44e}$$

$$y_A = 1 - \frac{2A}{1+A^2}.$$

$$E_k = \frac{r_K}{2},$$

$$|\Psi_k\rangle = \frac{1}{\sqrt{2}}|G\rangle - \frac{1}{\sqrt{2}}|D\rangle = \frac{1}{\sqrt{2}}|a\bar{b}\rangle + \frac{1}{\sqrt{2}}|b\bar{a}\rangle \tag{44f}$$

$$E_f = 1 + \sqrt{r_h^2 + \left(\frac{r_K}{2}\right)^2}$$

$$|\Psi_f\rangle = \frac{1}{\sqrt{2(1+B^2)}}\left(|G\rangle + \sqrt{2}B|S\rangle + |D\rangle\right) \tag{44g}$$

$$= \frac{1}{\sqrt{2(1+B^2)}}\left\{(1+B)|a\bar{a}\rangle + (1-B)|b\bar{b}\rangle\right\}.$$

Here,

$$A \equiv \frac{r_K + \sqrt{4r_h^2 + r_K^2}}{2r_h}(>0) \text{ and } B \equiv \frac{r_K - \sqrt{4r_h^2 + r_K^2}}{2r_h}(<0). \tag{44h}$$

For $r_h^2 < 1 - r_K$, states g and k are pure neutral (diradical) and ionic, while for $r_h^2 > 1 - r_K$, they are pure ionic and neutral (diradical), respectively. Namely, for $y_S = 1$, the diradical character y_A is abruptly reduced from 1 to $1 - \frac{2A}{1+A^2}$ when turning from $r_h^2 < 1 - r_K$ to $r_h^2 > 1 - r_K$. In the case of $r_h^2 = 1 - r_K$ at $y_S \to 1$, where $E_g = E_k$, y_A asymptotically approaches $1 - \frac{\sqrt{1+2A^2}}{1+A^2}$ since the neutral (Eq. 44a) and ionic (Eq. 44e) components contribute to the wavefunction equivalently. At the same r_h value, the y_A is shown to decrease in the intermediate y_S region with increasing r_K up to $1 - r_h^2$ (≥ 0), while further increase of r_K is found to increase y_A again as seen from Eqs. 44e and 44h. Namely, the increase in r_K operates similarly to asymmetricity r_h for $r_h^2 < 1 - r_K$. This is also exemplified by the decrease of critical r_h value ($r_h{}_c$), at which the exchange of the dominant configurations (neutral/ionic) in state g and k occurs, with increasing r_K until $1 - r_{hc}^2$ (≥ 0) (see Eq. 44d), which is shown in r_K dependence of P_N and P_I for g and k states (Fig. 9).

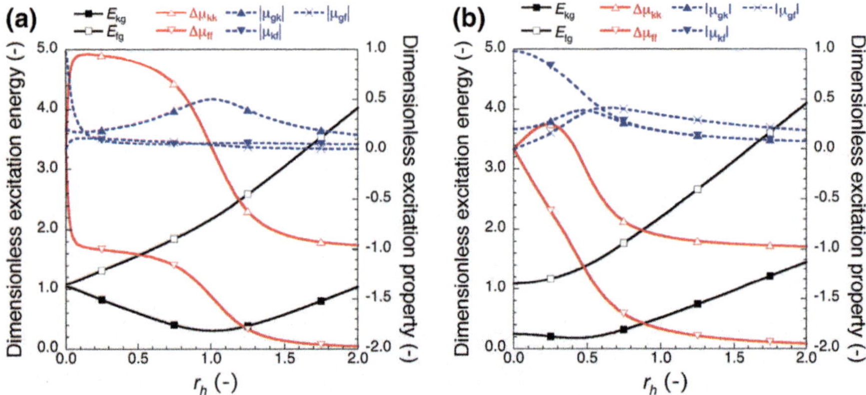

Fig. 10 r_h dependence of the dimensionless excitation energies (E_{ij}), dimensionless dipole moment differences ($\Delta\mu_{ii}$) and dimensionless transition moment amplitudes ($|\mu_{ij}|$) at $y_S = 0.6$ for $r_K = 0.0$ (**a**) and 0.8 (**b**)

The variations in r_h dependence of the dimensionless excitation energies E_{kg} and E_{fg} (for a fixed y_S) with increasing r_K are shown in Fig. 10. It is shown that E_{kg} and E_{fg} decrease and increase, respectively, with increasing r_K for $r_h^2 < 1 - r_K$, while that they increase with increasing r_K for $r_h^2 > 1 - r_K$ (see also Eqs. 44a–44h). The increase of r_K is found to move the behaviors around $r_h = 1.0$ of the excitation energies and transition moments to the lower r_h region due to the displacement of the critical point $r_{h\,c}$ as shown in Fig. 9. Also, the increase of r_K is turned out to decrease E_{kg} and $|\Delta\mu_{ii}|$ ($i = $ k, f), but increase $|\mu_{kf}|$ before $\sim r_{h\,c}$ as predicted from the analytical expressions of excitation energies and wavefunctions for $y_S = 1$ (Eqs. 44a–44c). Indeed, the asymmetric distributions represented by the relative contributions of $|a\bar{a}\rangle$ and $|b\bar{b}\rangle$ are shown to decrease with increasing r_K at the same r_h, e.g., $|a\bar{a}\rangle : |b\bar{b}\rangle = 1{:}0$ for $r_K = 0$ vs. $1 + $ A:1–A for $r_K \neq 0$ (see Eqs. 44a–44c), the feature of which decreases $|\Delta\mu_{ii}|$ ($i = $ k, f) and increases $|\mu_{kf}|$.

For the ground-state singlet–triplet energy gap, $E_{gT}(\equiv E_g - E_T)$ (see Fig. 11), it is found (a) that the increase of y_S causes the decrease of E_{gT} for $r_K = 0$, (b) that the increase in r_K stabilizes the triplet state, and (c) that for a given y_S the increase of r_h leads to the increase of the r_K value giving a triplet ground state. As seen from Fig. 11, the singlet ground state r_K (antiferromagnetic) region is broad in small y_S region and further broadens to larger y_S values with increasing r_h. For $r_h > 1$, the singlet ground state region is found to be widely extended over the whole y_S–r_K region.

Fig. 11 Dimensionless E_{gT} contours on the y_S-r_K plane for $r_h = 0.0$ (**a**), 0.8 (**b**) and 1.4 (**c**). The variation from *cold to warm color* indicates that from negative to positive E_{gT} values. The *black solid line* E_{gT} contours range from -3.0 to 3.0 with division 0.2 and 0.0 contour is shown by a *black dashed-dotted line*. The *black dashed lines* represent the iso-y_A lines

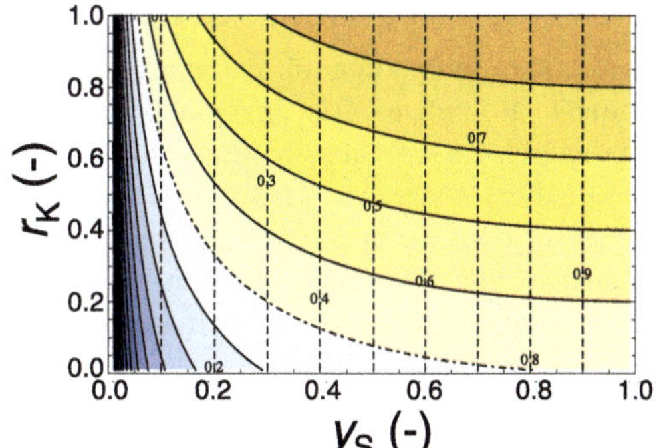

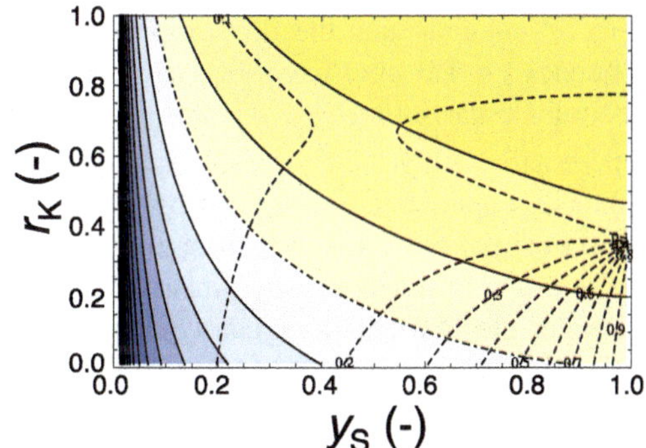

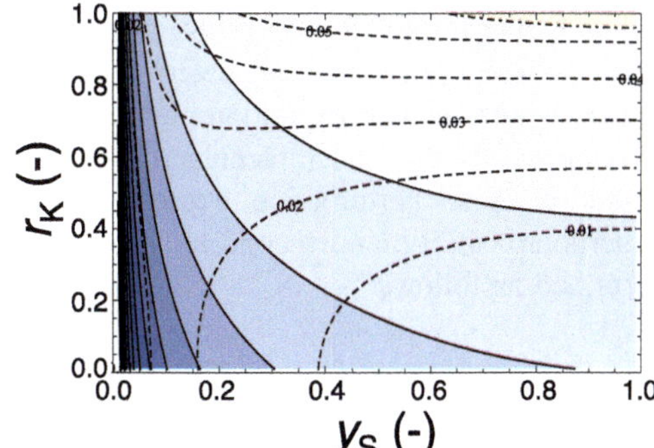

5 Relationship between Open-Shell Character and Optical Response Properties

The unique properties of excitation energies, transition moments, and dipole moment differences for open-shell singlet systems, i.e., their dependence on diradical character, cause a strong correlation of the optical response properties to the diradical character. This is understood by the perturbation analysis of those optical response properties. In general, the microscopic polarization p, which is defined by the difference between the induced dipole moment μ and permanent dipole moment μ_0, is expanded by using the applied electric field F [40–42]:

$$
\begin{aligned}
p^i &= \mu^i - \mu_0^i \\
&= \sum_j \alpha_{ij} F^j(\omega_1) + \sum_{jk} \beta_{ijk} F^j(\omega_1)F^k(\omega_2) + \sum_{jkl} \gamma_{ijkl} F^j(\omega_1)F^k(\omega_2)F^l(\omega_3) + \cdots
\end{aligned}
$$

$$(45)$$

Here, $F^i(\omega_l)$ indicates the ith component ($i = x, y, z$) of local electric field with frequency ω_l. The coefficient of each term indicates the optical responsibility of the nth order polarization: α_{ij}, β_{ijk} and γ_{ijkl} are referred to as the polarizability, first hyperpolarizability, and second hyperpolarizability, respectively. These optical response properties are described by the electronic states of the atom/molecule and environmental effects, and their signs and amplitudes determines the characteristic of microscopic linear and nonlinear optical properties at the molecular scale. For example, the real and imaginary parts of α_{ij} describe the linear polarization and optical absorption, respectively, while those of γ_{ijkl} are the off- and on-resonant third-order NLO properties, respectively, where the former and the latter typical phenomena are third-harmonic generation (THG), and two-photon absorption (TPA), respectively. As seen from Eq. 45, even-ordered coefficients such as β_{ijk} vanish when the system has centrosymmetry, while odd-ordered coefficients such as γ_{ijkl} generally have non-zero values regardless of the symmetry. The amplitude and sign of these coefficients are determined by the time-dependent perturbation formulae, which include excitation energies, transition moments and dipole moment differences, so that the molecular design for efficient NLO has been performed based on these perturbation expressions. For example, the polarizability, first hyperpolarizability, and second hyperpolarizability in the static limit ($\omega_i = 0$) are described as follows:

$$
\alpha_{ii} = 2 \sum_{n \neq 0} \frac{(\mu_{0n}^i)^2}{E_{n0}},
\tag{46}
$$

$$
\beta_{iii} = 3 \sum_{n \neq 0} \frac{(\mu_{0n}^i)^2 \Delta\mu_{nn}^i}{E_{n0}^2},
\tag{47}
$$

$$\gamma_{iiii} = 4\left\{ \sum_{n \neq 0} \frac{(\mu^i_{0m})^2 (\bar{\mu}^i_{mm})^2}{E^3_{m0}} \right.$$

$$\left. - \sum_{n,m \neq 0} \frac{(\mu^i_{0m})^2 (\mu^i_{n0})^2}{E_{m0} E^2_{n0}} + 2 \sum_{m \neq n} \frac{\mu^i_{0m} \Delta\mu^i_{mm} \mu^i_{mn} \mu^i_{n0}}{E^2_{m0} E_{n0}} + \sum_{m \neq n} \frac{\mu^i_{0m} \mu^i_{mn} \mu^i_{nq} \mu^i_{q0}}{E^2_{m0} E_{n0}} \right\} \quad (48)$$

Here, E_{n0} indicates the excitation energy of the nth excited state; μ^i_{mn} indicates the transition moment between the mth and nth states; $\Delta\mu^i_{mm}$ indicates the dipole moment difference between the mth excited state and the ground state (0). Applying these expressions to three singlet state model {g, k, f} for the symmetric two-site diradical model and using Eqs. 29–32, we obtain the analytical expressions of these response properties as functions of diradical character y. For symmetric systems, the terms including dipole moment differences are vanished due to $\Delta\mu^i_{mm} = 0$ in Eq. 48, and takes the form (where the component index "i" is omitted for simplicity):

$$\gamma = -4 \frac{(\mu_{gk})^4}{(E_{kg})^3} + 4 \frac{(\mu_{gk})^2 (\mu_{kf})^2}{(E_{kg})^2 E_{fg}} \quad (49)$$

and can be expressed as a function of effective bond order $q = 1-y$ [4, 7]:

$$\frac{\gamma}{(R^4_{BA}/U^3)} = \frac{\gamma^{II}}{(R^4_{BA}/U^3)} + \frac{\gamma^{III-2}}{(R^4_{BA}/U^3)}$$

$$= -\frac{8q^4}{\left(1 + \sqrt{1-q^2}\right)^2 \left(1 - 2r_K + \frac{1}{\sqrt{1-q^2}}\right)^3} + \frac{4q^2}{\left(1 - 2r_K + \frac{1}{\sqrt{1-q^2}}\right)^2 \frac{1}{\sqrt{1-q^2}}}. \quad (50)$$

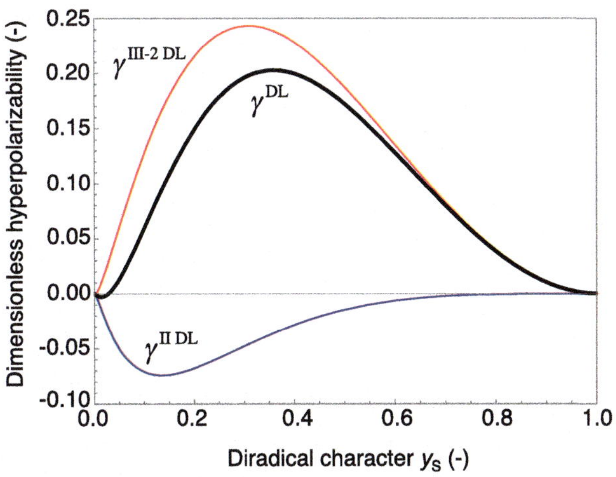

Fig. 12 Diradical character dependence of $\gamma^{DL}(= \gamma/(R^4_{BA}/U^3))$, $\gamma^{IIDL}(= \gamma^{II}/(R^4_{BA}/U^3))$ and $\gamma^{III-2DL}(= \gamma^{III-2}/(R^4_{BA}/U^3))$ in the case of $r_K = 0$

The first and the second terms, which are referred to as type II and III-2 virtual excitation processes [43, 44], respectively, are shown to be negative and positive contributions to total γ values, respectively. For $r_K = 0$ (usual case for open-shell molecules with singlet ground states), the variations of dimensionless total γ (γ^{DL}), as well as type II and III contributions ($\gamma^{\text{II DL}}$ and $\gamma^{\text{III}-2\,\text{DL}}$) as a function of y are shown in Fig. 12. It is found that $\gamma^{\text{II DL}}$ has a negative extremum in the small y region, while $\gamma^{\text{III}-2\,\text{DL}}$ has a positive extremum in the intermediate y region. Since the extremum amplitude of $\gamma^{\text{III}-2\text{DL}}$ is shown to be much larger than that of γ^{IIDL}, the variation of total γ^{DL} with y is found to be governed by that of $\gamma^{\text{III}-2\text{DL}}$ and gives positive values in the whole y region. This behavior of $\gamma^{\text{III}-2\text{DL}}$ is understood by the variation in the numerator $(\mu_{\text{gk}}^{\text{DL}})^2(\mu_{\text{kf}}^{\text{DL}})^2$ and denominator $(E_{\text{kg}}^{\text{DL}})^2 E_{\text{fg}}^{\text{DL}}$ in the second term of Eq. 50 as a function of y: the denominator and numerator approach infinity and a finite value, respectively, as $y \to 0$, leading to $\gamma^{\text{III}-2\text{DL}} \to 0$, while they do a finite value and 0, respectively, as $y \to 1$, leading to $\gamma^{\text{III}-2\text{DL}} \to 0$ again (see also Fig. 7b). Although both the denominator and numerator decrease with increasing y from 0 to 1, the denominator decreases more rapidly in the small y region than the numerator, which is the origin of the extremum of $\gamma^{\text{III}-2\text{DL}}$ (~ 0.306) in the intermediate y region (~ 0.243). The $\left|\gamma^{\text{III}-2\text{DL}}\right| > \left|\gamma^{\text{IIDL}}\right|$ except for $y \sim 0$ is understood by the fact that the numerator in the first term of Eq. 50, $(\mu_{\text{gk}}^{\text{DL}})^4$, decreases more rapidly than that in the second term of Eq. 50, $(\mu_{\text{gk}}^{\text{DL}})^2(\mu_{\text{kf}}^{\text{DL}})^2$ (see also Fig. 7b). As shown in Fig. 7b $(\mu_{\text{gk}}^{\text{DL}})^2$ and $(\mu_{\text{kf}}^{\text{DL}})^2$ show decrease and increase, respectively, with increasing y, so that $(\mu_{\text{gk}}^{\text{DL}})^2(\mu_{\text{kf}}^{\text{DL}})^2$ keeps a larger value than $(\mu_{\text{gk}}^{\text{DL}})^4$ in the whole y region. In summary, it turns out that the γ values of open-shell singlet systems with intermediate diradical character tend to be significantly larger than those of closed-shell and pure diradcial systems. For asymmetric systems, we also have revealed remarkable enhancements of $|\gamma|$ and $|\beta|$ values in the intermediate diradical/ionic character region [39, 45]. These theoretical predictions pioneer a novel class of highly-efficient second- and third-order NLO substances, i.e., open-shell NLO systems, which outstrip traditional closed^{shell} NLO systems.

6 Relationship between Open-Shell Character, Aromaticity, and Response Property

6.1 Indenofluorenes

The relationship between open-shell character and other traditional chemical concepts like aromaticity is useful for deeper understanding of the open-shell singlet electronic structures, as well as for its application to constructing design guidelines for highly efficient functional molecular systems. Indeed, the chemical and physical tuning schemes of the diradical character have been obtained by revealing the relationships between the open-shell character and the traditional chemical concepts/indices that most chemists are familiar with [16, 18]. Among the chemical

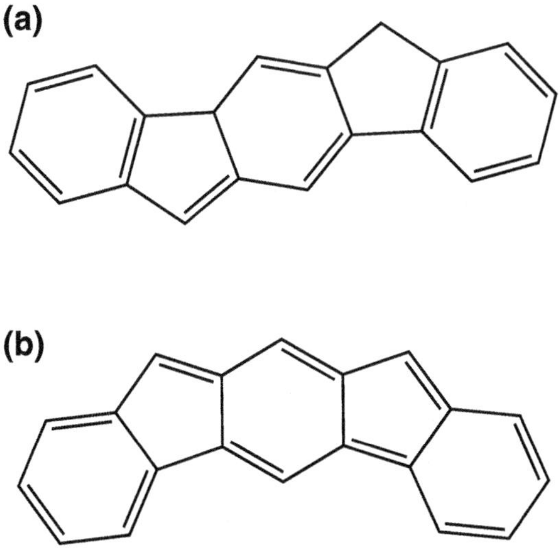

Fig. 13 Molecular frameworks of *para*- and *meta*-type indenofluorenes (**a** and **b**, respectively)

concepts/indices, "aromaticity" is one of the most essential and intuitive concepts relating to open-shell character for the chemists [46–48] since it is well-known that anti-aromatic systems have small energy gaps between the HOMO and the LUMO [47], which tend to increase the diradical character as shown in Eq. 27. Also, to clarify the structure–property relationship based on the diradical character and aromaticity, we have to reveal the spatial correlation between the open-shell character and aromaticity. In this section, we show the spatial correlation between the open-shell character, aromaticity, and the second hyperpolarizability (the third-order NLO response property at the molecular scale) by focusing on *para*- and *meta*-indenofluorenes (Fig. 13), which are π-conjugated fused-ring systems with alternating structures composed of three six-membered and two five-membered rings synthesized by Haley's and Tobe's groups [49–52]. Apparently, these are 20π electron systems, so that they are regarded as anti-aromatic analogues of pentacene [49]. On the other hand, these systems exhibit pro-aromatic quinodimethane framework in the central region, which is predicted to exhibit open-shell singlet character [50, 52, 53]. Thus, as shown in Sect. 5, these systems will be appropriate model systems for clarifying the relationships of spatial contributions between the open-shell character, the aromaticity, and the second hyperpolarizability.

6.2 Structure, Odd Electron Density, Magnetic Shielding Tensor, and Hyperpolarizability Density

The geometries of *para*- and *meta*-type indenofluorenes are optimized with the U(R)B3LYP/6-311 + G** method under the symmetry constraints of C_{2h} for *para*, and C_{2v} for *meta* systems. The diradical character y, unpaired(odd)-electron density, the magnetic shielding tensor component $-\sigma_{yy}$ and the second hyperpolarizabilities γ are evaluated using the long-range corrected (LC) density functional theory (DFT) method, LC-UBLYP (range separating parameter $\mu = 0.33$ bohr^{-1}) method, with the 6–311 + G** basis set. Within the single determinantal UDFT scheme, the diradical character is defined as the occupation number of the LUNO of the unrestricted wavefunctions n_{LUNO}:

$$y = n_{\mathrm{LUNO}} = 2 - n_{\mathrm{HONO}}. \tag{51}$$

Note here that the spin-projection scheme (see Eqs. 14, 15) is not applied in this case since the LC-UBLYP ($\mu = 0.33$) method is generally found to have smaller spin contamination than UHF and is found to reproduce well the diradical character and γ values at the strong correlated level of theory like UCCSD(T) (see Sect. 7.1). The spatial contribution of diradical character is clarified using the odd-electron density ρ_{odd} at position $\mathbf{r}$, which is calculated using the frontier natural orbitals $\phi_{\mathrm{HONO}}(\mathbf{r})$ and $\phi_{\mathrm{LUNO}}(\mathbf{r})$ as follows [31]

$$\rho_{\mathrm{odd}}(\mathbf{r}) = n_{\mathrm{LUNO}} \left(|\phi_{\mathrm{HONO}}(\mathbf{r})|^2 + |\phi_{\mathrm{LUNO}}(\mathbf{r})|^2 \right) \tag{52}$$

This contributes to the diradical character y as expressed by

$$y = \frac{1}{2} \int d\mathbf{r}\rho_{\mathrm{odd}}(\mathbf{r}) = \frac{1}{2} \int d\mathbf{r} \left[n_{\mathrm{LUNO}} \left(|\phi_{\mathrm{HONO}}(\mathbf{r})|^2 + |\phi_{\mathrm{LUNO}}(\mathbf{r})|^2 \right) \right] \tag{53}$$

In order to estimate the aromaticity, we here employ the magnetic shielding tensor calculated using the gauge invariant atomic orbital (GIAO) method [54]. Since the examined systems have planar structures (z–x plane), the magnetic shielding tensor ($-\sigma_{yy}$) is evaluated 1 Å above the center of each ring, which primarily reflects the contribution of the π-electron ring current. Namely, more negative $-\sigma_{yy}$ values indicate more aromaticity, while more positive values more antiaromaticity due to the diatropic (paratropic) ring current in aromatic (antiaromatic) ring. The longitudinal component of the static second hyperpolarizabilities γ_{zzzz}, which is along the spin polarization direction and thus reflects the polarization of odd electrons, is calculated by the forth-order differentiation of the total energy according to the static electric field, finite field (FF) method [55]. The spatial contribution of electrons to the γ_{zzzz} can be analyzed by using γ_{zzzz} density $\rho_{zzz}^{(3)}(r)$, which is defined as [44]

$$\rho_{zzz}^{(3)}(r) = \left. \frac{\partial^3 \rho(r)}{\partial F_z^3} \right|_{\mathbf{F}=0} \tag{54}$$

Here, $\rho(r)$ is the total electron density at the position r, and F_z is the z component of the external electric field $\mathbf{F}$. This γ_{zzzz} value is obtained by γ_{zzzz} density as

$$\gamma_{zzzz} = -\frac{1}{3!} \int r_z \rho_{zzz}^{(3)}(r) dr \tag{55}$$

This relationship indicates that a pair of positive and negative γ_{zzzz} densities with large amplitudes, separated by a large distance, contribute to the increase of $|\gamma_{zzzz}|$ values, and that the sign of the contribution is determined by the direction of the arrow drawn from positive to negative γ_{zzzz} density: when the direction of the arrow

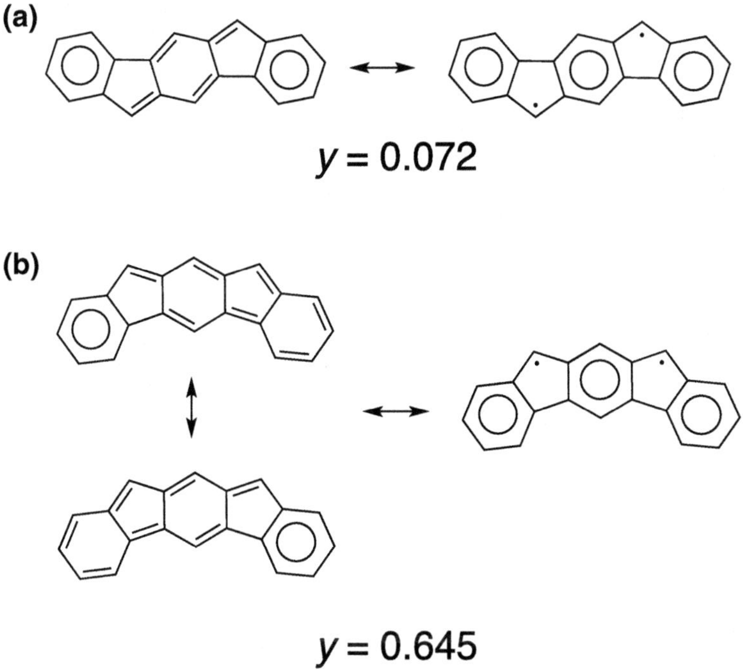

(a)

$$y = 0.072$$

(b)

$$y = 0.645$$

Fig. 14 Resonance structures with Clar's sextets (indicated by the delocalized benzene-ring forms) for *para-* (**a**) and *meta-* (**b**) type indenofluorenes. Diradical character y of each indenofluorene system is calculated at LC-UBLYP/6-311 + G**//U(R)B3LYP/6-311 + G** level of theory

coincides with (is opposite to) that of the coordinate axis, the contribution is positive (negative) in sign.

6.3 Diradical Character and Local Aromaticity of Indenofluorenes

The open-shell singlet character and resonance structure are shown in Fig. 14 for each indenofluorene system. It is found that the system involving *para*-quinodimethane framework, referred to as *para*, exhibits negligible diradical character (<0.1), while the system involving *meta*-quinodimethane framework, referred to as *meta*, shows a larger value ($y = 0.645$) [56]. This indicates that the *para* system is classified into nearly closed-shell systems, while the *meta* system is classified into intermediate singlet diradical system. These features are qualitatively understood based on their resonance structures with Clar's sextets rule. Namely, the open-shell resonance structures for both systems exhibit a larger number of Clar's sextets, (three benzene rings) than the closed-shell structures. This feature originates form the existence of the pro-aromatic quinodimethane structure in these systems. It is also noted that the *meta* system exhibits a smaller number of Clar's sextets in the closed-shell form than *para* systems, i.e., one for *meta* and two for *para*, the difference of which indicates the relatively larger stability of the open-shell

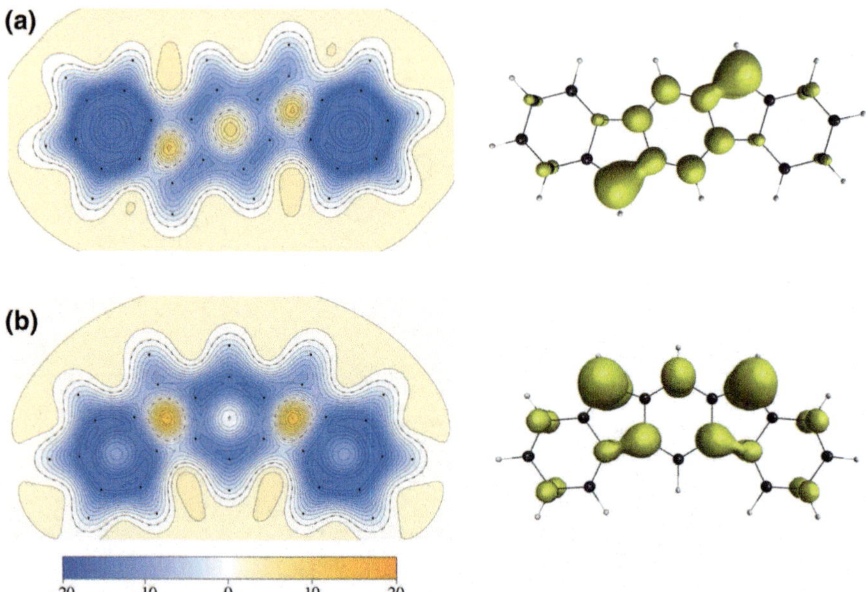

Fig. 15 $-\sigma_{yy}$ Maps (*left*) and odd-electron densities (*right*) of *para* (**a**) and *meta* (**b**) systems in the singlet states calculated at the LC-UBLYP/6-311 + G** level of theory (contour values of 0.0004 a.u. (*para*) and 0.004 a.u. (*meta*) for the odd-electron density distributions)

resonance structure in *meta* system than in *para* system, resulting in the larger diradical character of *meta* system than that of *para* system.

The local aromaticity is clarified using the magnetic shielding tensor component ($-\sigma_{yy}$) 1 Å above the center of each six- and five-membered ring plane (Fig. 15) [56]. For these systems, the middle three rings, the six-membered ring together with the adjoining two five-membered rings exhibit positive $-\sigma_{yy}$ (anti-aromatic), while the terminal benzene rings exhibit negative $-\sigma_{yy}$ (aromatic). On the other hand, it is found that the $-\sigma_{yy}$ values of the terminal benzene rings exhibit larger negative (aromaticity) (–20 ppm) for the *para* system than for the *meta* system (–11.1 ppm), while that of the anti-aromatic central six-membered ring is larger positive (anti-aromatic) (9.4 ppm) for the *para* system than for the *meta* system (0.2 ppm), which represents much reduced anti-aromatic or non-aromatic central six-membered ring. Considering the diradical characters $y = 0.072$ for *para* and 0.645 for *meta* systems, it is found that the difference in the local aromaticity between the central and the terminal rings is much smaller in the intermediate diradical *meta* system ($|-\sigma_{yy}$ (central) $+ \sigma_{yy}$(terminal)$| = 11.3$ ppm) than in the nearly closed-shell *para* system ($|-\sigma_{yy}$(central) $+ \sigma_{yy}$(terminal)$| = 29.4$ ppm). This feature is understood by comparing the number of the Clar's sextets in the resonance structures (see Fig. 14). The diradical resonance structures are shown to exhibit the Clar's sextets at all the six-membered rings, which contribute to the aromaticity at all the six-membered rings. In contrast, the closed-shell resonance structure of the *para* system exhibits the Clar's sextets at both the two terminal rings, while the *meta* system does the Clar's sextet at only one of the two terminal rings. This implies that the terminal six-membered rings of the nearly closed-shell *para* system exhibit fully Clar's sextet aromatic nature for all the resonance structures, while those of *meta* system do less aromatic nature due to the both contribution of the fully aromatic nature in the diradical resonance structure and the half in the closed-shell resonance structure. Similarly, the fact that the *meta* system has a larger contribution of the diradical resonance structure is found to lead to much reduced anti-aromatic or non-aromatic nature of the central six-membered ring, which exhibits a Clar's sextet at the central benzene ring in the diradical resonance structure. Figure 15 also shows the spatial distributions of $-\sigma_{yy}$ with color contours, where the blue and yellow contours represent aromatic (with negative $-\sigma_{yy}$) and anti-aromatic (with positive $-\sigma_{yy}$) regions, respectively. The central benzene ring together with the adjoining two five-membered rings for *para* system shows yellow contours (local anti-aromatic nature), while that of *meta* system does almost white contours (local non-aromatic nature). Such spatial features of $-\sigma_{yy}$ maps give more detailed spatial contribution features of the local aromaticities in the indenofluorene series. The spatial correlation between the local aromaticity and the diradical character is clarified by examining the maps of odd (unpaired)-electron density distribution (Fig. 15). Large odd-electron densities are shown to be generally distributed around the zigzag-edge region of the five-membered rings. Since this feature is consistent with that in the diradical resonance structures, the odd-electron density maps also substantiate Clar's sextet rule in these molecules. As seen from the odd-electron densities of the six-membered rings, the *para* system exhibits odd-electron densities more significantly distributed at the central benzene rings than at the terminal ones, while the *meta* system shows odd-electron densities more delocalized over both the central and terminal benzene rings. Although this distribution difference is not straightforwardly understood from the resonance structures, the primary odd-electron density distribution region well corresponds to the local anti- or weaker aromatic ones of the six-membered rings: the difference in the local aromaticity between the six-membered rings is more distinct in the *para* system than in the *meta* system. This indicates that for each indenofluorene system, the six-membered rings with larger odd-electron densities exhibit relatively anti-aromatic nature. This spatial correlation between the odd-electron density and local aromaticity is understood by the fact that the emergence of odd-electron density in the aromatic ring implies the partial destruction of the fully π-delocalization over the ring, resulting in the reduction of aromaticity or in the emergence of anti-aromaticity.

In order to confirm further the correlation between odd-electron density and local aromaticity, let us consider the triplet states of the *para* and *meta* systems since these triplet states correspond to the pure diradical states. It is found that unlike the corresponding singlet systems (which exhibit anti-aromaticity in the central benzene rings; Fig. 15), all the benzene rings exhibit similar aromatic nature in both systems (Fig. 16), the feature of which is particularly different in the central benzene ring from that of the singlet *para* system [56]. This is understood by the fact that the triplet states are described as pure diradical resonance structures for both the systems, which are stabilized by all the benzene rings with Clar's sextet form, whereas the singlet systems have contribution of closed-shell resonance structures,

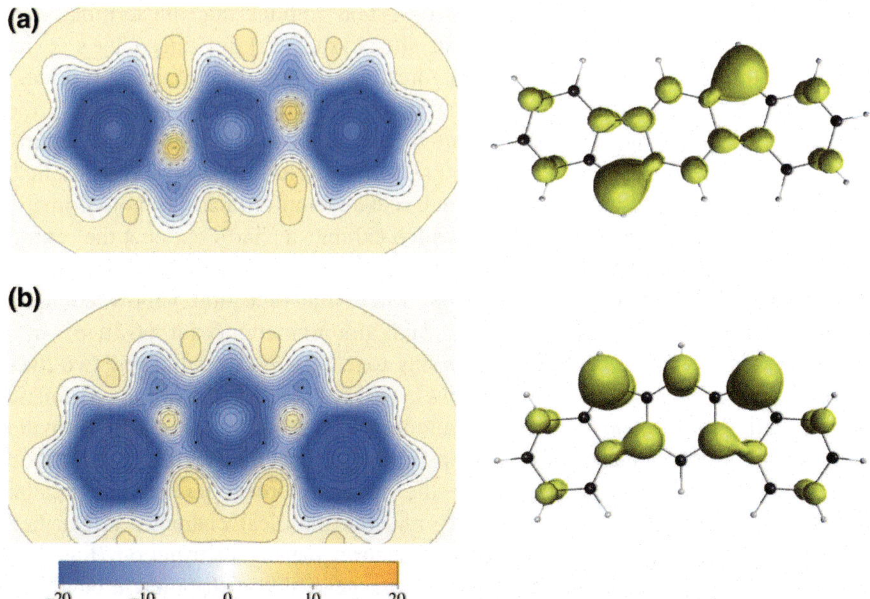

Fig. 16 $-\sigma_{yy}$ Maps (*left*) and odd-electron densities (*right*) of *para* (**a**) and *meta* (**b**) systems in the triplet states calculated at the LC-UBLYP/6-311 + G** level of theory (contour value of 0.004 a.u. for the odd-electron density distributions)

which in particular reduce the aromaticity or increase the anti-aromaticity in the central six-membered ring (see Fig. 14). Furthermore, the feature of odd-electron density distribution of *para* system in the triplet state differs from that in the singlet state (see Figs. 15, 16): (a) the odd-electron density amplitudes are much larger than those in the triplet state, and (b) the odd electron densities are negligible in the terminal benzene rings in the singlet state, while those are also observed in the terminal benzene rings in addition to the central one. Accordingly, by changing from singlet to triplet state, the central benzene ring drastically change from anti-aromatic to aromatic, and the terminal rings slightly reduce the aromaticity. These results indicate that there exists a correlation between the difference in the local aromatic nature and that in the amplitudes of odd-electron densities of the six-membered rings [56].

6.4 Second Hyperpolarizaibilities of Indenofluorene systems

We here show the impact of the diradical character and the aromaticity on the molecular functionality, i.e., the longitudinal components of second hyperpolarizabilities γ, γ_{zzzz}, which are the dominant components for these indenofluorene systems. Figure 17 shows the γ_{zzzz} densities together with the γ_{zzzz} values in singlet states. As expected from our y–γ correlation, the *meta* system with the intermediate diradical character exhibits larger γ_{zzzz} values than the *para* system with much smaller diradical character. As seen from Figs. 5 and 16, the relative amplitudes of

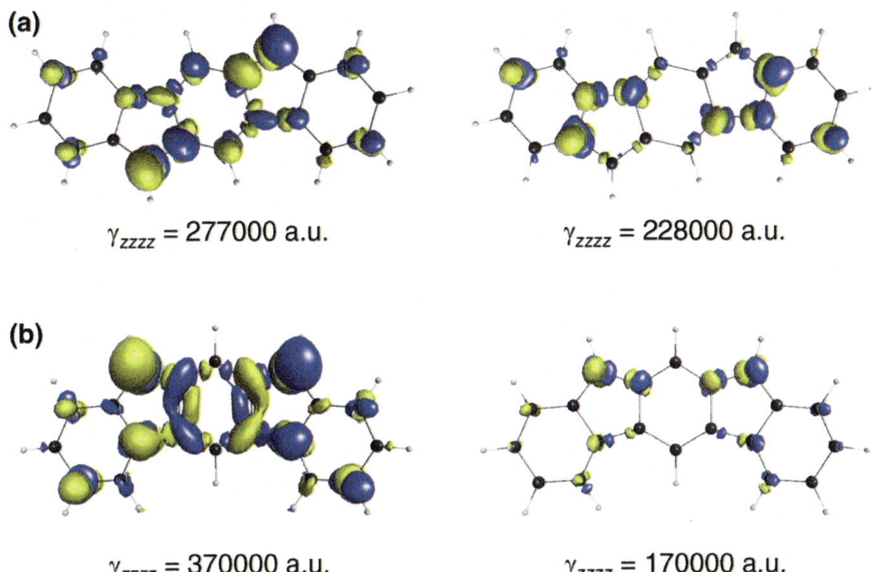

Fig. 17 γ_{zzzz} Density maps (contour value of 1000 a.u.) for *para* (**a**) and *meta* (**b**) systems in the singlet (*left*) and triplet (*right*) states calculated at the LC-UBLYP/6-311 + G** level of theory. The γ_{zzzz} values are also shown

γ_{zzzz} densities qualitatively accord with those of the odd electron densities. Indeed, for the *para* system with slight diradical character, the γ_{zzzz} density is primarily localized in the quinodimethane framework, while for *meta* system with intermediate dirdadical character, that is not localized in the quinodimethane region, but spreads over the molecule. These results demonstrate the spatial-distribution similarity between the γ_{zzzz} density, odd-electron density and $-\sigma_{yy}$ maps, especially around the six-membered ring regions.

Next, we consider spin state dependences of these quantities. For the *para* system with slight dirdadical character, the γ_{zzzz} amplitude in the triplet state is somewhat smaller than that in the singlet state, $\gamma_{zzzz}^{triplet}/\gamma_{zzzz}^{singlet} = 0.83$, while for the *meta* system with the intermediate diradical character the γ_{zzzz} amplitude in the triplet state shows significant reduction (only a half amplitude, $\gamma_{zzzz}^{triplet}/\gamma_{zzzz}^{singlet} = 0.46$) of the singlet state γ_{zzzz} value. As seen from the γ_{zzzz} density distribution maps for both systems (Fig. 17), the γ_{zzzz} density distribution features are significantly different between the triplet and singlet states: the γ_{zzzz} density distributions in the triplet states tend to spread over all the molecular frameworks, not localized at the central quinodimethane region. In particular, in the triplet states, the γ_{zzzz} densities are shown to be significantly reduced as compared to the singlet states around the zigzag edge region of the five-membered rings, where the large odd-electrons are distributed. These features indicate that in the triplet states, the spatial contribution to the γ_{zzzz} is not so correlated to the odd-electron density distributions unlike the intermediate open-shell singlet systems. This is understood by the Pauli effects [57]: the triplet diradical electrons are prohibited to be delocalized due to the Pauli principle and thus do not contribute to the enhancement of γ_{zzzz}.

In summary, it is found that there exists strong correlation in the absolute values and the spatial distributions between the diradical character, local aromaticity and γ values. In this regard, the diradical character and the odd electron density distribution are qualitatively predicted with Clar's sextets rule: the relative stability of the open-shell resonance structures varies based on the number of the Clar's sextets in the closed-shell and the open-shell resonance structures. Also, the indenofluorene frameworks can vary the relative stability of the resonance structures, which is accompanied by the change of the geometry around the anti-aromatic five-membered rings.

7 Diradical Character and Optical Response Properties Calculated Using Broken-Symmetry Density Functional Theory Methods

7.1 Functional Dependence of Diradical Character for Polycyclic Hydrocarbons

As shown in Eq. 51, diradical character y is usually defined by the occupation number of LUNO. This definition can be applied to highly correlated methods such as spin-unrestricted coupled-cluster with single, double, and perturbative triple excitations (UCCSD(T)), multi-reference Møller-Plesset perturbation (MRMP), and full CI. On the other hand, when applied to broken-symmetry methods, spin contamination effects must be considered for reproducing the highly correlated diradical character, e.g., UCCSD(T) y value. The spin contamination effects on the

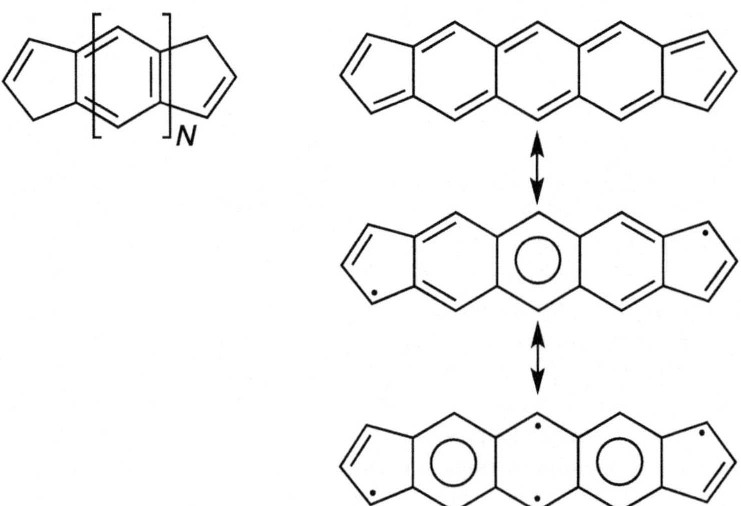

Fig. 18 Structure of dicyclopenta-fused oligoacenes (DPA[N], $N = 0$–3) together with the resonance structures for $N = 3$

diradical character is explicitly defined at UHF level of theory, so that approximate spin-projection schemes, for example, perfect-pairing type spin-projected HF (PUHF), Eq. 12, are applied to obtaining semi-quantitatively correct diradical character. Indeed, the PUHF y values are known to reproduce those at full CI level of theory. On the other hand, for the BS DFT (UDFT) case, it is known that spin contamination effects are smaller than those at UHF level of theory, and depend on the exchange–correlation (xc)-functionals. In order to clarify the functional dependences of the diradical character and response property, polarizability, which reflects the description of excitation energies and transition properties, we examine these quantities for dicyclopenta-fused oligoacenes (DPA[N], $N = 0$–3) (Fig. 18) optimized by UB3LYP/6-3111G* method under constraint of D_{2h} symmetry. These molecules are known to be anti-aromatic $4n\pi$ systems with a wide range of open-shell characters depending on the oligomer size [25, 47]. Indeed, as seen from the resonance structures, DPA involves two types of spin polarizations, which are along the longitudinal (between the terminal five-membered rings) and lateral (which is observed in the middle region of oligoacenes) directions, respectively. However, in the considered oligomer size, the main open-shell character is described by the diradical character, which corresponds to the spin polarization between the terminal five-membered rings.

In general, conventional DFT methods, e.g., BLYP and B3LYP methods, are known to provide highly accurate results with less computational effort for geometry, electronic structure, reaction, etc., which demonstrates great successes in closed-shell based molecules in chemistry. On the other hand, the BLYP and B3LYP methods are known to have several drawbacks originating from their local character, e.g., underestimated band gap energy and too large (non)linear optical properties [58], and no description of weak-interaction, undershot charge-transfer excited state energy, etc. The local character of conventional xc-functionals in the DFT can be improved by the long-range corrected (LC-)DFT, e.g., LC-BLYP, method [59–61]. In such methods, the range-separation of the DFT exchange functional is realized by dividing the electron repulsion operator ($1/r_{12}$) into long- and short-range parts (the range separation is controlled by the range-separating parameter μ), is found to improve the description of NLO response properties for extended π-conjugated molecular systems [62–65] and several open-shell singlet molecules [66]. In the case with a larger (smaller) range-separating parameter μ, the fraction of the HF exchange is larger (smaller) at a given r_{12}. The μ^{-1} represents the delocalization length [68].

Figure 19 shows the diradical character y at the UCCSD, LC-UBLYP, CAM-UBLYP, UBHandHLYP, and UB3LYP levels of approximation versus y at PUHF level of approximation for DPA[N] ($N = 0, 1, 2, 3$). It is found that the UCCSD and LCUBLYP ($\mu = 0.47$) methods give slightly larger y values at each size of DPA[N] than PUHF y value, while the UBHandHLYP gives a slightly smaller y value than PUHF y value at each N. The LC-UBLYP ($\mu = 0.33$) and CAM-UB3LYP methods are shown to reproduce the PUHF y values at large N values ($N = 3, 2$), while they are shown to result in smaller y values than the PUHF y values at small N values ($N = 2, 0$), e.g., about a half of the PUHF y value at $N = 0$. The B3LYP is found to give significantly undershot y values than the PUHF

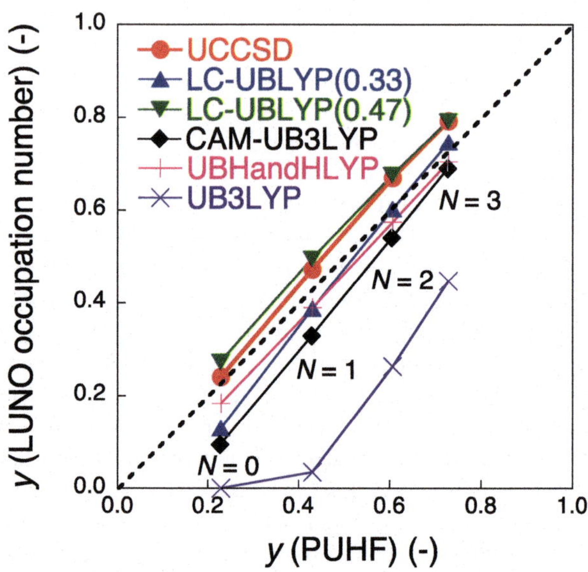

Fig. 19 Diradical character y at the UCCSD and UDFT levels of approximation versus y at PUHF level of approximation for DPA[N] ($N = 0, 1, 2, 3$). The 6-31G basis set is used

Table 1 Longitudinal polarizability α [a.u.] for DPA[N] up to $N = 2$

Method	DPA[0]	DPA [1]	DPA [2]
UCCSD	117	225	352
LC-UBLYP(0.33)	124	245	395
LC-UBLYP(0.47)	115	227	366
UB3LYP	129	273	470
UBHandHLYP	121	242	399
UBLYP	128	270	491
CAM-UB3LYP	129	270	453

y values in the whole y(PUHF) region. In summary, the inclusion of the HF exchange is important for well reproducing y values at PUHF and strong-correlated UCCSD levels of theory. The UBHandHLYP, LC-UBLYP, and CAM-UB3LYP methods are found to work well for evaluating diradical character of open-shell singlet systems. For polarizabilities α, it is found from Table 1 that the BHandHLYP and LC-UBLYP ($\mu = 0.33$, 0.47) methods well reproduce the UCCSD α values at least up to $N = 2$, while that the UBLYP, UB3LYP, and CAM-UB3LYP methods overestimate the UCCSD α values. The overestimation is shown to be more emphasized for larger size systems.

7.2 Approximate Spin-Projection Scheme of Diradical Character and Optical Response Properties

7.2.1 Calculation Methods and Model System

Although UDFT methods with several xc-functionals are found to reproduce semi-quantitatively the y–γ correlation obtained from strong-correlated quantum chemistry calculations [5, 6, 66, 69], the spin contamination effects involved in the broken-symmetry (BS) schemes often cause incorrect results on optimized molecular structures and magnetic properties of open-shell systems [29, 33, 34]. In this section, we show an approximate spin-projection scheme within the spin-unrestricted single-determinantal framework [31, 70]. This is based on a correction of the occupation numbers (diradical characters) and is expected to improve the description of the odd electron density [30, 71, 72] and the evaluation of the (hyper)polarizabilities of delocalized open-shell singlet molecules. The performance of this scheme is demonstrated by the static polarizability (α) and second hyperpolarizability (γ) of a typical open-shell singlet system, i.e., the p-quinodimethane (PQM) model using several xc-functionals are employed, ranging from the pure DFT BLYP, the hybrid B3LYP, and BHandHLYP functionals to the LC-UBLYP functional.

First, we briefly describe the approximate spin-projected method based on the NOs and occupation numbers obtained by spin-unrestricted single determinantal schemes. The chemicophysical properties like the optical responses of N-electron systems are calculated using the one-electron reduced density:

$$d(\mathbf{r}) = \sum_{i=0}^{N/2-1} [n_{\mathrm{HONO}-i}\phi^*_{\mathrm{HONO}-i}(\mathbf{r})\phi_{\mathrm{HONO}-i}(\mathbf{r}) + n_{\mathrm{LUNO}+i}\phi^*_{\mathrm{LUNO}+i}(\mathbf{r})\phi_{\mathrm{LUNO}+i}(\mathbf{r})],$$

$$(56)$$

where $\{\phi_k(\mathbf{r})\}$ is the kth NO with occupation number $\{n_k\}$ ($n_{\mathrm{HONO}-i}$ + $n_{\mathrm{LUNO}+i}$ = 2). Here, approximately removing spin contamination effects is performed by the perfect-pairing spin-projection scheme [28, 29] on the occupation numbers n_k (see also Eqs. 14, 15):

$$n^{\mathrm{ASP}}_{\mathrm{HONO}-i} = \frac{(n_{\mathrm{HONO}-i})^2}{1+(T_i)^2} \equiv 2 - y^{\mathrm{ASP}}_i, \text{ and } n^{\mathrm{ASP}}_{\mathrm{LUNO}+i} = \frac{(n_{\mathrm{LUNO}+i})^2}{1+(T_i)^2} \equiv y^{\mathrm{ASP}}_i \qquad (57)$$

T_i, the overlap between the corresponding orbital pairs, is expressed by

$$T_i = \frac{n_{\mathrm{HONO}-i} - n_{\mathrm{LUNO}+i}}{2}. \qquad (58)$$

Here, y^{ASP}_i is the spin-projected diradical character [0 (closed-shell) $\leq y^{\mathrm{ASP}}_i \leq 1$ (pure diradical)] defined by Yamaguchi [28, 29]. Using Eqs. 57 and 58, the approximate spin-projected (ASP) polarizability ($\alpha^{\mathrm{ASP}}_{ij}$) and the ASP second hyperpolarizability ($\gamma^{\mathrm{ASP}}_{ijkl}$) ($i, j, k, l = x, y, z$) are expressed by [31]

$$\alpha_{ij}^{\text{ASP}} = -\int r_i d_j^{\text{ASP}(1)}(\mathbf{r}) d\mathbf{r} \tag{59}$$

and

$$\gamma_{ijkl}^{\text{ASP}} = -\frac{1}{3!}\int r_i d_{jkl}^{\text{ASP}(3)}(\mathbf{r}) d\mathbf{r} \tag{60}$$

Here, $d_j^{\text{ASP}(1)}(\mathbf{r})$ and $d_{jkl}^{\text{ASP}(3)}(\mathbf{r})$ indicate the ASP polarizability and the ASP second hyperpolarizability densities [44], respectively:

$$d_j^{\text{ASP}(1)}(\mathbf{r}) = \left.\frac{\partial d^{\text{ASP}}(\mathbf{r})}{\partial F_j}\right|_{\mathbf{F}=0} \tag{61}$$

and

$$d_{jkl}^{\text{ASP}(3)}(\mathbf{r}) = \left.\frac{\partial^3 d^{\text{ASP}}(\mathbf{r})}{\partial F_j \partial F_k \partial F_l}\right|_{\mathbf{F}=0} \tag{62}$$

Here, ASP one-electron reduced density is expressed, using Eqs. 56 and 57, as [31]

$$d^{\text{ASP}}(\mathbf{r}) = \sum_{i=0}^{N/2-1} \left[n_{\text{HONO}-i}^{\text{ASP}} \phi_{\text{HONO}-i}^{*}(\mathbf{r})\phi_{\text{HONO}-i}(\mathbf{r}) + n_{\text{LUNO}+i}^{\text{ASP}} \phi_{\text{LUNO}+i}^{*}(\mathbf{r})\phi_{\text{LUNO}+i}(\mathbf{r}) \right]. \tag{63}$$

The PQM model, which is a prototypical open-shell singlet model molecule, has a contribution from mixing two resonance forms—the diradical (open-shell) and quinoid (closed-shell) forms (see Fig. 20). The optimized structure with D_{2h} symmetry ($R_1 = 1.351$ Å, $R_2 = 1.460$ Å, and $R_3 = 1.346$ Å) at UB3LYP/6-311G* level of approximation coincides with that at RB3LYP level, which implies that the equilibrium PQM exhibits the quinoid-like structure instead of diradical one. We consider a model with stretching R_1 from 1.35 to 1.7 Å under the $R_2 = R_3 = 1.4$ Å constraint PQM, which can display a wide range of diradical character (y) since the

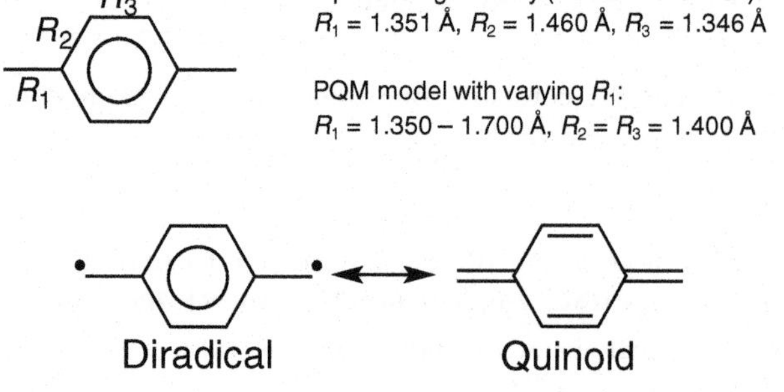

Fig. 20 p-Quinodimethane (PQM) model together with the resonance structures

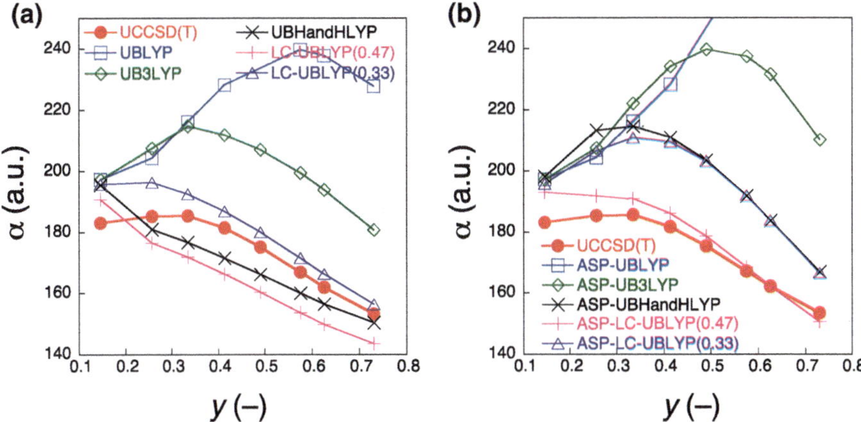

Fig. 21 Diradical character versus α [a.u.] of the PQM model calculated by the non-spin-projected (**a**) and the approximate spin-projected (ASP) (**b**) UDFT methods in comparison with the UCCSD(T) reference method

stretching R_1 mainly causes the π bond breaking. The diradical character y at PUHF/6-31G* + $p(\zeta = 0.0523)$ level ranges from 0.146 to 0.731 [6]. Here, we clarify the spin-projection effect on the diradical character dependence of α and γ using the several UDFT methods with 6-31G* + p basis set. The UBLYP, UB3LYP, and UBHandHLYP xc-functionals, as well as the long-range-corrected, LC-UBLYP ($\mu = 0.33$ and 0.47 bohr^{-1}), functional are employed for evaluating the longitudinal components of the static α ($\equiv \alpha_{xx}$) and γ ($\equiv \gamma_{xxxx}$). The non-spin-projected and ASP results are compared with the strong-correlated UCCSD(T) results.

7.2.2 Polarizability

First, we consider α values. Figure 21 displays the evolution of α of PQM as a function of y calculated by the non-spin projected (a) and ASP (b) UDFT methods, as well as the reference the UCCSD(T) method [70]. It is found that the UCCSD(T) α value slightly increases with y, attains a maximum ($\alpha_{max} = 185.4$ - a.u.) around $y = 0.34$, and then decreases. Although the UBLYP functional without spin projection reproduces qualitatively the diradical character dependence of α, α_{max} is significantly ($\sim 29\%$) overshot, while the y value giving γ_{max} (y_{max}) is shifted toward a larger y value ($y_{max} = 0.58$). It turns out that the increase of fraction of HF exchange reduces the overshot α_{max} and moves y_{max} to smaller values, e.g., $\alpha_{max} = 214.9$ a.u. at $y_{max} = 0.34$ (UB3LYP) vs. $\alpha_{max} = 239.6$ a.u. at $y_{max} = 0.58$ (UBLYP). In the case of using UBHandHLYP, which involves 50% of HF exchange, α_{max} is shown to be located at the initial $y = 0.146$ value, while α is shown to decrease monotonically with y. The differences in α between the UCCSD(T) and UBHandHLYP are, however, found to be relatively small in the whole y region. It is found that the y dependence and the amplitude of α at LC-UBLYP ($\mu = 0.47$) level are very similar to those at UBHandHLYP level though the LC-UBLYP ($\mu = 0.47$) α is slightly smaller (within $\sim 5\%$) than UBHandHLYP

α at each y. The α and y_{max} are found to increase as decreasing the range-separating parameter μ like the case of decreasing the fraction of HF exchange in the global hybrid functionals. It turns out that the α values at LC-UBLYP ($\mu = 0.33$) level somewhat overshoot those at UCCSD(T) level in the small y region (< 0.34), while that they well reproduce those for $y > 0.34$. In summary, among the non-spin-projected UDFT methods, the LC-UBLYP ($\mu = 0.33$) method is found to best reproduce the amplitude and the variation of α as a function of y in the whole y value region.

The approximate spin projection (Fig. 21b) is shown to cause an increase of α_{max} and a shift of y_{max} to larger y, e.g., $(y_{max}, \alpha_{max}) = (0.34, 214.9$ a.u.$)$ (UB3LYP) vs. $(0.49, 239.8$ a.u.$)$ (ASP-UB3LYP). Thus, the ASP-UBLYP, -UB3LYP, -UBHandHLYP, and –LC-UBLYP ($\mu = 0.33$) results are shown to go further apart from the reference UCCSD(T) result. It is found that the best agreement among the present methods is achieved by the ASP-LC-UBLYP ($\mu = 0.47$) method, which decreases (enhances) the overshot (undershot) α in the small (intermediate and large) y region, and then semi-quantitatively reproduces the UCCSD(T) y–α curve, though the y_{max} is slightly smaller than that at UCCSD(T) level of approximation.

As seen from these results, the static correlation and the spin contamination effects in the UDFT treatments sensitively depend on the fraction of HF exchange as well as on its range of application, i.e., μ value in the LC-UBBLYP method. Namely, we should be careful about the application of the spin-projection corrections to the UDFT method in the case of calculating the polarizability. It is found that the LC-UBLYP ($\mu = 0.33$) functional in the non-spin-projection scheme and the LC-UBLYP ($\mu = 0.47$) functional in the ASP scheme well reproduce the UCCSD(T) y-α curve.

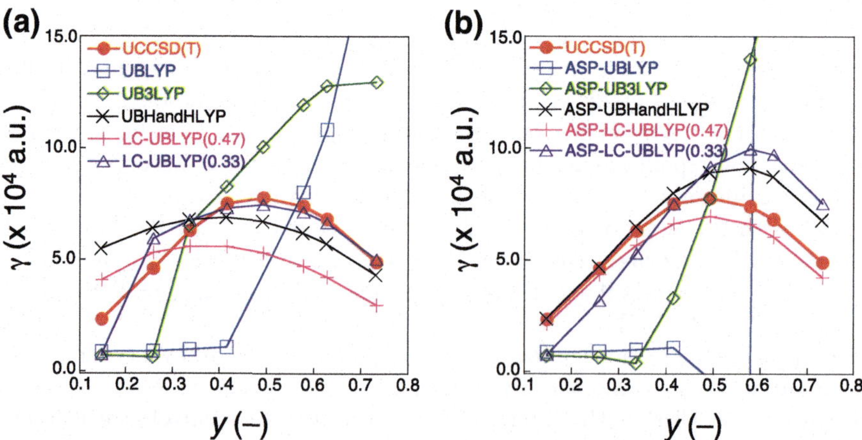

Fig. 22 Diradical character versus γ [a.u.] of the PQM model calculated by the non-spin-projected (**a**) and the approximate spin-projected (ASP) (**b**) UDFT methods in comparison with the UCCSD(T) reference method

7.2.3 Second Hyperpolarizability

First, we show the non-spin-projected y–γ curves of PQM calculated with hybrid and LC functionals [70]. The UCCSD(T) y–γ curves show a bell-shape variation with a maximum γ (γ_{max}) = 77500 a.u. around y_{max} = 0.49. As seen from Fig. 22a, for $y < 0.41$, the UBLYP results coincide with the RBLYP results, which give much smaller γ amplitudes, while for larger y, the UBLYP results give significantly overshot behavior. The increase in fractions of HF exchange in xc-functionals moves the y value at which the abrupt increase in γ occurs toward a smaller y value [70]. It is shown that the UBHandHLYP functional closely reproduces the UCCSD(T) γ variation for $y > 0.4$, while it overshoots the UCCSD(T) γ values in the small y value region ($y < \sim 0.34$). The LC-UBLYP ($\mu = 0.47$) functional is found to provide a similar curve to UBHandHLYP, though the γ values are found to be on average 13500 a.u. smaller. As a result, we find that LC-UBLYP ($\mu = 0.33$) functional best reproduces the UCCSD(T) y–γ curve, in particular for intermediate and large y values though it reduces to the spin-restricted solution at $y = 0.146$. In summary, it tuns out that tuning the HF exchange fraction in the hybrid UDFT methods or the range-separating parameter in the LC-UBLYP method can improve, to some degree, the agreement with the UCCSD(T) results, though the overshot behavior in the small y value region cannot be fully corrected.

Let us consider the performance of ASP scheme. As seen from Fig. 22b, the significant improvement of γ is achieved in the small y value region. Namely, the ASP is shown to correct the overshot behavior of γ in the small y region, as well as the slightly undershot curve of γ in the intermediate and large y value region for DFT results (Fig. 22a, b). As a result, the ASP-UBLYP($\mu = 0.47$) is found to best reproduce both the whole y-γ curve and the (y_{max}, γ_{max}) values [=(0.49, 77500 a.u.) using UCCSD(T)]: (y_{max}, γ_{max}) = (0.49, 69400 a.u.) [LC-UBLYP ($\mu = 0.47$)].

7.2.4 (Hyper)polarizability Densities Using the ASP-LC-UBLYP ($\mu = 0.47$) Method

In order to clarify the details of the ASP effects on the spatial electronic density, α and γ, we analyze the electronic density distributions as well as α and γ density distributions obtained by the ASP and non-spin-projected (NSP) UBLYP ($\mu = 0.47$) xc-functional [70].

Figure 23a shows the electron density differences $[\Delta d(\mathbf{r}) \equiv d^{ASP}(\mathbf{r}) - d^{NSP}(\mathbf{r})]$ for $y = 0.257$ (small), 0.491 (intermediate) and 0.731 (large). It is found that the increase (yellow) and decrease (blue) patterns of $\Delta d(\mathbf{r})$ are the same for any y value, while that the amplitudes significantly decrease with y. The increase of y (bond weakness), in other words, a decrease of quinoid character, is shown to be caused by a stretching of the terminal CC bonds, so that variations of $d(\mathbf{r})$ in these bonding regions are expected by spin projection. Namely, the positive $\Delta d(\mathbf{r})$ in the R_1 and R_3 regions and the negative one in the R_2 regions indicate that the spin contamination (primarily due to the mixing of the triplet component) emphasizes

(a)

$\Delta d(\boldsymbol{r})$

(b)

$d^{\mathrm{ASP}(1)}(\boldsymbol{r})$

$\Delta d^{(1)}(\boldsymbol{r})$

(c)

$d^{\mathrm{ASP}(3)}(\boldsymbol{r})$

$\Delta d^{(3)}(\boldsymbol{r})$

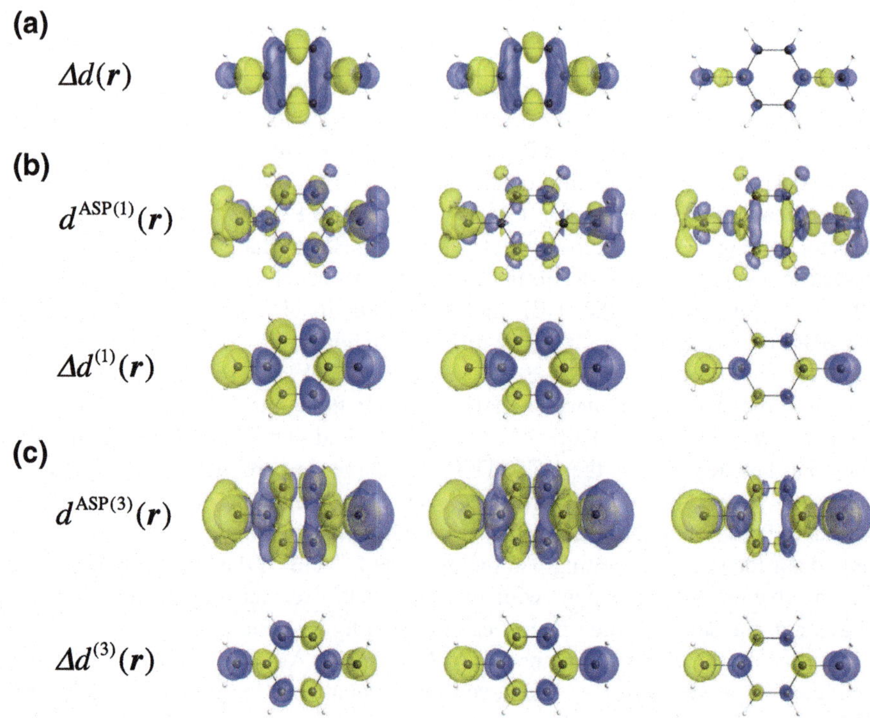

Fig. 23 Electron density differences $[\Delta d(\mathbf{r}) \equiv d^{\mathrm{ASP}}(\mathbf{r}) - d^{\mathrm{NSP}}(\mathbf{r})]$ **(a)**, α densities $[d^{\mathrm{ASP}(1)}(\mathbf{r})]$ and the differences between ASP and NSP α densities $[\Delta d^{(1)}(\mathbf{r}) \equiv d^{\mathrm{ASP}(1)}(\mathbf{r}) - d^{\mathrm{NSP}(1)}(\mathbf{r})]$ **(b)**, and γ densities $[d^{\mathrm{ASP}(3)}(\mathbf{r})]$ and the ASP-NSP differences $[\Delta d^{(3)}(\mathbf{r}) \equiv d^{\mathrm{ASP}(3)}(\mathbf{r}) - d^{\mathrm{NSP}(3)}(\mathbf{r})]$ **(c)** at $y = 0.257$ (*left*), 0.491 (*center*) and 0.731 (*right*) at LC-UBLYP ($\mu = 0.47$)/6-31G* $+ p(\zeta = 0.0523)$ level of approximation. The *yellow and blue surfaces* indicate positive and negative densities, respectively, with iso-surfaces: ± 0.0003 a.u. for $\Delta \rho(\mathbf{r})$, ± 0.1 a.u. for $d^{\mathrm{ASP}(1)}(\mathbf{r})$, ± 0.01 a.u. for $\Delta d^{(1)}(\mathbf{r})$, and ± 100 a.u. for $d^{\mathrm{ASP}(3)}(\mathbf{r})$ and $\Delta d^{(3)}(\mathbf{r})$

the bond dissociation nature, while the ASP favors the quinoid character, which corrects this excess bond dissociation features. Also, the reduction of the $\Delta d(\mathbf{r})$ amplitude in the large y value region in spite of an increased spin contamination is predicted to be caused by the fact that in the bond dissociation limit region the electron density distributions in the singlet state become very similar to those in the triplet state.

The α densities $(d^{\mathrm{ASP}(1)}(\mathbf{r}))$ and their differences $[\Delta d^{(1)}(\mathbf{r}) \equiv d^{\mathrm{ASP}(1)}(\mathbf{r}) - d^{\mathrm{NSP}(1)}(\mathbf{r})]$ are shown in Fig. 23b. We observe similar distribution pattern of α density for any y value, where the π-electrons distributed on both-end C atoms provide dominant positive contribution to α, while the σ-electrons do negative contributions with smaller amplitudes. The π-electron α density amplitudes decrease with y in agreement with α-y relationship (Fig. 21b). The α density differences $\Delta d^{(1)}(\mathbf{r})$ are primarily localized on the C atoms with alternant sign, which exhibit the same pattern for any y value and increase α amplitudes. The $\Delta d^{(1)}(\mathbf{r})$ is small for large y, as shown in the difference between the ASP- and NSP-LC-UBLYP ($\mu = 0.47$) results (see Fig. 21).

Figure 23c shows the γ densities ($d^{\mathrm{ASP}(3)}(\mathbf{r})$) and their differences [$\Delta d^{(3)}(\mathbf{r}) \equiv d^{\mathrm{ASP}(3)}(\mathbf{r}) - d^{\mathrm{NSP}(3)}(\mathbf{r})$]. Similar to the case of α, the distribution pattern is shown to be the same for any y value, and the primary positive contribution to γ is found to come from the π-electrons distributed on both-end C atoms. On the other hand, the positive and negative contributions in the middle benzene ring are shown to cancel with each other significantly. It is found that the amplitudes of the π-electron γ densities attain a maximum at $y = 0.491$ and then significantly decrease towards $y = 0.731$, the feature of which corresponds to the bell-shape variation of γ with y (see Fig. 22b). It turns out that the γ density differences $\Delta d^{(3)}(\mathbf{r})$ are primarily distributed on all C atoms with alternant sign, while that the sign of the $\Delta d^{(3)}(\mathbf{r})$ pattern is inverted when going from at $y = 0.257$–0.491, the feature of which represents the sign change of the ASP correction to γ from low to large y region (see Fig. 22).

7.2.5 Summary

As shown in this section, the degree of reproducibility of the UCCSD(T) (hyper)polarizabilities significantly depends on the fraction of HF exchange in the xc-functional or on the range-separating parameter μ in the LC-UBLYP functional. For the polarizability α, the LC-UBLYP ($\mu = 0.33$) and ASP-LC-UBLYP ($\mu = 0.47$) methods well reproduce α-y curve at the UCCSD(T) level in the whole y region. For the second hyperpolarizability γ, the ASP-LC-UBLYP ($\mu = 0.47$) method semi-quantitatively reproduces the y–γ curve in the whole y region, and the LC-UBLYP ($\mu = 0.33$) method also works well in the intermediate and large y regions.

8 Examples of Open-Shell Singlet Systems

In this section, we show several examples of open-shell singlet molecules based on polycyclic aromatic hydrocarbons (PAHs) together with their open-shell characters and hyperpolarizabilities. The architecture, edge shape, and size are found to significantly affect the diradical characters, the features of which are well understood based on the resonance structures with Clar's sextet rule and correspond to the aromaticity of the rings involved in those molecules.

8.1 Diphenalenyl Diradicaloids

As shown in Fig. 1b, diphenalenyl compound such as IDPL ($n = 1$) is known as a thermally stable diradicaloid because of contributions from both resonance structures, quinoid and benzenoid resonance forms [10–15, 17, 73]. Namely, the recovery of aromaticity in the central benzene ring in s-indacene part of IDPL leads to the diradical structures. As a closed-shell reference, we consider a similar size condensed-ring compound, PY2, which is composed of two pyrene moieties. From

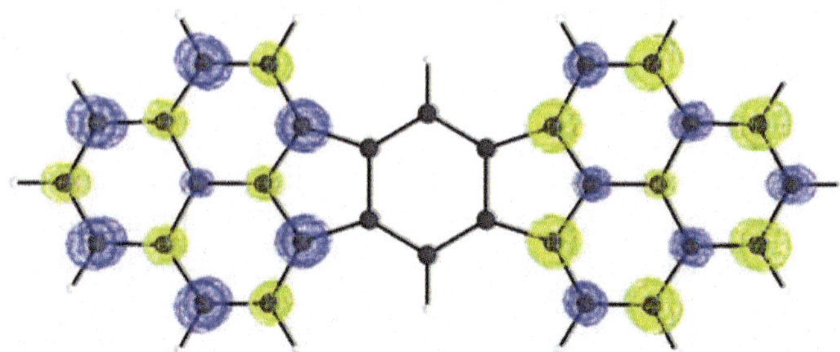

Fig. 24 Spin density distribution of IDPL at UBHandHLYP/6-31G* level of approximation. The *yellow* and *blue* meshes represent α and β spin densities with iso-surface 0.01 a.u., respectively. These spin densities represent spatial spin correlations between α and β spin

comparison of the geometries (optimized at RB3LYP/6-31G** level) of these two compounds, the larger bond length alternations in phenalenyl ring are observed for PY2, the feature of which well reflects the Kekulé structures [74]. This large π-electron delocalization in the phenarenyl rings contributes to stabilize the diradical form for IDPL. As expected from the resonance structures, the radical electrons are spin polarized between the both-end phenalenyl rings (see Fig. 24) and each up and down spin densities are delocalized in the phenalenyl ring though smaller spin polarization still exists in the phenalenyl ring. Although such spin density distribution is not observed in real singlet systems, this can be interpreted to indicate approximately the feature of spatial correlation between α and β spins. The diradical characters y values calculated from Eq. 16 using HONO and LUNO of UNOs at UHF/6-31G* level for IDPL and PY2 are 0.7461 (intermediate diradical character) and 0.0 (closed-shell), respectively, the feature of which is consistent with the features of the spin density distributions and the resonance structures [74]. The γ_{xxxx} (γ) values for IDPL and PY2 are 2383×10^3 a.u. and 194×10^3 a.u., respectively, at (U)BHandHLYP/6-31G* level of approximation. This significant enhancement of γ for IDPL as compared to that for PY2 (the ratio IDPL/PY2 = 12.3) is predicted to be caused by the intermediate diradical character ($y = 0.7461$) for IDPL in contrast to the closed-shell PY2 ($y = 0.0$) [74]. As seen from Fig. 25 (the γ density distributions), for both systems, main contributions come from π-electrons, whose contributions have opposite sign to those of σ-electrons. For intermediate diradicaloid, IDPL, we observe extended positive and negative γ densities distributed on the left and right phenalenyl ring regions, respectively, which cause a dominant positive contribution to γ though small opposite (negative) contributions appear in the central region. Judging from the fact that the sites with dominant γ density distributions on phenalenyl rings exhibit major spin density distributions, the spin-polarized π-electrons concerning left- and right-hand phenalenyl rings are the origin of the enhancement of γ values for these diradicaloids. Contrary to the dominant extended γ densities with positive contributions for IDPL, the γ density amplitudes for PY2 are smaller, and positive and negative γ densities appear alternately in the bond-length alternated (Kekulé structured) region though both-end benzene ring regions provide positive

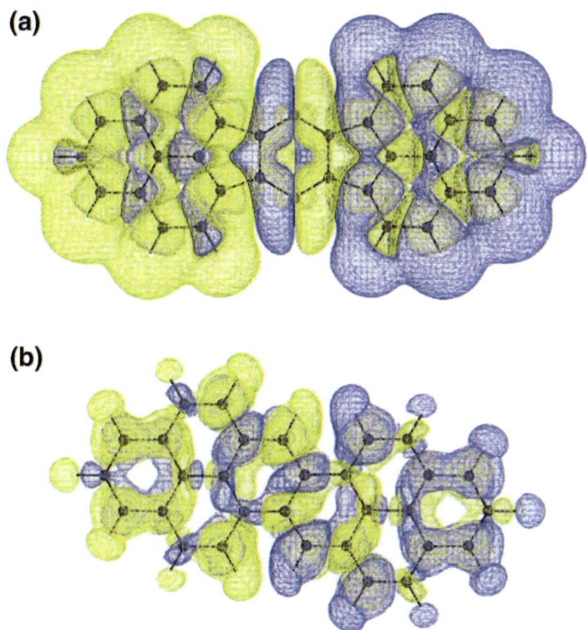

Fig. 25 γ density distributions for IDPL (**a**) and PY2 (**b**) at the BHandHLYP level. The *yellow and blue meshes* represent positive and negative γ densities with iso-surfaces with ±100 a.u

contribution to γ. This alternate change of sign of γ densities for PY2 significantly cancels the positive contribution to γ. In summary, the large enhancement of γ density amplitudes and spatially well-separated positive and negative γ densities on both end phenalenyl regions for IDPL turn out to be the origin of the significant enhancement of γ value for diradical molecules with intermediate diradical character as compared to closed-shell π-conjugated systems.

Another type of diphenalenyl diradicaloids, there are linked-type phenalenyl radical compounds, where two phenalenyl radical rings are bridged with acetylene or vinylene linkers. As examples, we consider acetylene-linked compounds, 1,2-bis(phenalen-1-ylidene)ethene (**1S**), and 1,2-bis(phenalen-2-yl)ethyne (**2S**), which are different in their linked positions from each other (see Fig. 26) [75]. As a closed-shell reference, 1,2-bis(pyren-4-yl)ethyne (**3S**) is also considered. Also, spin state dependences of γ are also investigated for **1S** and **2S**. The diradical characters of these singlet compounds are listed in Table 2 together with their γ values. It is found that **1S** (singlet) has a n intermediate diradical character, while **2S** (singlet) nearly pure diradicla character. Compound **3S** has a smaller diradical character and are regarded as a closed-shell systems as expected. The pure diradical character for **2S** (singlet) is caused by the linked position where no distribution (node line) of HSOMO of the phenalenyl ring. The ratios of γ, **1S** (singlet, $y = 0.6525$)/**3S** ($y = 0.1915$) is 4.4 and **1S** (singlet, $y = 0.6525$)/**2S** (singlet, $y = 0.9993$) is 7.5. This feature is in agreement with the y–γ correlation, which states that singlet diradical systems with an intermediate diradical character tend to exhibit larger γ

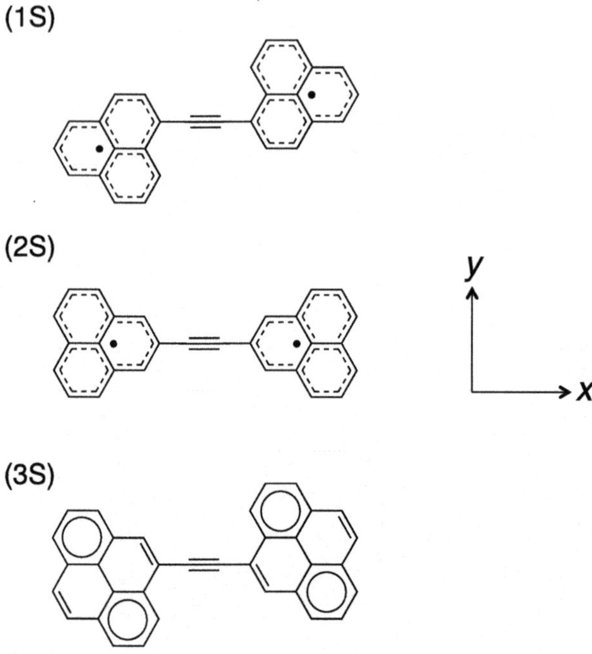

Fig. 26 Structures of 1,2-bis(phenalen-1-ylidene)ethene (**1S**) and 1,2-bis(phenalen-2-yl)ethyne (**2S**), as well as the closed-shell reference 1,2-bis(pyren-4-yl)ethyne (**3S**). The molecular geometries are optimized at (U)B3LYP/6-31G** level of approximation under the constraint of C_{2h} symmetry. The coordinate axis is also shown

Table 2 Diradical character y calculated at PUHF/6-31G** level and longitudinal components of second hyperpolarizabilities $\gamma(\gamma_{xxxx})$ for **1S–3S** (Fig. 26) calculated by the BHandHLYP/6-31G* method [75]

System	y	$\gamma\ [\times 10^3\ \text{a.u.}]$
1S (singlet)	0.6525	1568
2S (singlet)	0.9993	209
3S	0.1915	360
1S (triplet)	–	461
2S (triplet)	–	687

value than the closed-shell systems with a similar π-conjugation length. The spin state dependence is clarified by comparing singlet and triplet state results for **1S** and **2S**. In the intermediate diradical character region, the change from singlet (the lowest spin) to the triplet (the highest spin) state causes a significant reduction of γ, while in the pure diradical character region, such significant reduction is not observed. This significant reduction of γ in the triplet state in the intermediate diradical character region can be understood by the fact that in the triplet state the radical electron is localized in each atom due to the Pauli effect, resulting in suppressing the polarization over the two atoms. This feature is also similar to the

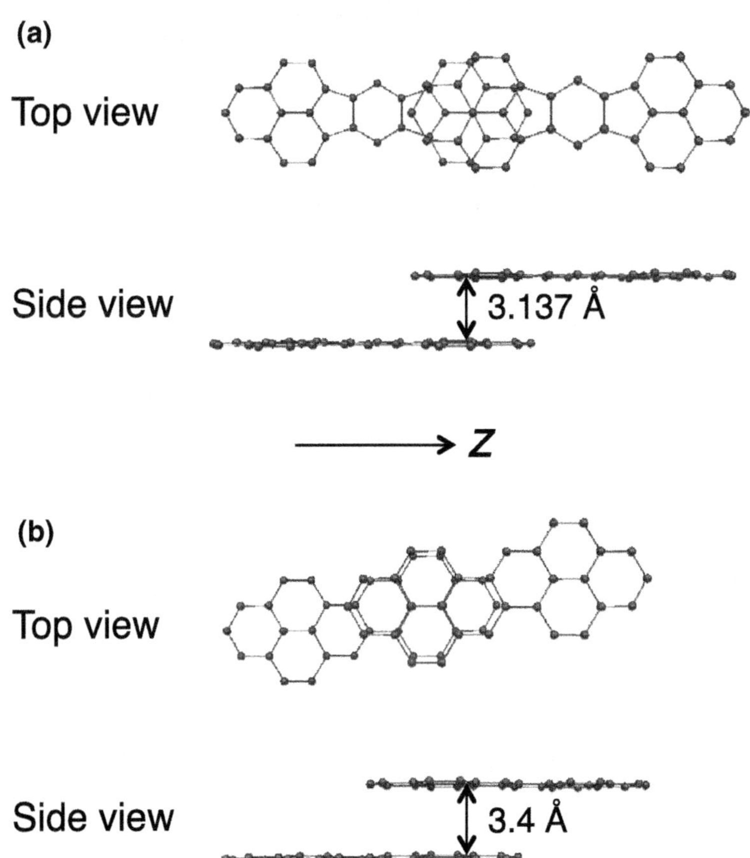

Fig. 27 Structures of IDPL dimer (R = H) (**a**) extracted from a single crystal (R = Ph) [10] and PY2 dimer (**b**). The average distance between cofacial phenalenyl rings of IDPL dimer is 3.137 Å, while that of PY2 dimer is fixed to be a typical van der Waals distance 3.4 Å. The structure of PY2 monomer is optimized by the RB3LYP/6-31G** method. The coordinate axis is also shown

Table 3 Longitudinal γ (γ_{zzzz}) values [$\times 10^3$ a.u.] per monomer (γ/N, N: the number of monomers) for IDPL monomer and dimer as well as PY2 monomer and dimer models (see Fig. 27) calculated by the BHandHLYP/6-31G* method [76]

N	IDPL dimer	PY2
1	2284[$]$	194[$]$
2	4539[$]$	224[$]$

[$] The γ values of IDPL systems are calculated by the UBHandHLYP/6-31G* method, while those of PY2 systems are done by the RBHandHLYP/6-31G* method

case of the pure diradical state, where the radical electron is well localized on each atom, and is hard to contribute to polarization over the two atoms. Such remarkable spin state dependence of γ is also a characteristic for diradicaloids with intermediate diradicla character.

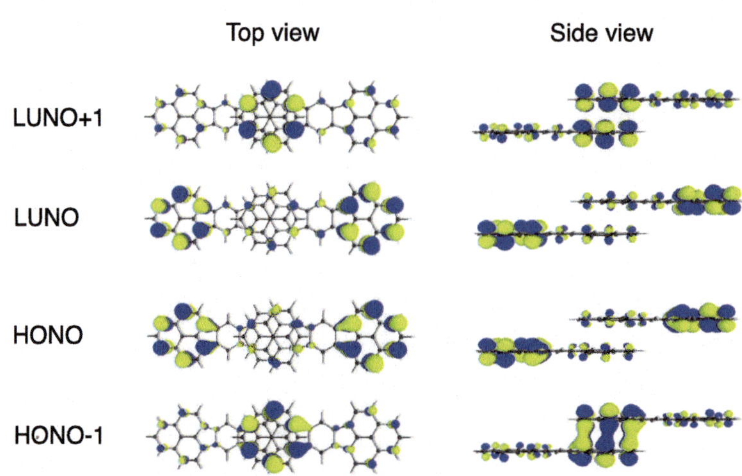

Fig. 28 UNO frontier orbitals (HONO-1, HONO, LUNO, and LUNO + 1) of the IDPL dimer model. The *yellow and blue surfaces* represent positive and negative MOs with iso-surfaces with ±0.025 a.u., respectively

The IDPL is known to make a needle-like one-dimensional (1D) crystal, where the monomer is arranged in a slipped stack form (see Fig. 27) [10]. This exhibits an unusually short π–π distance (3.137 Å) less than a typical van der Waals distance (3.4 Å), large conductivity, as well as an absorption peak shifted extraordinarily to the low-energy region [10]. These features stem from the resonance structures of intra- and inter-molecular interactions of the unpaired electrons in 1D chain. In this section, we show the effects of intra- and inter-molecular interactions on the longitudinal γ of IDPL and of its dimer in relation to their average diradical character. We examine a symmetric dimer model using the structure of IDPL monomer optimized by the UB3LYP/6-31G** method and the experimental interplanar distance of 3.137 Å [76]. As a reference, closed-shell dimer composed of PY2 (interplanar distance = 3.4 Å) is also considered. Table 3 lists the longitudinal BHandHLYP/6-31G* γ values per monomer (γ/N, N: the number of monomers) of IDPL monomer and dimer as well as PY2 monomer and dimer models. The interaction-induced increase ratio $r = \gamma(\text{dimer})/[2 \times \gamma(\text{monomer})]$ is found to be larger for IDPL system ($r = 1.99$) than for PY2 system ($r = 1.15$). This significant increase is predicted to be caused by the covalent-like intermolecular interaction with open-shell singlet nature and intramolecular intermediate diradical interaction. Figure 28 shows that HONO and LUNO exhibit dominant distributions at the both-end phenalenyl rings in the dimer, while the HONO-1 and LUNO + 1 have dominant distributions at the cofacial phenalenyl rings in the middle region of the dimer. Since the occupation numbers of HONO (HONO-1) and LUNO (LUNO + 1) are related to the diradical character y_0 (y_1), the primary diradical interaction for y_0 (0.898) occurs between both end phenalenyl rings in the dimer, while that concerning y_1 (0.770) is between cofacial phenalenyl rings in the middle region of the dimer. This relative amplitude of the diradical characters is understood by the fact that a pair of radicals with larger intersite distance gives larger y values.

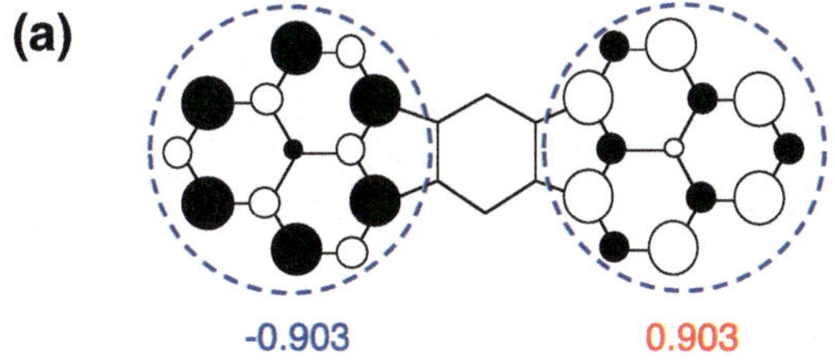

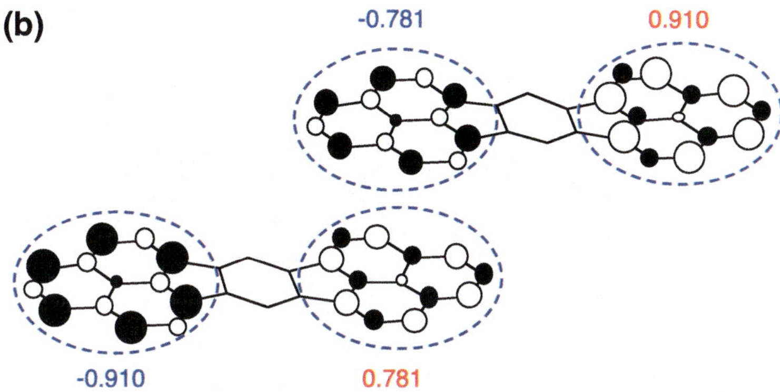

Fig. 29 Mulliken spin densities of IDPL monomer and dimer at UBHandHLYP/6-31G* level of approximation. The *white and black circles* represent α and β spin densities, respectively

In order to investigate further the spatial contribution of the diradical character, let us consider the spin polarization using Mulliken spin density distributions of the monomer and dimer of IDPL. From Fig. 29, both the monomer and the dimer, the primary α and β spin density distributions are separated into right- and left-hand side phenalenyl ring regions, respectively, though spin polarizations are observed in phenalenyl rings. For the dimer case, the spin density distributions on each cofacial phenalenyl ring (0.781) is smaller than that of the monomer (0.903), while that of the end-phenalenyl rings (0.910) is larger. This demonstrates a strong covalent-like intermolecular interaction, which leads to the smaller y_1 (0.508) concerning the cofacial phenalenyl rings of the dimer than the y_0 (0.770) of the monomer. Finally, we clarify the origin of the significant interaction-induced γ increase in IDPL dimer by using the γ density analysis (Fig. 30). As shown in Fig. 25, the large γ(monomer) value stems from the extended positive and negative π-electron γ densities well-separated on the left- and right-hand side phenalenyl rings, respectively. This separation is also observed for each monomer building the dimer, while the $\rho^{(3)}(\mathbf{r})$ amplitude on the cofacial phenalenyl rings get smaller, leading therefore to positive and negative $\rho^{(3)}(\mathbf{r})$ difference on the left and right monomers, respectively. This feature corresponds to the suppression of the spin polarization between cofacial phenalenyl rings (Fig. 29) and the decrease in diradical character y_1 as compared to y_0 of the monomer. As a result, the strong covalent-like interaction (with

        Springer

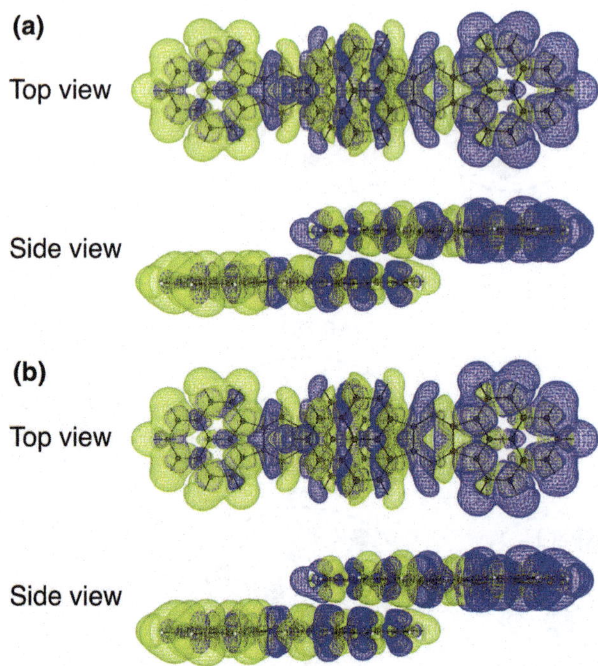

Fig. 30 γ (γ_{zzzz}) density distributions of the dimer (**a**) as well as γ density difference ($\rho_{\text{diff}}^{(3)}(\mathbf{r}) = \rho_{\text{int}}^{(3)}(\mathbf{r}) - \rho_{\text{non-int}}^{(3)}(\mathbf{r})$) (**b**) for the IDPL dimer model. The *yellow and blue meshes* represent positive and negative densities with iso-surface ± 500 a.u., respectively

intermediate diradical character) between the unpaired electrons of the cofacial phenalenyl rings provides a significant interaction-induced increase of γ for the dimer, which is exemplified by the field-induced virtual charge transfer between both end phenalenyl rings of the dimer (Fig. 30). Recently, such covalent-like interaction between phenalneyl radicals in the real 1D column of π-stacked phenalenyl aggregate is also found to exhibit strongly enhanced γ amplitude in the stacking direction, and is predicted to very large macroscopic γ value, which is comparable to that of polyacetylene [77].

8.2 Graphene Nanoflakes (GNFs)

8.2.1 Rectangular GNFs

First, we consider rectangular graphene nanoflake PAH[X,Y], where X and Y denote the number of fused rings in the zigzag and armchair edges, respectively. It is well known that the open-shell character strongly depends on the edge shape and architectures of PAHs [1, 31, 78–103]. For example, in rectangular GNFs having zigzag and armchair edges, the spin polarization or odd electron density distributions appear only on the zigzag edges, while not on the armchair edges. This suggests that there is a strong correlation between the edge shape, diradical character y and γ values. Namely, in square PAH[3,3] (Fig. 31), the γ_{yyyy} (diagonal

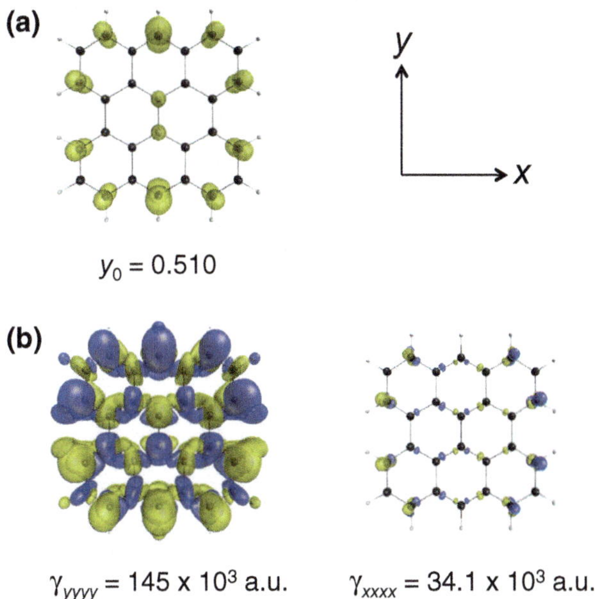

$y_0 = 0.510$

$\gamma_{yyyy} = 145 \times 10^3$ a.u. $\gamma_{xxxx} = 34.1 \times 10^3$ a.u.

Fig. 31 Odd electron (**a**) and γ (γ_{xxxx} and γ_{yyyy}) density (**b**) distributions of PAH[3,3] (singlet) at UBHandHLYP/6-31G* level of approximation. y_0 (PUHF/6-31G*) and γ values are shown with Cartesian axes. *Yellow* and *blue meshes* indicate positive and negative densities (with contours of 0.003 a.u. for odd electron density and ± 200 a.u. for γ density), respectively

component along its armchair edge) is expected to be larger than γ_{xxxx} (diagonal component along its zigzag edge) due to the spin polarization between the mutually facing zigzag edges. In order to verify this prediction, we calculated the γ components and their density distributions at UBHandHLYP/6-31G* level in its singlet and triplet states [93]. As seen from Fig. 31, PAH[3,3] has odd electron density distributions on the zigzag edges and an intermediate diradical character ($y_0 = 0.510$), the feature of which leads to a relative enhancement of γ_{xxxx} (34.1×10^3 a.u.) as compared to γ_{yyyy} (145×10^3 a.u.). These results are in qualitative agreement with our prediction. Namely, the γ is enhanced for the direction joining the radical sites. It is also found that a significant reduction of γ_{yyyy} occurs by changing the spin state from the singlet ($\gamma_{yyyy} = 145 \times 10^3$ a.u.) to the triplet ($\gamma_{yyyy} = 35.0 \times 10^3$ a.u.) state due to the Pauli effect, while a negligible difference in γ_{xxxx}. This is also an evidence that PAH[X,Y] belongs to the intermediate dirdadical character based NLO systems. Such edge shape effects are also observed in hexagonal GNFs (HGNFs) of similar size: the zigzag-edged form (Z-HGNF) presents intermediate tetraradical characters ($y_0 = y_1 = 0.410$) in contrast to its armchair-edged analogue (A-HGNF) (closed-shell) [94]. This difference is understood by the primary contributing resonance structures together with Clar's sextet rule (see Fig. 32): for Z-HGNF, the number of Clar's sextets in the closed-shell form (12), diradical form (12) and tetraradical form (12) are equal to each other, which implies that diradical and tetraradical contributions also exist with an equal weight to the closed-shell contribution, while for A-HGNF, only the

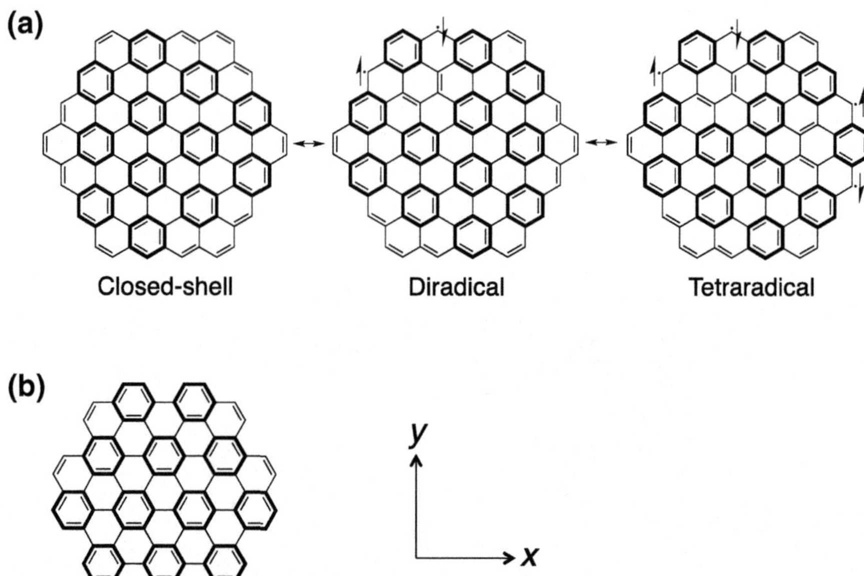

Fig. 32 Primary resonance structures Clar's sextets (*solid benzene rings*) in resonance structures of Z-HGNF (**a**) and of A-HGNF (**b**)

closed-shell form contribution exist due to the largest number of Clar's sextets. Reflecting these open-shell characters of these systems, the $\gamma_{xxxx} = \gamma_{yyyy}$ values of Z-HGNF (139×10^4 a.u.) are shown to be more than three times as large as those of A-HGNF (41.7×10^4 a.u.).

The size dependences of the diradical characters y_i ($i = 0, 1$) for rectangular GNFs PAH[X,Y], as well as the γ_{xxxx} and γ_{yyyy} values are shown in Table 4 [95, 96]. The y_0 and y_1 values are shown to increase with X with keeping Y constant and also with Y with keeping X constant. Also, the increase of y_0 value precedes that in y_1 value, i.e., the y_1 begins to significantly increase after the y_0 is close to 1. This feature indicates that for zigzag-edged rectangular GNFs, the extension of π-conjugation enhances the open-shell character, leading to a multiradical state beyond the diradical state. For example, PAH[7,7] presents the significant multiple diradical characters, $y_0 = 1.000$ and $y_1 = 0.899$, which implies that this system exhibits a nearly pure tetraradical singlet nature. Next, we consider the multiradical character (y_0, y_1) effects on the γ_{xxxx} and γ_{yyyy} values (see Table 4; Fig. 33). The variations of these quantities for PAH[X,Y] with $Y = 1, 3, 5,$ and 7 are plotted as functions of X [95]. For all the systems, the γ_{yyyy} values exhibit non-monotonic X dependences, though the γ_{xxxx} monotonically increase with X. The maximum γ_{yyyy} value for each family (PAH[X,1], PAH[X,3], PAH[X,5], and PAH[X,7] ($1 \leq X \leq 7$)) is obtained at the intermediate y_0 value, i.e., $y_0 = 0.559$ (PAH[1,6]), $y_0 = 0.510$ (PAH[3,3]), $y_0 = 0.372$ (PAH[2,5]), and $y_0 = 0.487$ (PAH[2,7]). In addition, we observe the γ enhancement in the intermediate diradical character y_0

Table 4 Diradical characters y_0 and y_1 [-] (at PUHF/6-31G* level) and γ_{xxxx} and γ_{yyyy} values [$\times 10^4$ a.u.] (at UBHandHLYP/6-31G* level) for PAH[X,Y] ($1 \leq X \leq 7$, $1 \leq Y \leq 7$) in the singlet states [95, 96]

X	Y	y_0	y_1	γ_{yyyy}	γ_{xxxx}
1	1	0.000	0.000	0.032	0.032
2	1	0.050	0.016	0.305	0.090
3	1	0.149	0.022	1.52	0.160
4	1	0.282	0.034	7.48	0.561
5	1	0.419	0.068	19.8	0.730
6	1	0.559	0.118	47.3	0.748
7	1	0.696	0.183	99.3	0.667
1	3	0.037	0.012	0.051	1.73
2	3	0.217	0.022	0.409	4.30
3	3	0.510	0.053	3.41	14.5
4	3	0.806	0.137	15.0	7.33
5	3	0.922	0.259	41.2	5.51
6	3	0.972	0.407	79.3	7.33
7	3	0.995	0.560	123	9.29
1	5	0.069	0.014	0.068	11.5
2	5	0.372	0.057	0.928	80.3
3	5	0.808	0.100	5.36	50.0
4	5	0.953	0.233	21.9	25.2
5	5	0.989	0.420	49.9	40.5
6	5	0.999	0.623	82.5	57.2
7	5	0.999	0.790	152	56.0
1	7	0.094	0.026	0.086	37.3
2	7	0.487	0.114	1.13	316
3	7	0.921	0.152	7.08	102
4	7	0.999	0.304	27.5	88.8
5	7	0.999	0.534	53.1	197
6	7	1.000	0.763	100	215
7	7	1.000	0.899	220	147

region, as well as in the intermediate second diradical character y_1 region. The PAHs with $Y \geq 3$ exhibit the second γ_{yyyy} peak with increasing X, which occurs at $(y_0, y_1) = (\sim 1, 0.623)$ for PAH[5,6] and $(\sim 1, 0.763)$ for PAH[6,7] and $(\sim 1, 0.560)$ for PAH[3,7]. These features indicate that the y–γ correlation holds for y_0, as well as for y_1, although the γ_{yyyy} peak at intermediate y_1 value is smaller than that at intermediate y_0 value. This reduced enhancement for intermediate y_1 value is predicted to be associated with the corresponding larger excitation energy (originating form the smaller HOMO$-1 -$ LUMO $+ 1$ gap) than that concerning the intermediate y_0. Such multiradical character effect on γ is furthermore clarified by using the odd electron and γ density analyses. The odd electron densities for y_0 and y_1, and γ_{yyyy} density distributions of PAH[3,3] and PAH[6,7] are show in Fig. 34. For both systems, π-electrons give primary positive and negative γ_{yyyy} densities. The density distributions are well separated to around the bottom- and top-edges, respectively, and they rapidly decrease in amplitude toward the center

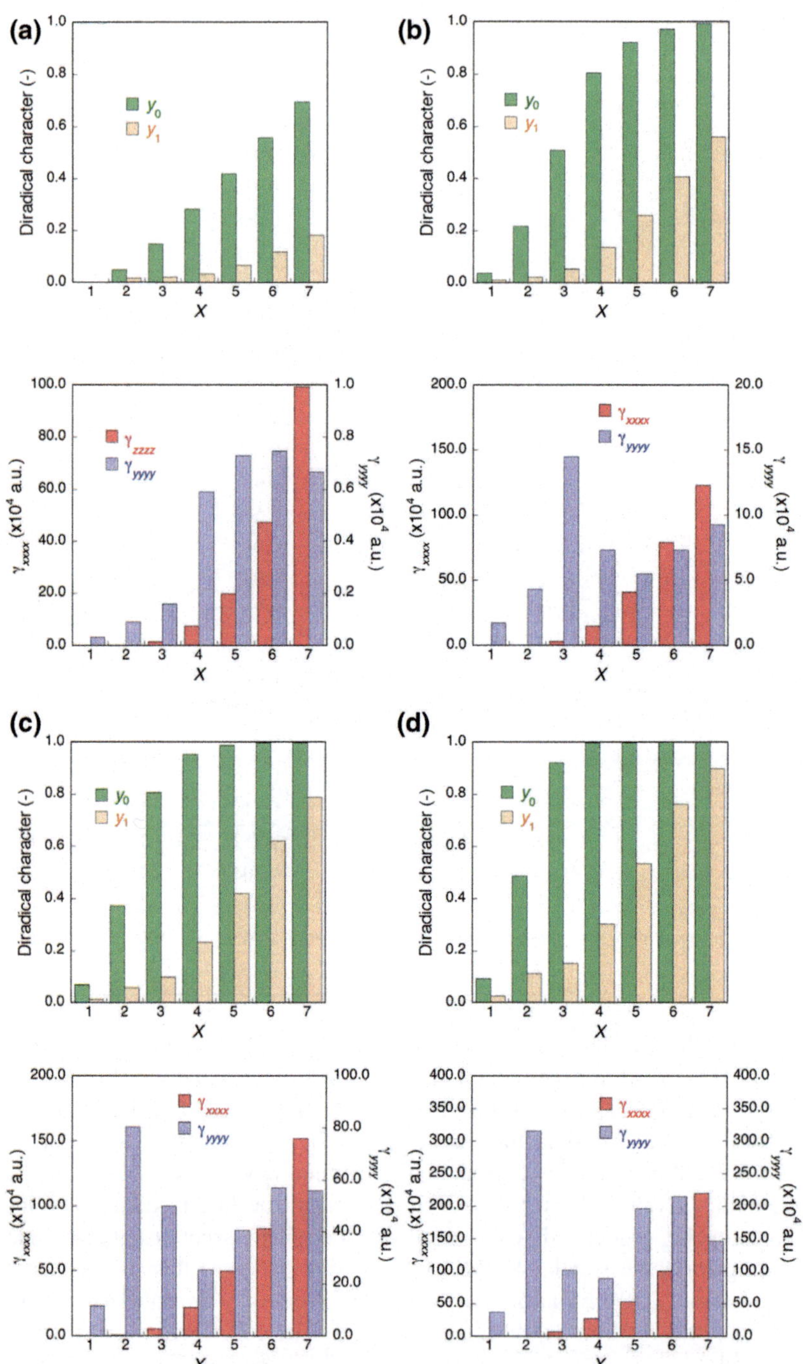

Fig. 33 Variation in diradical character y_i ($i = 0, 1$) (at PUHF/6-31G* level) and γ (γ_{xxxx} and γ_{yyyy}) (at ASP-UBHandHLYP/6-31G* level) for PAH[X,Y] with $Y = 1$ (**a**), 3 (**b**), 5 (**c**), and 7 (**d**) as a function of X

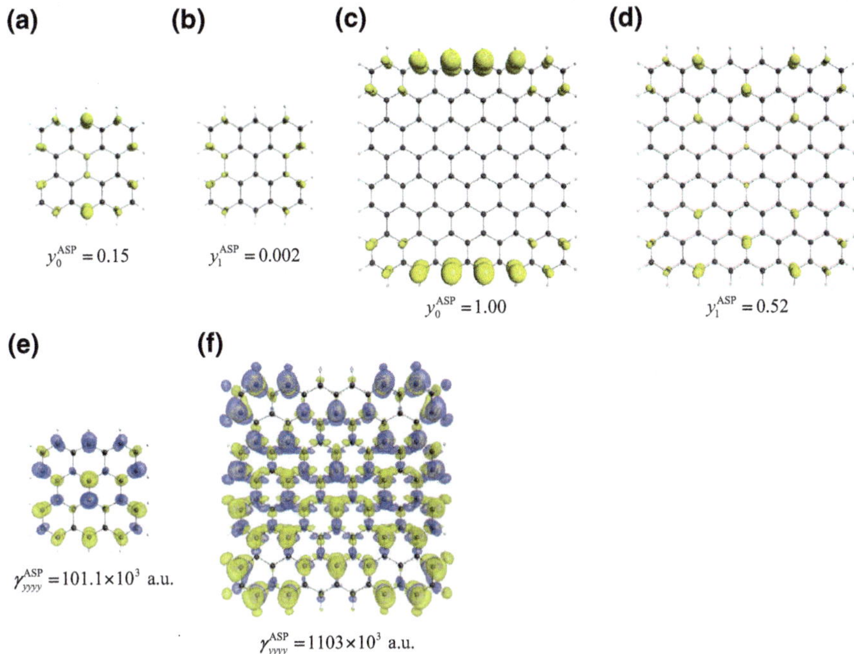

Fig. 34 Odd electron density [$D_{y_0}^{\text{odd}}(\mathbf{r})$ and $D_{y_1}^{\text{odd}}(\mathbf{r})$] distributions (at ASP-LC-UBLYP ($\mu = 0.47$)/6-31G* level) with the iso-surfaces of 0.0015 a.u. for $D_{y_0}^{\text{odd}}(\mathbf{r})$ of PAH[3,3] (**a**), 0.0003 a.u. for $D_{y_1}^{\text{odd}}(\mathbf{r})$ of PAH[3,3] (**b**), and 0.0025 a.u. for $D_{y_0}^{\text{odd}}(\mathbf{r})$ (**c**) and $D_{y_1}^{\text{odd}}(\mathbf{r})$ (**d**) of PAH[6,7]. γ_{yyyy} density [$\rho_{yyy}^{(3)}(\mathbf{r})$] distributions (at ASP-LC-UBLYP ($\mu = 0.47$)/6-31G* level) (**e**, **f**), where the *yellow* and *blue meshes* represent positive and negative densities with iso-surface $\pm$ 300 a.u., respectively

region. For PAH[3,3], the γ_{yyyy} densities are primarily distributed in the middle region of the both zigzag edges, which coincide with the region with large odd electron density distributions for y_0. In contrast, the primary amplitudes of γ_{yyyy} densities for PAH[6,7] are located in the end zigzag-edge region, i.e., four-corner phenalenyl blocks, and this distribution feature is in good agreement with the odd electron density distribution for y_1. These results correspond to the fact that γ_{yyyy} of PAH[3,3] is determined by the intermediate y_0 value, while that of PAH[6,7] by the intermediate y_1 value since the $y_0 \sim 1$ reduces the HONO–LUNO contribution. In summary, the first and the second γ_{yyyy} peaks for PAH[X,Y], which appear at intermediate y_0 and y_1 values, respectively, are evidences of the multiradical effect (tetraradical in this case) on γ.

8.2.2 Rhombic and Bow-Tie GNFs

Another structural dependences of open-shell character is introduced for two types of GNFs composed of two phenalenyl rings, rhombic (a) and bow-tie (b) [97], where the two phenalenyl units are linked in a different manner, i.e., linked via their sides (a) and vertices (b) (see Fig. 35). Although both systems have open-shell singlet states, the diradical character (at PUHF/6-31G* level) for rhombic GNF ($y_0 = 0.418$) is found to

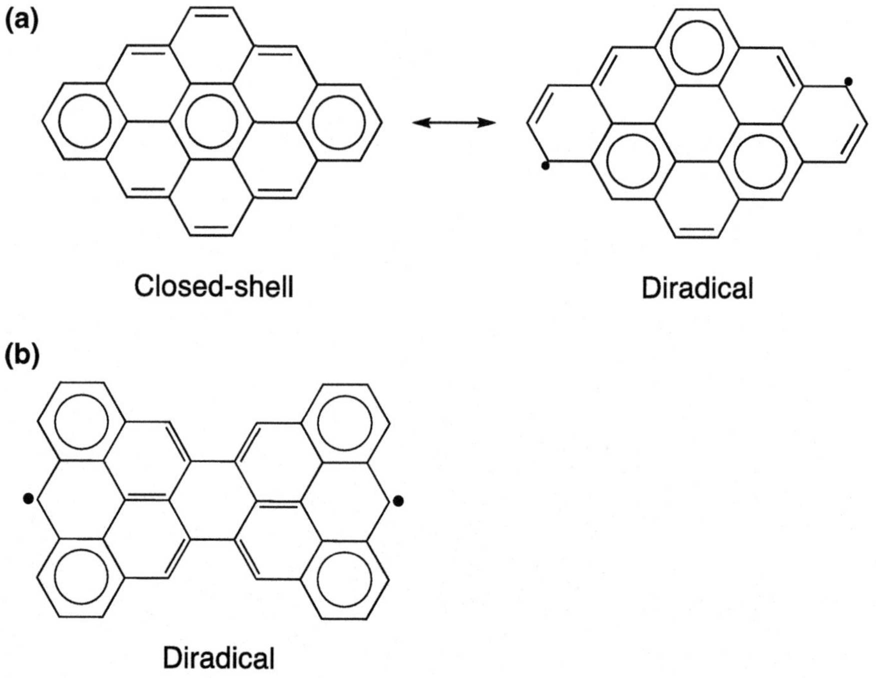

Fig. 35 Resonance forms of rhombic (**a**) and bow-tie (**b**) GNFs

be smaller than that for bow-tie GNF ($y_0 = 0.970$). This difference indicates that there are unique structural dependences of open-shell characters of GNFs based on the linked form, and this is different from the conventional simple π-conjugation size dependence of NLO properties. This structural dependence can be understood based on the resonance structures of these GNFs (Fig. 35): the rhombic GNF shows both closed-shell and diradical resonance forms, while there are no closed-shell forms for the bow-tie GNFs. It is also found that the number of Clar's sextets in the diradical form of rhombic GNF is the same as that in the closed-shell form, while that it is less than that of diradical bow-tie GNF. This predicts that for the rhombic GNF, the thermal stabilities of the closed-shell and diradical forms are similar to each other, resulting in its intermediate diradical character.

8.2.3 One-Dimensional (1D) GNFs Composed of Phenalenyl Radical Units

As another example of linked form dependence of open-shell character and γ, we consider two types of GNFs composed of phenalenyl radiclal units, alternately linked (AL) and nonalternately linked (NAL) systems shown in Fig. 36, i.e., singlet and highest spin states of 1D conjugated systems composed of the linked N phenalneyl units [$2 \leq N$ (even number) ≤ 10] [98, 99]. We first examine the two-unit systems. As seen from the resonance structures with Clar's sextet rule, the AL system has both contributions from closed-shell and diradical forms, while the NAL system from only diradical form. Indeed, from Lieb's theorem [104] and

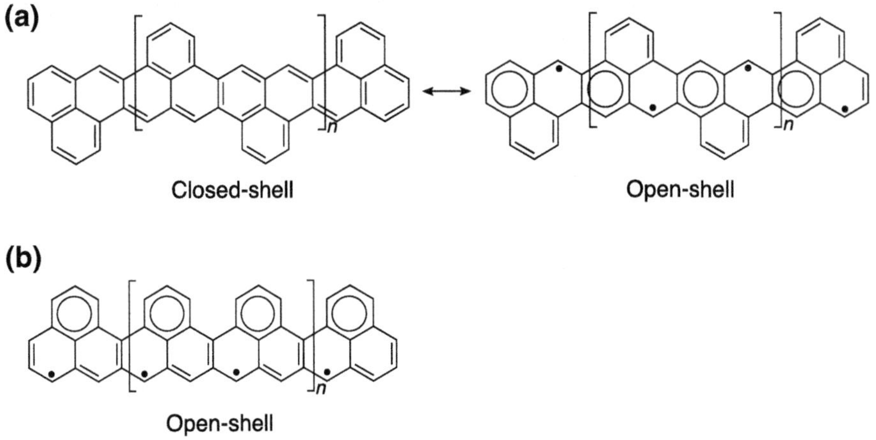

(a)

Closed-shell

Open-shell

(b)

Open-shell

Fig. 36 Structures of alternately linked (AL) (**a**) and nonalternately linked (NAL) (**b**) systems together with their primary resonance structures

Ovchinnikov's rule [105], the AL (NAL) system has the singlet (the highest) spin state as the ground state. Also, this difference is understood by the difference in the HOMO–LUMO gap, which is predicted by the interaction between the phenalenyl SOMOs [99]. As a result, the two-unit AL (2-AL) and 2-NAL exhibit $y_0 = 0.101$ and 0.919, respectively, at LC-UBLYP/6-31G* level of approximation. Reflecting these diradical characters, in the singlet states, the γ_{xxxx} of 2-AL (30.7 × 10^4 a.u.) is more than twice as large as that of 2-NAL (13.4 × 10^4 a.u.). Furthermore, when changing from singlet to triplet, the both γ amplitudes are significantly reduced [γ_{xxxx}(triplet)/γ_{xxxx}(singlet) = 0.33 (2-AL) and 0.57 (2-NAL)], and the ratio of γ_{xxxx} amplitude of 2-AL to that of 2-NAL system is also reduced [2.29 (singlet) → 1.33 (triplet)].

Next, we consider the size dependence of y_i and γ_{xxxx} for N-AL and N-NAL systems [99]. As seen from Fig. 37, the diradical characters of AL systems show slower increases with N than NAL systems, while y_i values for both systems show a systematic increase, which appears in the order of increasing i. As a result, the relatively large size systems such as ten-unit AL system exhibit intermediate multiple diradical characters, i.e., intermediate multiradical nature, e.g., for 10-AL, $y_0 = 0.630$, $y_1 = 0.383$, $y_2 = 0.236$, $y_3 = 0.153$, $y_4 = 0.122$. In contrast, in NAL systems, the y_i values show abrupt transitions from almost 0 to almost 1 at every addition of a pair of units, due to the almost negligible orbital interaction. As a result, NAL systems are pure multiradical systems regardless of the number of units.

Figure 38 shows the size dependences of γ for AL and NAL systems in their singlet and highest spin states. Remarkable differences in the amplitudes and size dependences are observed between those systems. Namely, the AL systems show much larger γ values together with much stronger size-dependent enhancement than the corresponding NAL systems. In particular, the γ value of the 10-AL system in the singlet state (13300 × 10^4 a.u.) is more than 30 times as large as that of the analogous NAL system (392 × 10^4 a.u.) and the $\gamma(N = 10)/\gamma(N = 2)$ ratio for the singlet AL systems attains 433, which is more than ten times larger than that for the

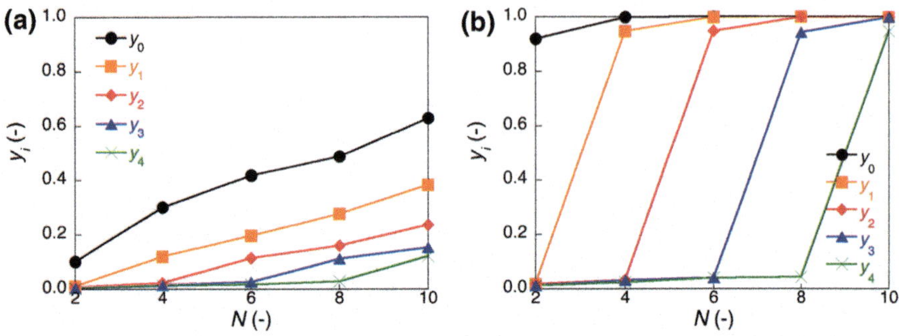

Fig. 37 Size dependences of the y_i values $(0 \leq i \leq 4)$ in N-AL (**a**) and N-NAL systems (**b**) $(2 \leq N \leq 10)$, where $N = 2n + 2$. The y_i values are calculated from the LC-UBLYP/6-31G* NO occupation numbers

Fig. 38 Size dependences of the longitudinal γ values in N-AL and N-NAL systems $(2 \leq N \leq 10)$ in their singlet and highest spin $[(N + 1)$-multiplet] states. The γ values are calculated using the LC-UBLYP/6-31G* method

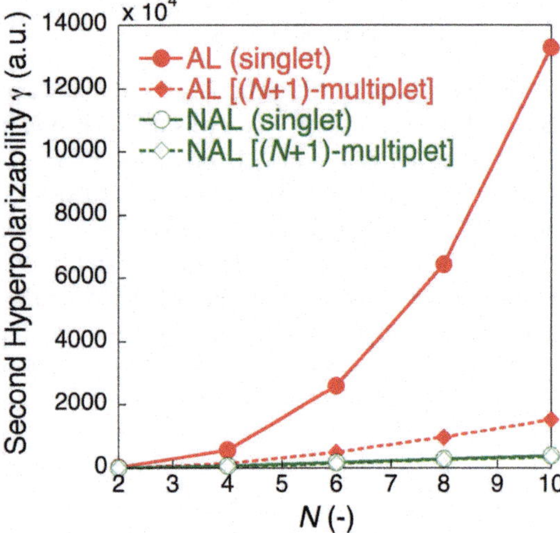

singlet NAL systems (38.4). In contrast, all AL systems exhibit a significant spin state dependence: the change from singlet to the highest spin states significantly reduce the γ_{xxxx} values, e.g., 88% reduction at 10-AL system, whereas such change is negligible for NAL systems, e.g., 7% reduction at 10-NAL. As a result, in the highest spin state, size dependence of γ is significantly reduced for NAL, e.g., $\gamma_{xxxx}(N = 10)/\gamma_{xxxx}(N = 2)$ attains 151 for AL versus 48 for NAL. Such spin state dependence indicates that the large γ_{xxxx} enhancement rate with N in singlet AL systems originates in their intermediate multiple diradical characters.

8.3 Asymmetric Open-Shell Singlet Systems

As shown in Sect. 4, asymmetric open-shell molecular systems have a potential for exhibiting further enhancement of amplitudes of hyperpolarizabilities by tuning the diradical character and asymmetricity [38, 39, 45, 106, 107]. On the other hand,

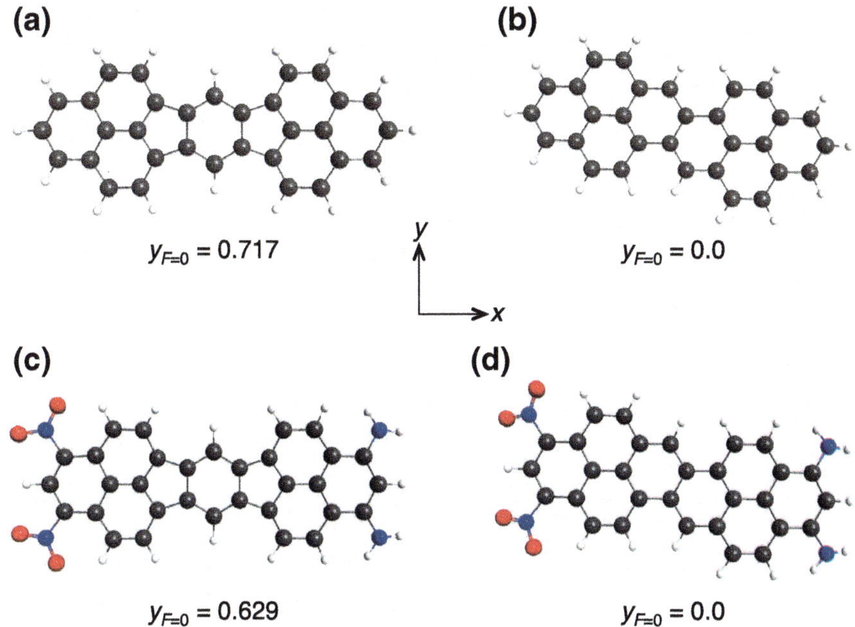

Fig. 39 Molecular structures (*grey* carbon, *blue* nitrogen, *red* oxygen, *white* hydrogen) of IDPL (**a**), PY2 (**b**), DA-IDPL [$(NO_2)_2$-IDPL-$(NH_2)_2$] (**c**) and DA-PY2 [$(NO_2)_2$-PY2-$(NH_2)_2$] (**d**) (in absence of static electric field) optimized by the (U)B3LYP/6-31G* method. The diradical characters ($y_{F=0}$) in absence of a field calculated by the LC-UBLYP/6-31G* method are also shown

there have been few realistic asymmetric systems with open-shell character, so that the molecular design and synthesis of such systems have been eagerly anticipated. Here, we show two types of asymmetric open-shell molecules, where the

Fig. 40 Static electric field F effect on γ_{xxxx} [a.u.] of IDPL and PY2 at LC-UBLYP/6-31G* level of theory

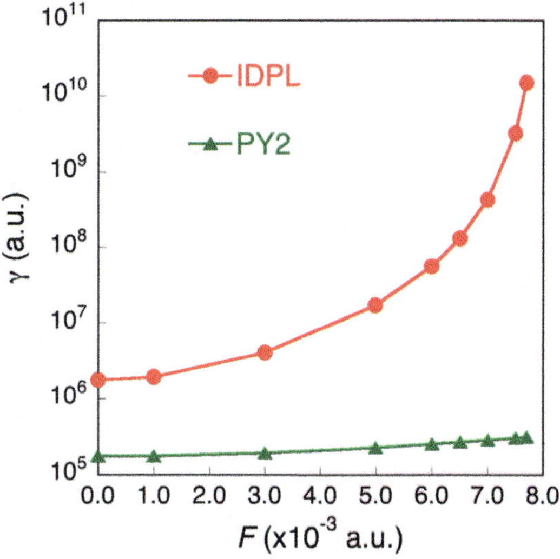

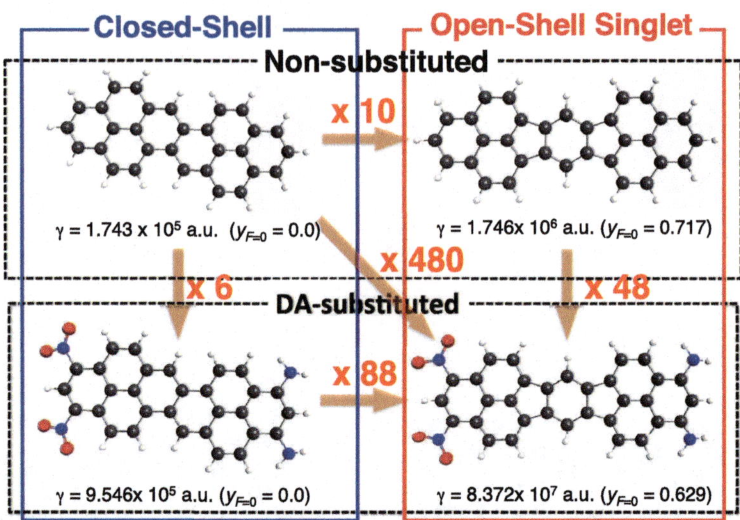

Fig. 41 Comparison of γ values and diradical characters between nonsubstituted PY2 (closes-shell) and IDPL (diradicaloid), and donor (NH_2)-acceptor (NO_2) disubstituted PY2 and IDPL

asymmetricity is induced by a static electric field application, and by donor/acceptor substitution.

First, we consider static electric field application to IDPL and PY2 (Fig. 39a, b) [106], the former and the latter of which are diradicaloid and closed-shell systems, respectively. The amplitude of the static electric field along the longitudinal (x) direction ranges from 0.0 to 0.0077 a.u. (0.0–0.4 V/Å). The diradical character in the absence of the static field, $y_{F=0} = 0.717$ (IDPL), is found to decrease as increasing the field amplitude, e.g., the y_F (IDPL) amplitude goes down to 0.293 at $F = 0.0077$ a.u. This is caused by the field-induced relative increase of the ionic component in the ground state of IDPL. Indeed, at $F = 0.0077$ a.u., the charge transfer (CT) occurs from the right- to the left-hand side of IDPL and then leads to the x component of the ground-state dipole moment ($\mu_{g\,x} = 9.88$ a.u.), which is significantly larger than that in the pyrene rings of PY2 ($\mu_{g\,x} = 5.76$ a.u.). On the other hand, the IDPL spin densities are reduced and asymmetrized. The evolution of γ of IDPL and PY2 for the static fields ranging from 0.0 to 0.0077 a.u. is shown in Fig. 40. The γ is much larger in IDPL than in PY2 in the whole F region. The γ values of both systems increase with F, e.g., $\gamma = 1.746 \times 10^6$ a.u. (IDPL) vs. 1.743×10^5 a.u. (PY2) at $F = 0.0$ a.u. and $\gamma = 1.456 \times 10^{10}$ a.u. (IDPL) vs. 3.107×10^5 a.u. (PY2) at $F = 0.0077$ a.u., and the enhancement ratio, γ(IDPL)/γ(PY2), increases with F, e.g., γ(IDPL)/γ(PY2) = 10 ($F = 0.0$ a.u.) vs. 4.7×10^4 ($F = 0.0077$ a.u.).

Second, we consider donor (NH_2)-acceptor (NO_2) substitution into PAHs, i.e., DA-IDPL [$(NO_2)_2$-IDPL-$(NH_2)_2$] and DA-PY2 [$(NO_2)_2$-PY2-$(NH_2)_2$] (Fig. 39c, d) [106]. Such chemical modification is expected to achieve a similar situation to the static field application. Indeed, these substituted systems exhibit CTs from the right- to the left-hand side though the amounts of CT of DA-IDPL (0.162) and DA-PY2

(0.153) are smaller than those [0.427(IDPL) and 0.270(PY2)] at $F = 0.0077$ a.u. From comparing the γ values between substituted (Fig. 41) and field-application (Fig. 40) systems, the two donor–acceptor pairs substitution cause an effect on γ comparable to a field of approximately 0.0060–0.0065 a.u.: the γ of DA-IDPL becomes 8.373×10^7 a.u., in comparison to $\gamma = 5.501 \times 10^7$–$1.278 \times 10^8$ a.u. at $F = 0.0060$–0.0065 a.u., respectively. However, electric field amplitudes of 0.0077 a.u. or smaller are not enough to reproduce the γ value of DA-PY2 (9.546×10^5 a.u.), which indicates that the relationship between the effects of D/A pairs and external electric field is not universal and depends on the nature of the linker. Nevertheless, the γ value of DA-IDPL (8.373×10^7 a.u.) is about 88 times larger than that of DA-PY2 (9.546×10^5 a.u.), the ratio of which is strongly enhanced as compared to the non-substituted case at $F = 0$: $\gamma(\text{IDPL})/\gamma(\text{PY2}) = 10$.

In summary, it is found that there is a gigantic enhancement of γ by applying an electric field (F) along the spin polarization direction to polycyclic aromatic diradicaloids with intermediate diradical character. Indeed, for IDPL, the enhancement with respect to the field-free case attains four orders of magnitude by applying an electric field of 0.0077 a.u., while a similar-size closed-shell analogue PY2 shows a weak field effect. Similar effects are achieved when substituting both end phenalenyl rings of IDPL by donor (NH_2)/acceptor (NO_2) groups. In this case, DA-IDPL also exhibits a γ value more than two orders of magnitude larger than in the reference closed-shell PY2. Furthermore, the diradical character in this open-shell singlet system is reduced due to either the application of an electric field or the substitution by donor/acceptor groups. This behavior is an advantage towards improved thermal stability. The present results demonstrate that the introduction of asymmetricity into the open-shell molecular systems provide a new design guideline for further enhancement/tuning of the NLO responses.

9 Experimental Estimation of Diradical Character

As shown in previous sections, the diradical character is not an observable index, but a chemical/physical index for bond nature and electron correlation. From the experimental side, the diradical nature is qualitatively estimated by using the molecular structure, e.g., quinoid vs. benzenoid forms, optical absorption spectrum, singlet–triplet energy gap, and so on. However, the quantitative evaluation scheme of the diradcial character based on the experimental measurements has not been proposed yet. We have firstly proposed an approximate evaluation scheme of the diradical character y using the relationships between the excitation energies and y value for two-site VCI diradical model, Eqs. 22, 23, 24b, 25b, and 27 [108]. The diradical character y can be expressed by [108]

$$y = 1 - \sqrt{1 - \left(\frac{{}^1E_{1u} - {}^3E_{1u}}{{}^1E_{2g} - {}^1E_{1g}}\right)^2} = 1 - \sqrt{1 - \left(\frac{\Delta E_{S(u)} - \Delta E_T}{\Delta E_{S(g)}}\right)^2} \quad (64)$$

where the energies in the first right-hand side (rhs) are concerned with the four electronic states shown in Sect. 3.1. $\Delta E_{S(g)} (\equiv {}^1E_{2g} - {}^1E_{1g})$, $\Delta E_{S(u)} (\equiv {}^1E_{1u} - {}^1E_{1g})$

Fig. 42 The diradical character (y^{expr}) deduced from experimetal results versus that theoretical one (y^{theor}) obtained from PUHF/6-31G** calculations for several compounds [108]

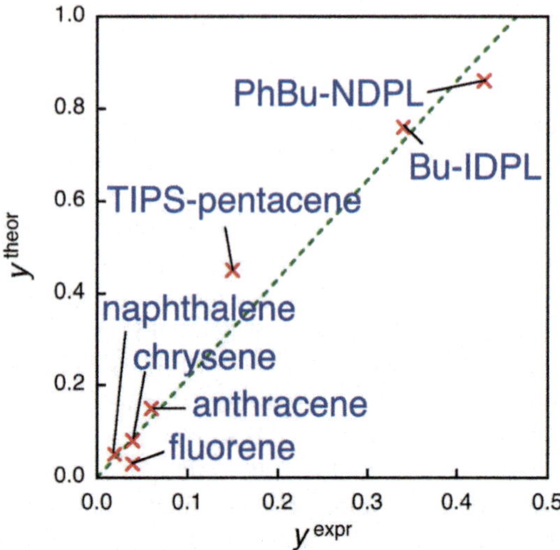

and $\Delta E_T (\equiv {}^3E_{1u} - {}^1E_{1g})$ in the second rhs represent the excitation energies of the higher singlet state of g symmetry (two-photon allowed excited state), of the lower singlet state with u symmetry (one-photon allowed excited state), and of the triplet state with u symmetry, respectively. Namely, $\Delta E_{S(u)}$ and $\Delta E_{S(g)}$ are obtained from the lowest-energy peaks of the one- and two-photon absorption spectra, respectively, while ΔE_T is obtained from phosphorescence and ESR measurement. Figure 42 shows the experimental y (y^{exp}) values and theoretically calculated ones (y^{theor}). A strong correlation between the experimental and theoretical results for all compounds is observed in spite of the difference in the scales originating from several factors, e.g., including the effects of the environment (solvation or crystal packing), inconsistencies among the different experimental methods, and the approximate nature of the VCI model [7]. As a result, the theoretical relationship between the diradical character and measurable quantities (Eq. 64) is found to provide semiquantitative estimates (after scaling) for the diradical characters

10 Summary

This chapter focus on the theoretical aspects of electronic structures of open-shell molecular systems including symmetric and asymmetric di/multi-radicaloids and open-shell molecular aggregates from the viewpoint of diradical character. Also, on the basis of the two-electron two-site valence configuration interaction model, the excitation energies and properties (dipole moment differences and transition moments), as well as nonlinear optical (NLO) responses are found to be strongly correlated to the diradical character. In particular, we found (a) that the first optically allowed excitation energy tends to decrease and then increase as increasing the dirdaical character and (b) that the second hyperpolarizability (third-order NLO property at the molecular scale) is maximized in the intermediate diradical character

Fig. 43 Diradical character based molecular design for functional molecules

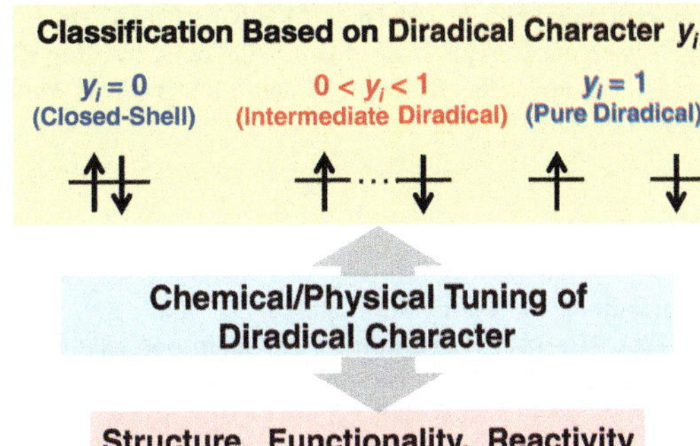

region. These properties are useful for building the design guidelines for tuning the optical absorption spectra and NLO properties including two-photon absorption. Several realistic molecules and aggregates are designed based on these principles, and their electronic structures, as well as functionalities are analyzed by first-principles quantum chemical calculations. Recently, some of our designed molecules and a new class of stable open-shell singlet systems have been synthesized and their unique properties and high functionalities are confirmed by experiments [3, 9, 12–15, 109–116]. Our design principle for functional systems based on the open-shell character is illustrated in Fig. 43, which indicates that each category classified based on the diradical character, i.e., closes-shell, intermediate open-shell, and pure open-shell regimes, characterizes the functionalities (structure, reactivity, and property). If we could clarify the relationship between the diradical character and these functionalities, we could construct design guidelines for controlling such functionalities. In this point, this strategy has advantage over conventional one because the diradical character is an chemical index for effective bond nature in the ground state and can be easily connected to conventional chemical concepts, e.g., resonance structure and aromaticity, which most chemists are familiar with. Indeed, the second hyperpolarizability mentioned in this chapter is described, in principle, by excitation energies, transition properties and dipole moment differences for all the electronic states including ground and excited states, which are very complicated and are hard to be used for constructing useful design guidelines. Our strategy can clarify the relationship between hyperpolarizability and diradical character, and thus succeeds in presenting familiar design guidelines for enhancing/tuning the hyperpolarizbaility by tuning the molecular architecture, symmetry, aromaticity, substitution of atom species and so on based on the resonance structures and Clar's sextet rule. Namely, such diradical character based method is useful for revealing the mechanism of the functionality and providing its simple physicochemical pictures, as well as for constructing practical molecular design guidelines for various functionalities including NLO properties, singlet fission and exciton migration [16, 18, 117, 118].

Acknowledgements This work has been supported by JSPS KAKENHI Grant Number JP25248007 in Scientific Research (A), Grant Number JP24109002 in Scientific Research on Innovative Areas "Stimuli-Responsive Chemical Species", Grant Number JP15H00999 in Scientific Research on Innovative Areas "π-System Figuration", and Grant Number JP26107004 in Scientific Research on Innovative Areas "Photosynergetics". This is also partly supported by King Khalid University through a grant RCAMS/ KKU/001-16 under the Research Center for Advanced Materials Science at King Khalid University, Kingdom of Saudi Arabia.

References

1. Bendikov M, Duong HM, Starkey K, Houk KN, Carter EA, Wudl F (2004) J Am Chem Soc 126:7416–7417 (**Erratum: J Am Chem Soc 126: 10493**)
2. Bettinger HF (2010) Pure Appl Chem 82:905–915
3. Sun Z, Ye Q, Chi C, Wu J (2012) Chem Soc Rev 41:7857–7889
4. Nakano M, Champagne B (2015) J Phys Chem Lett 6:3236–3256
5. Nakano M, Champagne B (2016) WIREs Comput Mol Sci 6:198–210
6. Nakano M, Kishi R, Nitta T, Kubo T, Nakasuji K, Kamada K, Ohta K, Champagne B, Botek E, Yamaguchi K (2005) J Phys Chem A 109:885–891
7. Nakano M, Kishi R, Ohta S, Takahashi H, Kubo T, Kamada K, Ohta K, Botek E, Champagne B (2007) Phys Rev Lett 99:033001-1–4
8. Nakano M, Yoneda K, Kishi R, Takahashi H, Kubo T, Kamada K, Ohta K, Botek E, Champagne B (2009) J Chem Phys 131:114316-1–7
9. Kamada K, Ohta K, Kubo T, Shimizu A, Morita Y, Nakasuji K, Kishi R, Ohta S, Furukawa S, Takahashi H, Nakano M (2007) Angew Chem Int Ed 46:3544–3546
10. Kubo T, Shimizu A, Sakamoto M, Uruichi M, Yakushi K, Nakano M, Shiomi D, Sato K, Takui T, Morita Y, Nakasuji K (2005) Angew Chem Int Ed 44:6564–6568
11. Lambert C (2011) Angew Chem Int Ed 50:1756–1758
12. Sun Z, Wu J (2012) J Mater Chem 22:4151–4160
13. Sun Z, Zeng Z, Wu J (2013) Chem-Asian J 8:2894–2904
14. Abe M (2013) Chem Rev 113:7011–7088
15. Sun Z, Zeng Z, Wu J (2014) Acc Chem Res 47:2582–2591
16. Nakano M (2014) Excitation Energies and Properties of Open-Shell Singlet Molecules, Springer International Publishing
17. Kubo T (2015) Chem Lett 44:111–122
18. Nakano M (2017) Chem Rec 17:27–62
19. Clar E (1972) The Aromatic Sextet. Wiley, New York
20. Hachmann J, Dorando JJ, Avilés M, Chan GK-L (2007) J Chem Phys 127:134309-1–9
21. Casanova D, Head-Gordon M (2009) Phys Chem Chem Phys 11:9779–9790
22. Hajgató B, Huzak M, Deleuze MS (2011) J Phys Chem A 115:9282–9293
23. Plasser F, Pasalic H, Gerzabek MH, Libisch F, Reiter R, Burgdörfer J, Müller T, Shepard R, Lischka H (2013) Angew Chem Int Ed 52:2581–2584
24. Chakraborty H, Shukla A (2013) J Phys Chem A 117:14220–14229
25. Motomura S, Nakano M, Fukui H, Yoneda K, Kubo T, Carion R, Champagne B (2011) Phys Chem Chem Phys 13:20575–20583
26. Fukuda K, Nagami T, Fujiyoshi J, Nakano M (2015) J Phys Chem A 119:10620–10627
27. Hayes EF, Siu AKQ (1971) J Am Chem Soc 93:2090–2091
28. Yamaguchi K (1975) Chem Phys Lett 33:330–335
29. Yamaguchi K (1990) In: Carbo R, Klobukowski M (eds) Self-Consistent Field: Theory and Applications. Elsevier, Amsterdam, The Netherlands, pp 727–828
30. Head-Gordon M (2003) Chem Phys Lett 372:508–511
31. Nakano M, Fukui H, Minami T, Yoneda K, Shigeta Y, Kishi R, Champagne B, Botek E, Kubo T, Ohta K, Kamada K (2011) Theoret Chem Acc 130:711–724 (**erratum (2011) 130:725**)
32. Nakano M, Champagne B, Botek E, Ohta K, Kamada K, Kubo T (2010) J Chem Phys 133:154302-1–15
33. Yamanaka S, Okumura M, Nakano M, Yamaguchi K (1994) J Mol Structure (Theochem) 310:205–218
34. Kitagawa Y, Saito T, Nakanishi Y, Kataoka Y, Matsui T, Kawakami T, Okumura M, Yamaguchi K (2009) J Phys Chem A 113:15041–15046
35. Calzado CJ, Cabrero J, Malrieu JP, Caballol R (2002) J Chem Phys 116:2728–2747

36. Minami T, Ito S, Nakano M (2013) J Phys Chem A 117:2000–2006
37. Kishi R, Nakano M (2011) J Phys Chem A 115:3565–3575
38. Nakano M, Champagne B (2013) J Chem Phys 138: 244306-1-13
39. Nakano M, Fukuda K, Champagne B (2016) J Phys Chem C 120:1193–1207
40. Special Issue on Optical Nonlinearities in Chemistry (1994) Burland D, ed. Chem Rev 94:1–278
41. Nalwa HS, Miyata S (eds) (1997) Nonlinear Optics of Organic Molecules and Polymers. CRC Press, Boca Raton, FL
42. Papadopoulos MG, Sadlej AJ, Leszczynski J (eds) (2006) Nonlinear optical properties of matter—from molecules to condensed phases. Springer, Dordrecht
43. Nakano M, Yamaguchi K (1993) Chem Phys Lett 206:285–292
44. Nakano M, Shigemoto I, Yamada S, Yamaguchi K (1995) J Chem Phys 103:4175–4191
45. Fukuda K, Suzuki Y, Matsui H, Nagami T, Kitagawa Y, Champagne B, Kamada K, Yamamoto Y, Nakano M (2017) Chem Phys Chem 18:142–148
46. Chen Z, Wannere CS, Corminboeuf C, Puchta R, Schleyer PVR (2005) Chem Rev 105:3842–3888
47. Kertesz M, Choi CH, Yang S (2005) Chem Rev 105:3448–3481
48. Rosenberg M, Dahlstrand C, Kilså L, Ottosson H (2014) Chem Rev 114:5379–5425
49. Chase DT, Rose BD, McClintock SP, Zakharov LN, Haley MM (2011) Angew Chem Int Ed 50:1127–1130
50. Shimizu A, Tobe Y (2011) Angew Chem Int Ed 50:6906–6910
51. Fix AG, Deal PE, Vonnegut CL, Rose BD, Zakharov LN, Haley MM (2013) Org Lett 15:1362–1365
52. Shimizu A, Kishi R, Nakano M, Shiomi D, Sato K, Takui T, Hisaki I, Miyata M, Tobe Y (2013) Angew Chem Int Ed 52:6076–6079
53. Thomas S, Kim K (2014) Phys Chem Chem Phys 16:24592–24597
54. Ruud L, Helgaker T, Bal KL, Jørgensen P, Jensen HJA (1993) J Chem Phys 99:3847–3859
55. Cohen HD, Roothaan CCJ (1965) J Chem Phys 43:S34–S39
56. Fukuda K, Nagami T, Fujiyoshi J, Nakano M (2015) J Phys Chem A 119:10620–10627
57. Nakano M, Nitta T, Yamaguchi K, Champagne B, Botek E (2004) J Phys Chem A 108:4105–4111
58. van Gisbergen SJA, Schipper PRT, Gritsenko OV, Baerends EJ, Snijders JG, Champagne B, Kirtman B (1999) Phys Rev Lett 83:694–697
59. Iikura H, Tsuneda T, Yanai T, Hirao K (2001) J Chem Phys 115:3540–3543
60. Tawada Y, Tsuneda T, Yanagisawa S, Yanai T, Hirao K (2004) J Chem Phys 120:8425–8433
61. Song J, Hirosawa T, Tsuneda T, Hirao K (2007) J Chem Phys 126:154105-1-7
62. Kamiya M, Sekino H, Tsuneda T, Hirao K (2005) J Chem Phys 122:234111–1–10
63. Sekino H, Maeda Y, Kamiya M, Hirao K (2007) J Chem Phys 126.014107-1-6
64. Peach MJG, Cohen AJ, Tozer DJ (2006) Phys Chem Chem Phys 8:4543–4549
65. Jacquemin D, Perpète EA, Medved M, Scalmani G, Frisch MJ, Kobayashi R, Adamo C (2007) J Chem Phys 126:191108-1-4
66. Kishi R, Bonness S, Yoneda K, Takahashi H, Nakano M, Botek E, Champagne B, Kubo T, Kamada K, Ohta K, Tsuneda T (2010) J Chem Phys 132:094107-1-11
67. Bonness S, Fukui H, Yoneda K, Kishi R, Champagne B, Botek E, Nakano M (2010) Chem Phys Lett 493:195–199
68. Refaely-Abramson S, Baer R, Kronik L (2011) Phys Rev B 84:075144–1–8
69. Kishi R, Nakano M, Ohta S, Takebe A, Nate M, Takahashi H, Kubo T, Kamada K, Ohta K, Botek E, Champagne B (2007) J Chem Theory Comput 3:1699–1707
70. Nakano M, Minami T, Fukui H, Yoneda K, Shigeta Y, Kishi R, Champagne B, Botek E (2010) Chem Phys Lett 501:140–145
71. Takatsuka K, Fueno T, Yamaguchi K (1978) Theor Chim Acta 48:175–183
72. Staroverov VN, Davidson ER (2000) Chem Phys Lett 330:161–168
73. Kubo T, Shimizu A, Uruichi M, Yakushi K, Nakano M, Shiomi D, Sato K, Takui T, Morita Y, Nakasuji K (2007) Org Lett 9:81–84
74. Nakano M, Kubo T, Kamada K, Ohta K, Kishi R, Ohta S, Nakagawa N, Takahashi H, Furukawa S, Morita Y, Nakatsuji K, Yamaguchi K (2006) Chem Phys Lett 418:142–147
75. Ohta S, Nakano M, Kubo T, Kamada K, Ohta K, Kishi R, Nakagawa N, Champagne B, Botek E, Takebe A, Umezaki S, Nate M, Takahashi H, Furukawa S, Morita Y, Nakasuji K, Yamaguchi K (2007) J Phys Chem A 111:3633–3641
76. Nakano M, Takebe A, Kishi R, Fukui H, Minami T, Kubota K, Takahashi H, Kubo T, Kamada K, Ohta K, Champagne B, Botek E (2008) Chem Phys Lett 454:97–104

77. Salustro S, Maschio L, Kirtman B, Rérat M, Dovesi R (2016) J Phys Chem A 120:6756–6761
78. Fujita M, Wakabayashi K, Nakada K, Kusakabe K (1996) J Phys Soc Jpn 65:1920–1923
79. Jiang DE, Dai S (2008) J Phys Chem A 112:332–335
80. Jiang DE, Sumpter BG, Dai S, J Chem Phys 127:124703-1–5
81. Chen Z, Jiang DE, Lu X, Bettinger HF, Dai S (2007) Schleyer PvR. Houk KN 9:5449–5452
82. Son YW, Cohen ML, Louie SG (2006) Phys Rev Lett 97:216803-1–4
83. Hod O, Barone V, Peralta JE, Scuseria GE (2007) Nano Lett 7:2295–2299
84. Hachmann J, Dorando JJ, Avilés M, Chan GKL (2007) J Chem Phys 127:134309-1–9
85. Pisani L, Chan JA, Montanari B, Harrison NM (2007) Phys Rev B 75:064418-1–9
86. Castro Neto AH, Guinea F, Peres HMR, Novoselov KS, Geim AK (2009) Rev Mod Phys 81:109–162
87. Yazyev OV, Wang WL, Meng S, Kaxiras E (2008) Nano Lett 8:766–
88. Wang WL, Meng S, Kaxiras E (2008) Nano Lett 8:241–245
89. Ezawa M (2008) Physica E 40:1421–1423
90. Ezawa M (2007) Phys Rev B 76: 245415-1–6
91. Fernándz-Rossier J, Palacios JJ (2007) Phys Rev Lett 99:177204-1–4
92. Dias JR (2008) Chem Phys Lett 467:200–203
93. Nakano M, Nagai H, Fukui H, Yoneda K, Kishi R, Takahashi H, Shimizu A, Kubo T, Kamada K, Ohta K, Champagne B, Botek E (2008) Chem Phys Lett 467:120–125
94. Nagai H, Nakano M, Yoneda K, Fukui H, Minami T, Bonness S, Kishi R, Takahashi H, Kubo T, Kamada K, Ohta K, Champagne B, Botek E (2009) Chem Phys Lett 477:355–359
95. Nagai H, Nakano M, Yoneda K, Kishi R, Takahashi H, Shimizu A, Kubo T, Kamada K, Ohta K, Botek E, Chanpagne B (2010) Chem Phys Lett 489:212–218
96. Yoneda K, Nakano M (2016) Chap 27 in Graphene Science Handbook edited by Mahmood A, Ali N, Milne WI, Ozkan CS, Mitura S, Gervasoni JL, CRC Press: pp 437–455
97. Yoneda K, Nakano M, Kishi R, Takahashi H, Shimizu A, Kubo T, Kamada K, Ohta K, Champagne B, Botek E (2009) Chem Phys Lett 480:278–283
98. Nakano M, Kishi R, Takebe A, Nate M, Takahashi H, Kubo T, Kamada K, Ohta K, Champagne B, Botek B (2007) Comput Lett 3:333–338
99. Yoneda K, Nakano M, Fukui H, Minami T, Shigeta Y, Kubo T, Botek E, Champagne B (2011) ChemPhysChem 12:1697–1707
100. Yoneda K, Nakano M, Inoue Y, Inui T, Fukuda K, Shigeta Y, Kubo T, Champagne B (2012) J Phys Chem C 116:17787–17795
101. Konishi A, Hirao Y, Matsumoto K, Kurata H, Kishi R, Shigeta Y, Nakano M, Tokunaga K, Kamada K, Kubo T (2013) J Am Chem Soc 135:1430–1437
102. Konishi A, Hirao Y, Kurata H, Kubo T, Nakano M, Kamada K (2014) Pure Appl Chem 86:497–505
103. Yoneda K, Matsui H, Fukuda K, Takamuku S, Kishi R, Nakano M (2014) Chem Phys Lett 595–596:220–225
104. Lieb EH (1989) Phys Rev Lett 62:1201–1204
105. Ovchinnikov AA (1978) Theor Chim Acta 47:297–304
106. Nakano M, Minami T, Yoneda K, Muhammad S, Kishi R, Shigeta Y, Kubo T, Rougier L, Champagne B, Kamada K, Ohta K (2011) J Phys Chem Lett 2:1094–1098
107. Fukuda K, Matsushita N, Minamida Y, Matsui H, Nagami T, Takamuku S, Kitagawa K, Nakano M (2017) Chemistry Select 2:2084–2087
108. Kamada K, Ohta K, Shimizu A, Kubo T, Kishi R, Takahashi H, Botek E, Champagne B, Nakano M (2010) J Phys Chem Lett 1:937–940
109. Kamada K, Fuku-en SI, Minamide S, Ohta K, Kishi R, Nakano M, Matsuzaki M, Okamoto H, Higashikawa H, Inoue K, Kojima S, Yamamoto Y (2013) J Am Chem Soc 135:232–241
110. Ishida M, Shin JY, Lim JM, Lee BS, Yoon MC, Koide T, Sessier JL, Osuka A, Kim D (2011) J Am Chem Soc 133:15533–15544
111. Li Y, Heng WK, Lee BS, Aratani N, Zafra JL, Bao N, Lee R, Sung YM, Sun Z, Huang KW, Webster RD, Navarrete JTL, Kim D, Osuka A, Casado J, Ding J, Wu J (2012) J Am Chem Soc 134:14913–14922
112. Zeng Z, Sung YW, Bao N, Tan D, Lee R, Zafra JL, Lee BS, Ishida M, Ding J, Navarrete JTL, Li Y, Zeng W, Kim D, Huang KW, Webster RD, Casado J, Wu J (2012) J Am Chem Soc 134:14513–14525
113. Zeng Z, Ishida M, Zafra JL, Zhu X, Sung YM, Bao N, Webster RD, Lee BS, Li RW, Zeng W, Li Y, Chi C, Navarrete JTL, Ding J, Casado J, Kim D, Wu J (2013) J Am Chem Soc 135:6363–6371

114. Kishida H, Hibino K, Nakamura A, Kato D, Abe J (2010) This Solid Films pp 1028–2030
115. Quah HS, Chen W, Schreyer MK, Yang H, Wong MW, Ji W, Vittal JJ (2015) Nat Comm 6:7954-1–7
116. Takauji K, Suizu R, Awaga K, Kishida H, Nakamura A (2014) J Phys Chem C 116:5241–5246
117. Minami T, Nakano M (2012) J Phys Chem Lett 3:145–150
118. Nakano M, Ito S, Nagami T, Kitagawa Y, Kubo T (2016) J Phys Chem C 120:22803–22815

Top Curr Chem (Z) (2017) 375:83
https://doi.org/10.1007/s41061-017-0171-2

REVIEW

Benzenoid Quinodimethanes

Akihito Konishi[1] · **Takashi Kubo**[2]

Received: 6 July 2017 / Accepted: 1 October 2017 / Published online: 17 October 2017
© Springer International Publishing AG 2017

Abstract Reactivity and physical properties of π-conjugated hydrocarbon systems depend predominantly on the topology of π-electrons array. Quinoidal conjugations serve as giving diradical character to molecules, leading to unique chemical behaviors. The simplest member of quinodimethanes are *o*-, *m*-, and *p*-quinodimethanes, which are very reactive due to diradical character and cannot be isolated under normal experimental conditions. However, chemical modifications, such as π-extension or introduction of substituent groups, of quinodimethanes imparts stabilities to quinodimethanes that can be handled under ambient conditions. This chapter offers an overview of reactivity and magnetic properties of benzenoid *o*-, *m*-, and *p*-quinodimethanes.

Keywords Benzenoid · Quinodimethane · Diradical · Clar sextet · Graphene nanoribbon

---

Chapter 2 was originally published as Konishi, A. & Kubo, T. Top Curr Chem (Z) (2017) 375: 83. https://doi.org/10.1007/s41061-017-0171-2.

✉ Takashi Kubo
kubo@chem.sci.osaka-u.ac.jp

[1] Department of Applied Chemistry, Graduate School of Engineering, Osaka University, 2-1 Yamada-oka, Suita, Osaka 565-0871, Japan

[2] Department of Chemistry, Graduate School of Science, Osaka University, Toyonaka, Osaka 560-0043, Japan

1 Introduction

Quinodimethanes (QDMs) and related compounds have attracted much attention of chemists in a wide range of research fields due to their unique chemical reactivity and physical properties. QDMs can be classified into two categories. The first category includes *o*- and *p*-quinodimethanes (*o*- and *p*-QDMs), whose spin multiplicity is singlet in the ground state (Fig. 1). These molecules can be drawn by the resonance hybrid of quinoidal Kekulé and diradical forms. The second category is characterized by open-shell triplet ground state and typified by *m*-quinodimethane (*m*-QDM).

Most of the interesting phenomenological features of QDMs arise from their peculiar electronic structures in the ground state. *m*-QDM is a non-Kekulé diradical with the triplet ground state and has been extensively used as an effective ferromagnetic coupler. In contrast to *m*-QDM, the understanding of the ground-state electronic structure for *o*- and *p*-QDMs has long been elusive. Although *o*- and *p*-QDMs can be represented as quinoidal (closed-shell singlet) structures, they inherently exhibit diradical reactivity such as intramolecular valence tautomerization and intermolecular polymerization. The manifestation of the diradical character originating from the quinoidal conjugation was experimentally demonstrated by structural and spectroscopic analyses for π-extended *p*- or *o*-QDMs like Chichibabin's hydrocarbon and pleiadene. Recent extensive studies on π-extended QDMs with sufficient stability in air have demonstrated that they possess prominent diradical character in the ground state and behave as open-shell singlet molecules (biradicaloids), which have proposed the possibility of biradicaloid compounds for the use of electronic, non-linear optical and magnetic materials [1–6]. Moreover, *o*- and *p*-QDMs contain a cross-conjugated π-electron system, which can exhibit redox reactions, and upon the redox events the π-conjugation style is altered from a quinoid to benzenoid structure [7]. These peculiar electronic structures have attracted much attention that QDMs are promising frameworks for constructing unique π-conjugated system. In fact, *p*-QDM derivatives have been extensively studied as electronic and optical materials and *o*-QDM derivatives are well-known versatile reagents in synthetic chemistry due to their high reactivity. These spacious properties of QDMs, including chemical reactivity, electronic and optical properties, and magnetic behaviors, have been constantly attracting our attention.

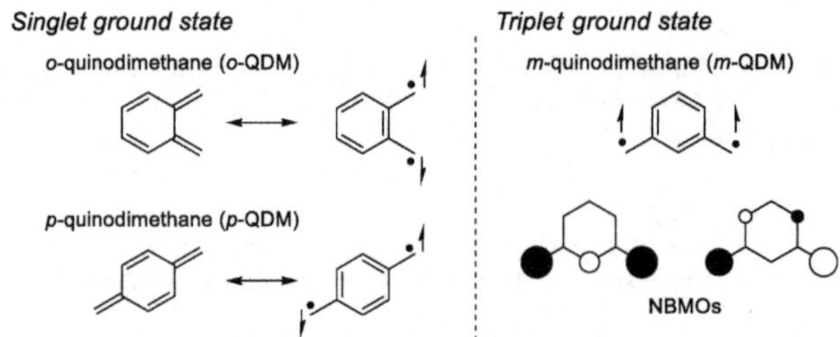

Fig. 1 Ground electronic structure of QDMs

In this chapter, we present the recent research progresses on electronic properties and chemical reactivity of benzenoid QDMs with diradical character. The ground state electronic structure of π-conjugated systems that incorporate a QDM scaffold is deeply influenced by the π-conjugation manner, varying from close-shell singlet to open-shell singlet or triplet diradical ground state, because a QDM inherently has diradical character induced by the recovery of aromaticity. Here, with the exception of some compounds, we would like to confine to benzenoid QDMs, which composed of all six-membered (benzene) rings and good models for the investigation of the origin of the diradical character of QDMs. The wide studies on the heteroanalogue compounds, such as π-extended tetracyanoquinodimethanes (TCNQs), viologenes, and oligothiophenes, have been discussed in other chapters, and will not be discussed here.

2 Brief History of QDMs

The generation of *p*-QDM **1** was first achieved in 1947 by the pyrolysis of *p*-xylene [8]. **1** is stable in gas phase, whereas labile in condensed phase to give rise to an insoluble material, poly(phenylenevinylene). Later it was found that the insoluble material also includes [2.2]paracyclophane [9] (Fig. 2). Exposure of the pyrolysis product to iodine in gas phase afforded *p*-xylylene diiodide. All these biradicaloid behaviors imply the open-shell character of **1** in the ground state, which attracted attention of theoreticians, but Coulson demonstrated, by using Hückel molecular orbital (HMO) calculation, that the ground state of **1** possesses a closed-shell quinoid form with very large free valence (that is, radical reactivity) at two terminal methylene carbons [10]. The quinoid ground state for **1** are also confirmed by ^{1}H NMR [11], UV–Vis [12], IR [12, 13], Raman [13, 14], photoelectron [15], and electron diffraction [16] measurements. On the other hand, it is unlikely that triplet diradical species causes the biradicaloid chemical behaviors, because a CASSCF(8,8)/6-31+G(d) calculation predicted the large singlet–triplet energy gap ($\Delta E_{S-T} = E_{singlet} - E_{triplet}$) of $-$ 159 kJ/mol [17]. The biradicaloid behaviors of **1** appears not to arise from the diradical electronic structure. As described below, the extension of the π-conjugated system of **1**, which started from the Thiele's [18] and Chichibabin's [19] hydrocarbons, has been exclusively investigated in terms of singlet diradical character and application to functional organic materials.

The chemistry of *o*-QDM **2** started from the interest in a reactive intermediate, and similar elusive nature, which derives from the bilateral character of quinoidal and diradical structure, has also been seen. In 1957, Cava and co-workers first suggested that an *o*-QDM derivative **2'** generates as a key intermediate in the Diels–

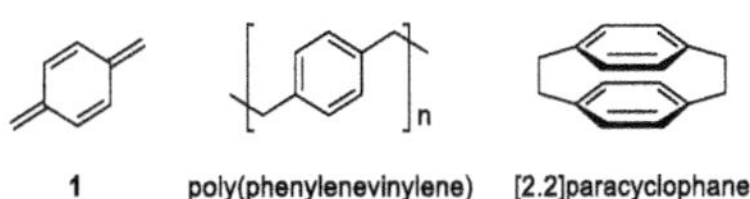

Fig. 2 *p*-QDM **1** and its polymeric and dimeric compounds

Alder reaction of α,α,α′,α′-tetrabromo-*o*-xylene to *trans*-1,2-dibromobenzocy-clobutene **3′** (Scheme 1) [20].

The reactivity of **2** as a diene was later proven by trapping experiments with a number of cyclic dienophiles. Since then, many methods for the generation of **2** and its heteroanalogues have been developed as building blocks in the synthesis of natural products and functional materials [21–23]. The generation of non-substituted **2** was first achieved in 1966 [24] as a reactive intermediate that readily dimerizes at − 150 °C to give the Diels–Alder spiro-dimer (Scheme 2a) [23, 25]. The direct evidence of the existence of **2** was confirmed by UV–Vis [26–30], fluorescence [26, 31–33], IR [28, 30], and Raman [28] measurements. In gas phase, **2** exists as an equilibrium mixture with its bicyclic valence tautomer benzocyclobutene **3** (Scheme 2b) [34, 35]. The thermodynamics and kinetics have been the subject of detailed studies. To confirm the contribution of the diradical species on the reaction pathway, Kametani et al. have estimated the stabilities for both structures of the intermediate by using MINDO/3, STO-3G, and 4-31G methods [36]. They concluded that the ring-open reaction of benzocyclobutene proceeds via a diradical intermediate by a stepwise cycloaddition, which is in agreement with the results reported by Trahanovsky who suggested that **2** dimerizes via a stepwise mechanism involving a diradical intermediate [37, 38]. Although no experimental evidence for a diradical structure of **2** in the ground state has been obtained, the diradical character would emerge on a reaction profile [26, 28, 39, 40]. As described below, the several stabilizations of **2** toward the developments of functional organic materials have been reported.

m-QDM **4** is a non-Kekulé diradical with non-disjoint type nonbonding molecular orbitals, which leads to triplet spin multiplicity in the ground state [41, 42]. Although a stable derivative of **4**, tetraphenyl-substituted Schlenk's hydrocarbon (see below) was synthesized in 1915 [43], the clean generation of the non-substituted **4** was challenging problem. In 1975, the generation of **4** was first detected by a sensitive fluorescence spectroscopy via UV irradiation of *m*-xylene [39]. However, **4** itself is a highly photo-labile species and therefore its yield is diminished by secondary photolysis, which makes it difficult to generate and detect **4** feasibly. In 2008, Neuhaus et al. accomplished the clean generation and detection of **4** from 1,3-bis(iodomethyl)benzene using flash vacuum pyrolysis (FVP) at 450–500 °C (Scheme 3a) [44]. **4** was trapped in argon matrices at 10 K and characterized by IR, UV–Vis, and ESR measurements. Irradiation of **4** results in the formation of three isomeric hydrocarbons, **5**, **6**, and **7**, which could be identified by the comparison of their IR spectra with the predictions given by DFT calculations (Scheme 3b). The triplet ground state was experimentally identified by Wright and Platz via ESR measurements [45]. The singlet–triplet energy gap ($\Delta E_{\text{S–T}}$) of **4** was

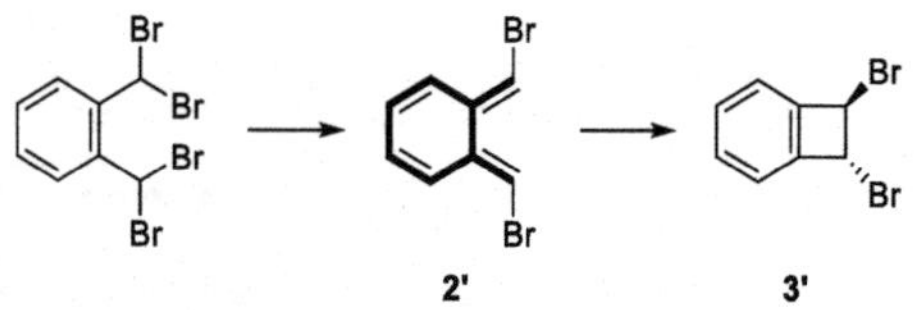

Scheme 1 Generation of the *o*-QDM derivative **2′** by Cava's group

72

Scheme 2 **a** Dimerization of *o*-QDM **2** and **b** equilibrium between *o*-QDM **2** and valence isomer **3** in gas phase

Scheme 3 **a** Generation and characterization of *m*-QDM **4** via FVP process by Neuhaus group, **b** photo reactivity of **4**, and **c** thermal reactivity of **4**

determined by Wenthold et al. to be $\sim +40.2$ kJ/mol using photoelectron spectroscopy [46], which is in good agreement with recent ab initio studies [47–49]. Detailed studies of the thermal reactivity of **4** have been carried out by Goodman and Berson [50–52]. It was found that **4**, prepared from the bis-methylenic hydrocarbon **8**, can readily react with conjugated dienes to form several adducts and afford a dimerized *m*-cyclophenes **9** (Scheme 3c).

3 π-Extended Quinodimethanes

3.1 Thiele's and Chichibabin's Hydrocarbons

The diradical nature of *p*-QDMs is deeply investigated through the study on the Thiele's **10** and Chichibabin's **11** hydrocarbons (Fig. 3a). In 1904, Thiele and Balhorn isolated **10** as the first isolable derivative of *p*-QDM [18]. Although **10** readily reacts with oxygen, a variety of physical measurements concluded that **10** has a singlet ground state [53–58]. In contrast to **10**, the examination of the Chichibabin's hydrocarbon **11** provoked a great controversy on spin multiplicity in the ground state: singlet, triplet, or double doublets. In 1907, Chichibabin isolated **11** as a blue-violet compound that reacts readily with oxygen to yield a polymeric peroxide [19]. Although magnetic susceptibility measurements indicated that **11** is diamagnetic [53], a solution ESR measurement suggested that $\sim 4\%$ of the molecule is in the paramagnetic triplet state at 300 K [59]. Another solution ESR

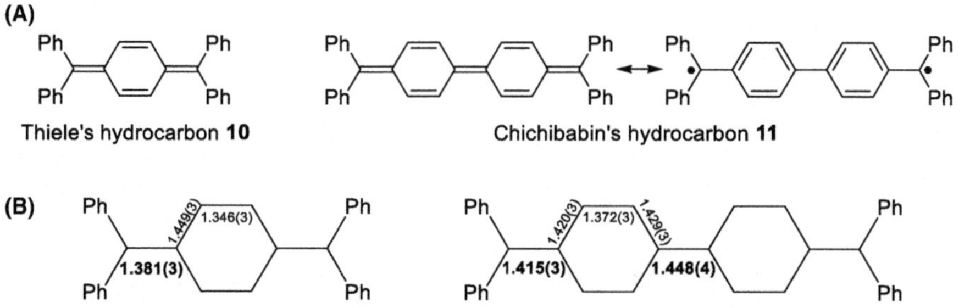

Fig. 3 **a** Thiele's **10** and Chichibabin's **11** hydrocarbons and **b** selected bond lengths of **10** and **11**

study demonstrated that the interaction between two unpaired electrons is weak due to the observation of sharp hyperfine lines in the spectrum [55, 60]. However, the weak coupling is inconsistent with the diamagnetic behavior or the small amount ($\sim$ 4%) of triplet species, evoking the issue "The Diradical Paradox" [61]. In 1964, this controversial issue was partially resolved by further ESR studies, from which it was concluded that the paramagnetic species are derived from a dimer or higher polymer of **11** [62], and in 1969 a hydrogen abstracted molecule was proposed as another candidate of the paramagnetic species on the basis of ENDOR studies by Brauer et al. [63]. Conclusive remarks were finally given by Montgomery et al. in 1986 [64]. In order to prove the spin multiplicity of **10** and **11** in the ground state, the X-ray crystal structures for **10** and **11** in a high quality were investigated. For **10**, the well-defined bond alternation is observed in its p-QDM framework (Fig. 3b). The bond lengths in the central benzene core (1.346 (3) and 1.449 (3) Å) substantially differ from that of the usual aromatic ring (1.39 Å). The *exo*-methylene double bonds are 1.381 (3) Å, which is slightly longer than those of typical olefins ($\sim$ 1.34 Å) but roughly comparable to the electron diffraction values for the non-substituted p-QDM [16]. On the other hand, the crystal structure of **11** gave unusual lengths of the *exo*-methylene double bonds (1.415 (3) and 1.448 (4) Å), which are far from the ordinal olefin double bonds and 0.03 Å longer than the corresponding bond in **10** (Fig. 3b). In addition, the bond length alternation in the biphenyl core also decreases (1.420 (3), 1.372 (3) and 1.429 (3) Å). From these structural features, it is concluded that **11** adopts a singlet quinoidal structure with an apparent manifestation of diradical character in the ground state. An ESR measurement of the pure **11**, which gave no signal at 77 K, also supported this conclusion.

3.2 Pleiadenes

Another important compound for the investigation of the singlet diradical character of QDMs is pleiadene **12**. In 1963, Cava and Schlessinger first generated **12** as a reactive intermediate by a pyrolysis of sulfone precursor (Fig. 4a) [65]. **12** could be trapped as a Diels–Alder adduct with *N*-phenylmaleimide but otherwise readily afforded a dimerize form [66]. These reactivities result from the *o*-quinoid structure, because the hydrocarbons **13** and **14** in the absence of the *o*-quinoidal moiety are

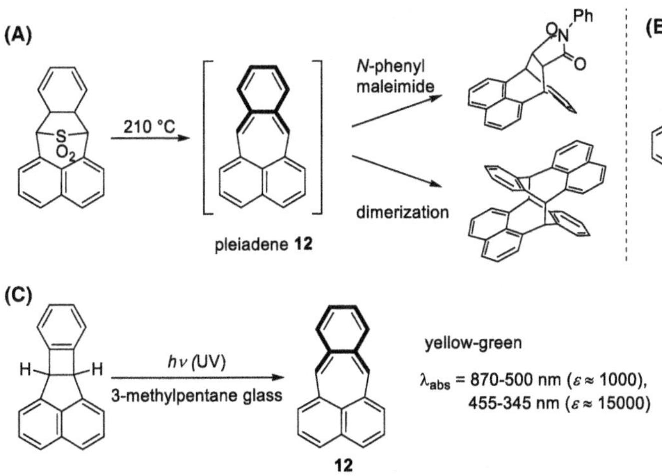

Fig. 4 **a** Cava's thermal generation of **12**, **b** stable hydrocarbons **13** and **14**, and **c** Michl's photogeneration of **12** with absorption bands in visible region

thermally stable and completely unaffected by heating with maleic anhydride at 80 °C for 8.5 days (Fig. 4b). In 1970, Michl and his co-workers successfully synthesized **12** in a milder condition, that is, photochemical generation in a low-temperature matrix [67]. Irradiation of a benzocyclobutene precursor in 3-methylpentane glass with UV light afforded a stable yellow-green **12**, which disappeared in 10–15 s at the glass-melting temperature (110 K) (Fig. 4c). The product was the dimer and the mechanism of the rapid reaction from two **12** to the dimer is quite interesting because it is formally a thermally forbidden [4 + 4]cycloaddition. A stepwise mechanism involving radical intermediates might be more plausible, but Michl proposed that the concerted mechanism is also a compelling path if the transition state bears biradicaloid character. From a quantum chemical calculation, **12** possesses a low-lying doubly excited configuration, which is closely related to singlet diradical character, and admixing of the ground state configuration with the doubly excited configuration would make the transition state lower in energy. The state dominated by the doubly excited configuration was found to be low-lying by using UV–Vis spectroscopy and a PPP-CI calculation [68, 69].

3.3 π-Extended *p*-QDMs and Related Quinones

As mentioned above, the Chichibabin's hydrocarbon **11** possesses appreciable singlet diradical character. The insertion of quinoid units into **11** would make the singlet diradical character enhanced because the recovery of aromatic stabilization

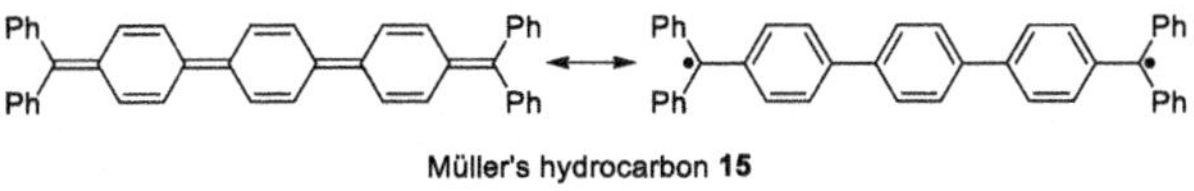

Scheme 4 Resonance formula of Müller's hydrocarbon **15**

in the diradical form increases. Indeed, tris-quinoid system, the Müller's hydrocarbon **15**, was synthesized [70] and characterized by ESR measurements (Scheme 4) [71]. **15** gave an ESR signal typical for triplet species ($|D| = 0.0035$ cm^{-1}, $|E| \leq 0.0002$ cm^{-1}). From the temperature dependency of signal intensities, **15** was concluded to have a singlet ground state, which is only ~ 4 kJ/mol stable than the lowest triplet excited state.

Benzannulation of the central quinoid moieties also makes a direct influence on the magnetic properties of **11**. Recently, Wu and his coworkers successfully synthesized and isolated the two types of tetrabenzannulated Chichibabin's hydrocarbons **16** and **17** (Scheme 5a) [72]. Their biradicaloid behaviors were thoroughly investigated by means of X-ray crystallographic analyses and spectroscopic measurements. The interesting feature of these compounds is summarized in that the electronic structure of the ground state is determined by the energy balance between a π-bond formation and an intramolecular structural strain. Due to the steric hindrance between two anthracene cores, **16** and **17** adopt not a planer form, but either an orthogonal diradical or a folded quinoid form as the most stable conformation. The orthogonal diradical **16a** was afforded by the reduction of a diol precursor with SnCl$_2$, which displayed a broad absorption spectrum with $\lambda_{max} = 834$ nm and a strong ESR signal at $g = 2.0029$. The orthogonal diradical **16a** gradually transformed into the quinoidal form **16b** with a half-life time of 415 min. The closed-shell singlet state of **16b** was confirmed by the X-ray crystallographic analysis, which showed that the *exo*-methylene double bonds (1.342 and 1.349 Å) between anthracene units are close to those in typical olefins. According to the DFT calculations at the UCAM-B3LYP/6-31G* level, the open-shell singlet state of **16a** is located above the closed-shell quinoidal singlet state of **16b** by 18.4 kJ/mol. In contrast to the stable closed-shell compound **16**, the fluorenyl-terminated derivative **17** was isolated as an only orthogonal diradical form **17a**. **17a** gave no ^{1}H NMR signals at even $- 100$ °C, and instead, an intense ESR signal at $g = 2.0027$ was observed at 153 K. From the temperature dependency of

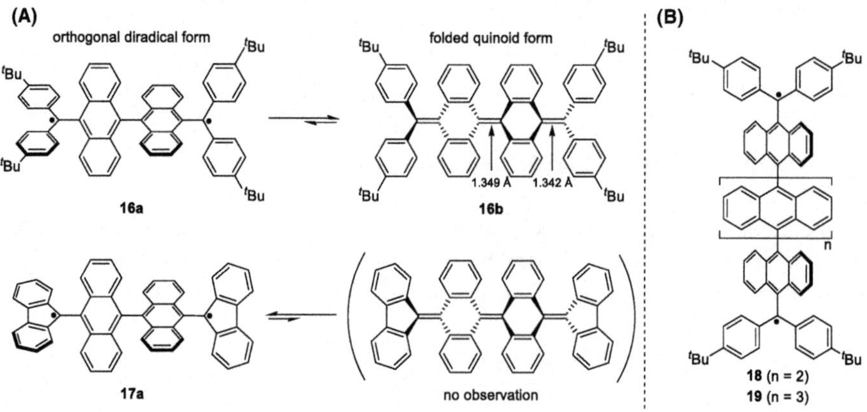

Scheme 5 **a** Thermal equilibrium of **16** and **17** between an orthogonal diradical and a folded quinoid form and **b** molecular structure **18** and **19**

magnetic susceptibility at 5–380 K, **17a** has very small ΔE_{S-T} of + 1.4 kJ/mol, suggesting that the ground state of **17a** is triplet, which is consistent with the DFT calculations (+ 2.5 kJ/mol). The difference in conformation between **16** and **17** is ascribable to the difference of steric strain around the terminal double bonds. The rigid fluorenyl moieties suffer from larger strain in the folded form, leading to the larger energy difference between the folded and the orthogonal forms. In following the same strategy for the molecular design, the more π-extended homologues of **16**, quinoidal 9,10-anthryl trimer **18** and tetramer **19**, were investigated (Scheme 5b) [73]. The synthesized **18** and **19** gave broad low-energy absorption bands centered at 813 nm and 801 nm, respectively. **18** and **19** showed clear ESR signals and the DFT calculations implied that **18** and **19** adopt an orthogonal diradical form like **17** due to the high steric repulsions between 9,10-anthryl units. The nearly isolated two unpaired electrons demonstrate that they appear to be better described as two individual radicals.

In terms of the electronic structure of biradicaloids, π-extended quinones are also attractive compounds because the incorporated quinomethide units can be represented as the resonance mixture of quinoidal and diradical forms (Fig. 5a). Biphenoquinone **20** and terphenoquinone **21** are air stable and diamagnetic compounds. The X-ray crystallographic analysis of **21** clearly showed a large bond length alternation in the quinoid framework of a nearly planar form and no detectable ESR signal in both solution and solid state was observed [74]. In contrast, larger homologues, quaterphenoquinone **22** [75, 76] and quinquephenoquinone **23** [77] gave clear ESR signals in solution and solid states, which implies that these compounds have appreciable diradical character. Because **22** and **23** appear to adopt predominantly a diamagnetic quinoid form in solid state by SQUID measurements, thermally excited triplet species may be responsible for the solid state ESR signals. On the other hand, in solution, the rotation of the quinoid moieties would result in decoupling of two unpaired electrons, giving rise to the ESR signals.

The replacement of quinoid framework with other π-conjugated systems largely affects the structure–property relationship of the quinone system (Fig. 6). Cansei et al. found that the ground state electronic structure of the bithiophene-inserted quinones **24** can be controlled by the substituents on the thiophene ring [78]. According to optical and magnetic measurements and quantum chemical calculations, the introduction of the electron-donating group (R = OMe or OC_5H_{11}) onto the thiophene ring largely stabilizes the quinoidal (closed-shell singlet) ground state, which is in sharp contrast to the non-substituted derivative (R=H) with prominent singlet diradical character ($\langle S^2 \rangle = 0.67$). Wu and his coworkers synthesized

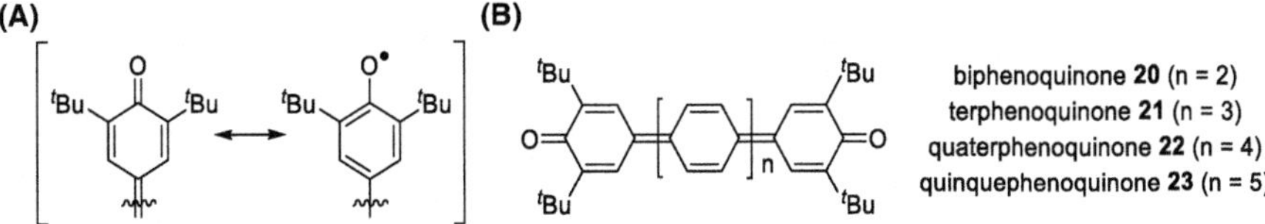

Fig. 5 **a** Resonance formula of quinomethide unit and **b** molecular structures of **20–23**

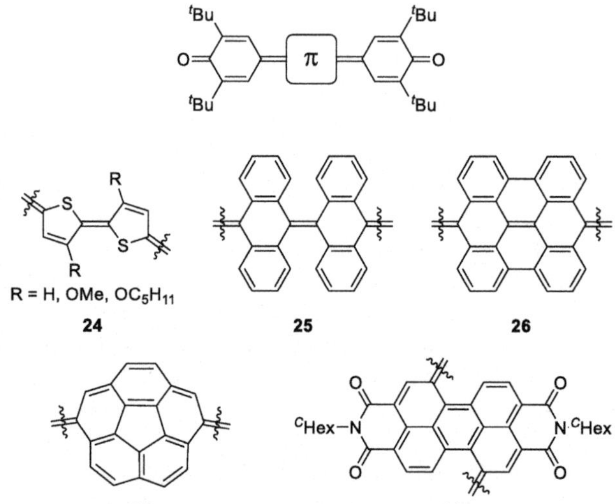

Fig. 6 π-Extended quinones **24–28**

anthracene- and bisanthene-inserted quinones, **25** and **26**, which showed folded and twisted quinoid forms, respectively [79]. Due to the rigid framework of bisanthene, **26** prefers a twisted form over a folded form. **25** and **26** showed sharp NMR signals even at elevated temperature, indicating the closed-shell ground state. In contrast, a corannulene-inserted quinone **27** showed diradical properties: a triplet ESR signal in a rigid glass and small ΔE_{S-T} of –3.4 kJ/mol [80]. A perylene bisimide-inserted quinone **28** also exhibited biradicaloid behaviors: a progressive signal sharping in the ^{1}H NMR measurement was observed upon cooling to − 93 °C [81]. Its ΔE_{S-T} was theoretically estimated to be − 3.9 kJ/mol.

Systematic investigations in terms of the relationship between the molecular geometry and singlet diradical character were carried out for thienoquinoids **29**, oligo(*N*-annulated perylene)quinodimethanes **30**, and π-extended *N*-annulated rylenes **31** (Fig. 7). For these compounds, the introduction of dicyanomethylidene at the terminal carbon stabilizes the radical character by the strong electron-withdrawing ability. Otsubo et al. prepared a series of dicyanomethylene-terminated thienoquinoids **29a–f** and revealed that **29e** and **29f** shows no NMR signal, instead, ESR signals are observed [82]. From the ESR signal intensity, the fractions of magnetic species were estimated to be 2.8% for **29e** and 29% for **29f**. According to the DFT calculations, thermally excited triplet species would give the observed ESR signals. Triplet populations estimated from theoretical ΔE_{S-T} (− 10.9 kJ/mol for **29e**, − 5.65 kJ/mol for **29f**) and a Boltzmann distribution are 3.6% for **29e**, and 24% for **29f** [83]. The contribution of the diradical form to the ground state was experimentally investigated using Raman spectroscopy [83]. Raman spectra measured at 123 K revealed that the small systems give quinoidal Raman bands, whereas the bands of the larger systems move into aromatic region. Contribution

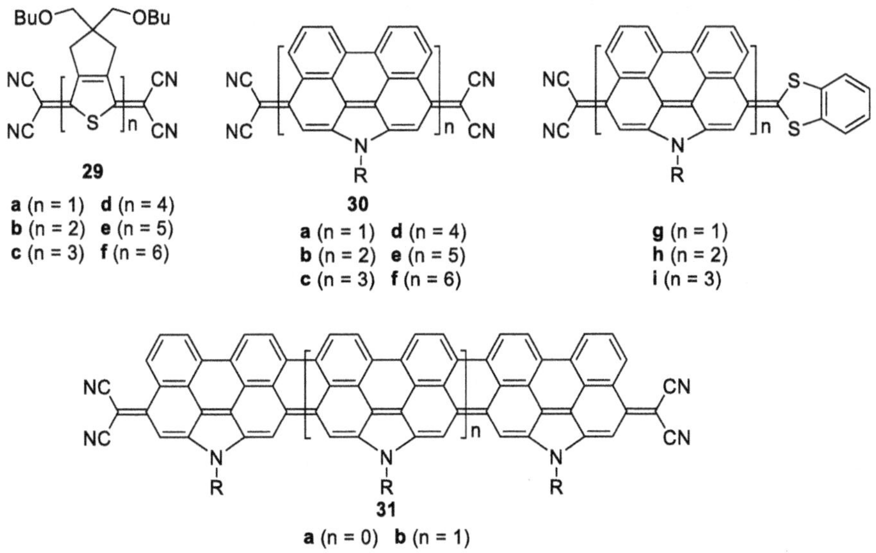

Fig. 7 π-Extended quinodimethanes 29–31

weight of the aromatic diradical form to the ground state increases with increasing the number of thienoquinoid unit due to a gain of large aromaticity in the diradical form. Wu and coworkers synthesized a series of the oligo(*N*-annulated perylene)quinodimethanes **30a–f** and investigated the electronic structure of the ground state [84]. **30** showed the tunable ground states, varying from a closed-shell quinoidal monomer **30a**, to a singlet diradical for dimer **30b**, trimer **30c** and tetramer **30d** and to a triplet diradical for pentamer **30e** and hexamer **30f**. The difference in the ground state electronic structure is ascribed to the balance between the cumulative steric repulsion among perylene units in the oligomer and the recovery of aromaticity of the quinoidal *N*-annulated perylene cores in the diradical form. Asymmetric push–pull substituted systems of **30g–i** were also investigated by the same group, in terms of the diradical and zwitterionic character of the ground state [85]. Detailed spectroscopic studies revealed that with the extension of the molecular chain length, the diradical character increased whereas the contribution of the zwitterionic form to ground state became smaller. In contrast to these twisted oligomeric perylenes **30**, extended rylenes **31** with highly planar fused structures

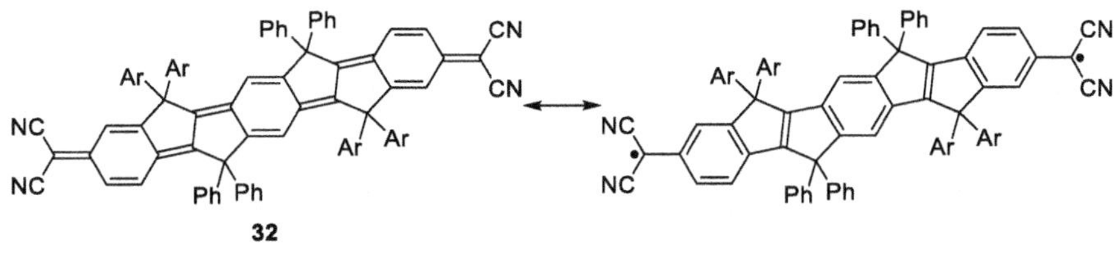

Fig. 8 Planar tri-*p*-quinodimethane system 32

gave a gradual shift of the diradical character with the extension of the rylene core size: the perylene derivative and quaterrylene derivative **31a** adopt quinoidal closed-singlet forms while the hexarylene derivative **31b** possesses an open-shell diradical singlet ground state, ΔE_{S-T} was experimentally determined to be $-$ 17.6 kJ/mol [86].

Tuji and Nakamura et al. reported the synthesis of the planar tri-*p*-quinodimethane system **32** with a distinct diradical character (Fig. 8) [87]. **32** was prepared via intramolecular cyclization of the bisacetylene precursor and obtained as an air- and heat-stable solid. The ^{1}H NMR signals of **32** gave the temperature dependent broadening, suggesting the presence of thermally excited triplet species. The singlet ground state of **32** was confirmed by SQUID measurements, which indicated that the lowest triplet state of **32** was located above the singlet ground state by 8.9 kJ/mol.

3.4 π-Extended *o*-QDMs and Related Compounds

Despite the synthetic versatilities of *o*-QDMs as reactive species, the synthesis of stable *o*-QDM derivatives are still challenging. In 1968, 7,7,8,8-tetraphenyl-*o*-QDM **33**, in which four benzene rings are attached to *exo*-methylene double bonds for kinetic stabilization, was reported by Quinkert et al. (Fig. 9a) [88]. In spite of the steric protection of four benzene rings, **33** is unstable, spontaneously to isomerize into 1,1,2,2-tetraphenylbenzocyclobutene (BCB) and 4a,10-dihydro-9,10,10-triphenylanthracene (DHA) by intramolecular cyclization. Suzuki and his coworkers found that the dibenzo-annulation of **33** surpassed the cyclization, because dibenzo-annulation would destabilize the corresponding BCB and DHA isomers by additional steric congestions [89, 90]. They successfully isolated the dibenzo-annulated derivatives **34** which possess a helical conformation due to steric repulsions between aryl groups. The detailed studies on the redox behaviors of **34** were carried out: **34** is readily oxidized into stable dication spices upon electron transfer and its helical molecular framework in both neutral and cation states exhibited chiroptical properties (Fig. 9b). Escudié et al. achieved to synthesize and characterize a *o*-QDM derivative **35** by a formal [2 + 3]cycloaddition of the germene Mes$_2$=GeR$_2$ with 1,4-naphthoquinone (Fig. 9c) [91]. **35** displayed the radical-addition of triplet oxygen upon exposure to air and the Diels–Alder reaction with a standard dienophile at room temperature.

Recently, Tobe et al. reported an air-persistent *o*-QDMs derivative, indenofluorene **36** (Fig. 10) [92]. Okada and his coworkers also achieved to synthesize an air-stable *o*-QDM derivative **37** (Fig. 10) [93]. The X-ray analysis of **37** showed a twisted π-framework along with distinct bond length alteration in the *o*-QDM core. **37** adopts a closed-shell quinoid structure, which is supported by the DFT calculations and the variable temperature ^{1}H NMR measurements.

3.5 Oxidized State of Oligo(*p*-phenylene)s and Cycloparaphenylenes (CPPs)

The interconversion of π-conjugated systems of QDMs between benzenoid and quinoid structures can be attained through chemical or electronic redox processes.

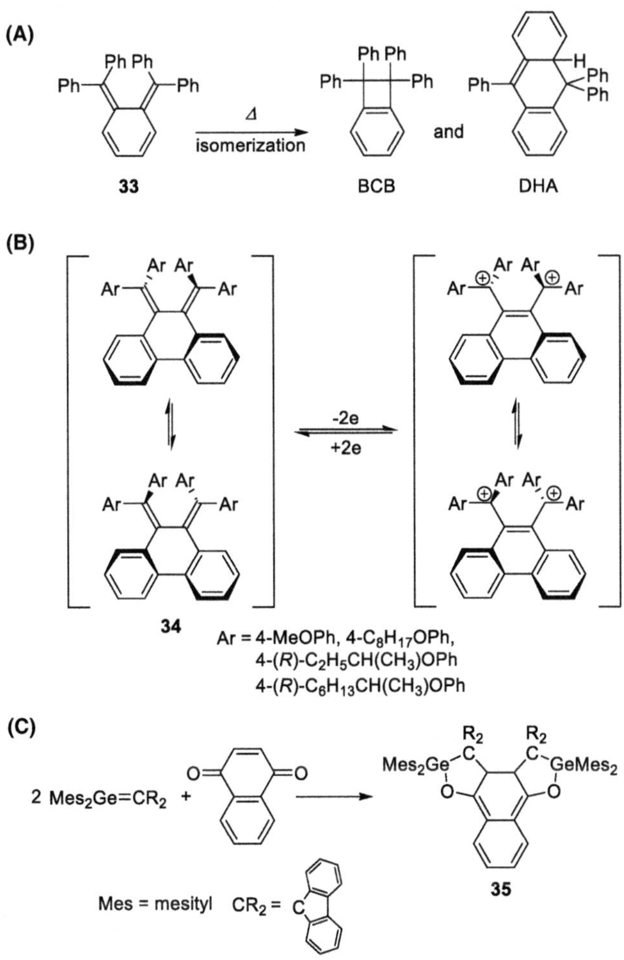

Fig. 9 π-Extended o-QDMs. **a** Intramolecular isomerization of **33**, **b** redox behaviors of **34**, and **c** synthesis of **35**

Fig. 10 Stable π-extended o-QDMs **36** and **37**

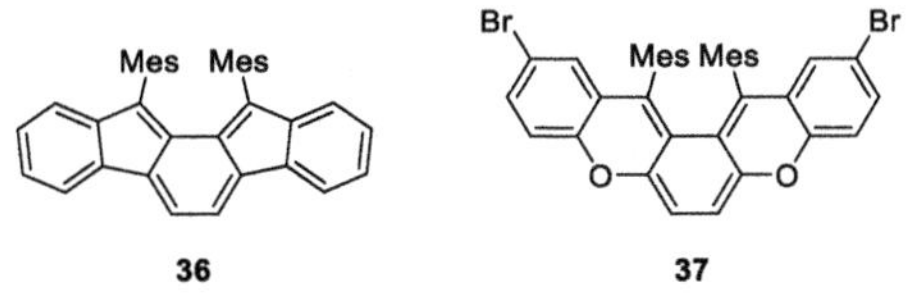

Rathore et al. investigated the oxidized state of oligo(p-phenylene)s, in which the radical cation was prepared by the chemical oxidation using nitronium hexachloroantimonate (NOSbCl$_6$) (Fig. 11a) [94, 95]. The X-ray analysis of the cation radical of the quaterphenylene derivative **38** provided the evidence for the

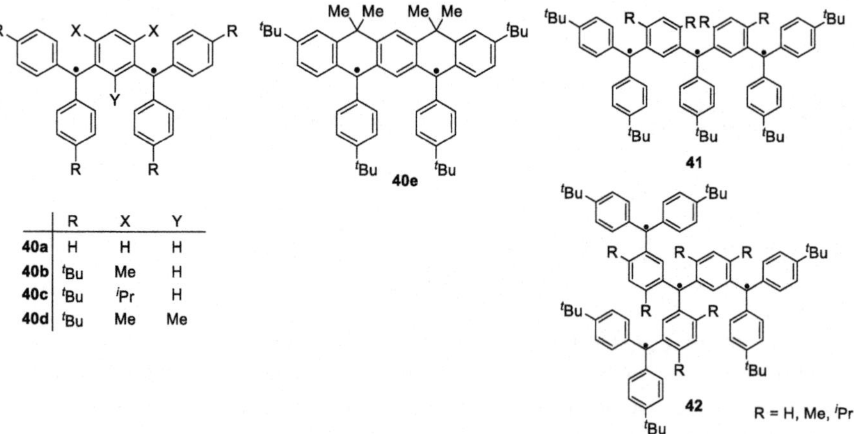

Fig. 11 **a** Synthesis of the cation radical of the quaterphenylene derivative $38^{\bullet+}$, **b** structures of [n]CPPs **39** (*left*), [n]CPPs$^{\bullet+}$ (*middle*), and [n]CPPs^{2+} (*right*), and **c** bond lengths of [8] CPPs (*left*) and [8] CPPs^{2+}(SbCl$_6^-$)$_2$ (right)

manifestation of a quinoidal form, which gave the distinct bond length alternation and an increase in coplanarity among *p*-phenylene units.

Cycloparaphenylenes (CPPs), which are end-free, cyclic vinylogues of oligo(*p*-phenylene)s, are good candidates for the investigation of the quinoidal character in the oxidized states of poly(*p*-phenylene)s (Fig. 11b). Yamago and coworkers successfully synthesized and isolated the radical cations and dications of [n]CPPs ($n = 5$–12) **39** as SbCl$_6$ salts [96]. ESR and UV–Vis–NIR measurements indicates the spin and charge are fully delocalized over the *p*-phenylene rings. The molecular structure of [8] CPPs^{2+}(SbCl$_6^-$)$_2$ determined by the X-ray analysis clearly showed a bond length alternation of benzene rings, which supported the quinoidal character increased in the oxidized state [97]. Circular delocalization of charge or spin on the CPP units leads to the curious of in-plane aromaticity. Molecular orbital and nucleus-independent chemical shift (NICS) analysis revealed that [n]CPPs^{2+} show in-plane aromaticity [98]. ^{1}H NMR measurements of [n]CPPs^{2+} ($n = 5$–12) apparently supported the effect of in-plane aromaticity: the characteristic singlet signals appeared with upfield shifts compared to those of neutral CPPs.

	R	X	Y
40a	H	H	H
40b	^{t}Bu	Me	H
40c	^{t}Bu	^{i}Pr	H
40d	^{t}Bu	Me	Me

Fig. 12 Molecular structures of the Schlenk's diradical **40a** and related compounds **40b–e**, **41** and **42**

3.6 π-Extended *m*-QDMs and Related Compounds

As describe above, *m*-QDM is used as a high-spin coupler to design organic magnetic materials. One of the most famous examples is the sterically crowded Schlenk's diradical **40a**, which was first synthesized in 1915 by Schlenk and four phenyl groups are substituted into its radical centers (Fig. 12) [43]. In 1970, **40a** was identified as a triplet ground state diradical by ESR measurement, in which an ESR signal typical for triplet species was observed ($|D| = 0.0079$ cm^{-1}, $|E| = 0.0005$ cm^{-1}) and the temperature-dependency of the ESR intensities followed the Curie's law [99]. Various substituted Schlenk's diradicals **40b–e** have been investigated by Rajca et al. [100, 101]. They designed and synthesized different alkyl and halogen substituted Schlenk's diradicals, and elucidated the most of them possesses triplet ground states by ESR and SQUID measurements. They also developed the related triradical **41** and tetraradical **42** systems incorporating several Schlenk's diradical units. The spin multiplicity in the ground state of **41** and **42** was determined to be quartet or quintet, respectively, by ESR and SQUID measurements [102–106]. The effect of the substituents on the exchange coupling constant (J) between two unpaired electrons on Schlenk's diradicals was theoretically investigated by Datta and coworkers [107]. According to the DFT calculations at the (BS)-UB3LYP/6-311++G(d,p) level, both mesomeric and inductive effects play a role in tuning of the exchange coupling constant, which corresponds to the singlet–triplet energy gap, for the substituted Schlenk's diradical derivatives. The introduction of groups with mesomerically or inductively electron-donating effects decreases the singlet–triplet energy gaps (ΔE_{S-T}) from the unsubstituted Schlenk's diradical, while mesomerically or inductively electron-withdrawing groups at the same position increases ΔE_{S-T}. These calculated results were explained in terms of the influence of the functional groups on the spin density distribution of the Schlenk's diradical.

The polycyclic aromatic hydrocarbons (PAHs) embedded with *m*-QDM units behave as Non-Kekulé hydrocarbons with high-spin multiplicity in the ground states. One of the most famous Non-Kekulé hydrocarbons is triangulene **43**, which

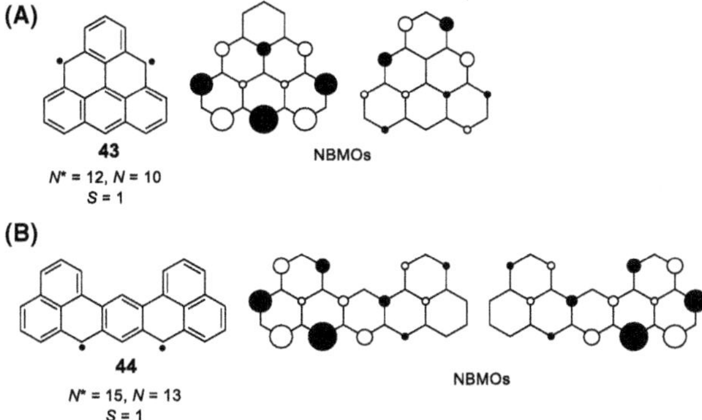

Fig. 13 Prediction of spin quantum number S and spin multiplicity in the ground state of **a** triangulene **43** and **b** **44** with non-bonding molecular orbitals (NBMOs)

is a triangular zigzag-edged PAH and predicted to have two unpaired electrons by Clar (Fig. 13) [108]. The Ovchinnikov's rule [109] and the non-disjoint nonbonding molecular orbital [110, 111] of **43** suggest that **43** possesses multi unpaired electrons localized over the periphery edges with a high spin multiplicity in the ground state, i.e. triplet ground state. Experimental determination of the spin multiplicity was carried out by Nakasuji et al. Tri-*tert*-butyl derivative of **43** gave an ESR signal typical for triplet species with threefold symmetry, and a Curie's plot of the signal intensity supports the triplet ground state [112]. Recently, the mesityl derivative of **44**, which is a heptazethrene isomer incorporating a *m*-xylene diradical unit, was prepared by Wu et al. and shows the triplet ground state confirmed by an electron spin transient nutation (ESTN) method [113]. These results of the triangular-ring fused non-Kekulé PAHs propose the adequate explanation of the ferromagnetic spin alignment along a zigzag edge in graphene nanoribbons (GNRs).

4 Acenes, Periacenes, Anthenes, and Graphene Nanoribbons

4.1 Acenes

Acenes, which are constituted only of linearly-fused benzene rings, have long been the subject of intense studies because of high reactivity and unique electronic properties associated with their π-conjugated topology. In particular, recent growing interests in organic electronic devices, such as field-effect transistors or solar cells, have been uncovering the peculiar electronic properties of acenes that include a biradicaloid behavior in the ground state and a singlet-fission deactivation process in the excited state. The unique properties can be demonstrated by the concept of the Clar's aromatic sextet (Clar sextet) that can well explain physical and chemical properties of most of polycyclic aromatic compounds (Fig. 14) [108, 114]. In the case of phenacenes with an angularly-fused system, multiple circles, which are the notation of the Clar sextets, can be drawn, which lead to the accumulation of the aromatic stabilization. In contrast, acenes can be represented as the form with only one circle and cumulative *o*-quinoidal segments, and furthermore the migration of the circle into the adjacent rings dilutes the effect of the aromatic stabilization. Therefore, acenes suffer from the dramatic decrease in stability and HOMO–LUMO energy gap with the extension of the molecular size.

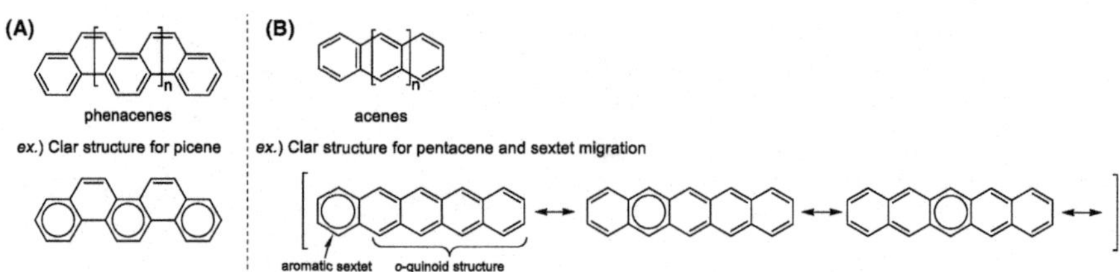

Fig. 14 Molecular structures and Clar structures of **a** phenancenes and **b** acenes

The small HOMO–LUMO gap in larger acenes would bring about approximately degenerate frontier orbitals, and lead to the emergence of an open-shell singlet ground state. The possibility that larger acenes possess diradical character has been first invoked by Bendikov et al. in 2004 [115]. Their DFT calculations using a broken symmetry UB3LYP/6-31G* predicted that the open-shell singlet state is the lowest state in acenes larger than hexacene and they behave as singlet diradicals, where singly occupied molecular orbitals (SOMOs) are distributed over their zigzag edges. The change in the electronic ground state can also be understood using Clar sextet. As shown in Scheme 6, by changing from closed-shell to open-shell singlet configuration, a π-bond is cleaved but this instability is partially compensated by the formation of an extra Clar sextet, which is facilitated for larger acenes than hexacene.

Non-substituted acenes larger than hexacene had not been isolated in pure form. In 2006, Neckers and co-workers successfully generated a non-substituted heptacene in a poly(methyl methacrylate) (PMMA) matrix by photo-decarbonylation of a dione precursor (Fig. 15a) [116]. Photo-decarbonylation of dione precursors facilitates the preparation of more larger non-substituted acenes. Bettinger and co-workers reported the synthesis and spectroscopic detection of non-substituted octacene and nonacene under the condition of matrix isolation [117]. Very recently, Bettinger and his group successfully obtained pure bulk heptacene [118]. The dissociation of heptacene-dimers through thermally induced cycloreversion in the solid state leads to bulk heptacene, which can persist in solid state (a half life time of several weeks at room temperature).

Full spectroscopic characterization of heptacene was first achieved by Anthony et al. in 2005 (Fig. 15b) [119]. The introduction of tris(trimethylsilyl)silyl acetylene group at 7,6-positions makes heptacene soluble and stable. The molecular structure was determined by the means of X-ray crystallographic analysis. Electronic absorption spectrum and cyclic voltammogram gave the HOMO–LUMO energy gaps of 1.36 eV (912 nm, estimated from the onset of the longest absorption band) and 1.30 eV (estimated from the difference between the first reduction and oxidation potentials). In contrast to the prediction of singlet open-shell ground state from the broken-symmetry UB3LYP calculation, ^{1}H and ^{13}C NMR signals of the substituted heptacene **45** are still sharp and no ESR signal is observed, indicating that **45** has singlet closed-shell ground state. The diradical character of acenes is experimentally demonstrated for the substituted nonacene derivative **46** [120]. The optical HOMO–LUMO gap was determined to be 1.2 eV (1033 nm), which is almost identical to the electrochemical HOMO–LUMO gap of 1.19 eV. These small HOMO–LUMO gaps of nonacene invoke the contribution of singlet diradical character to the ground state. Actually, ^{1}H NMR spectrum of **46** gave no signal in

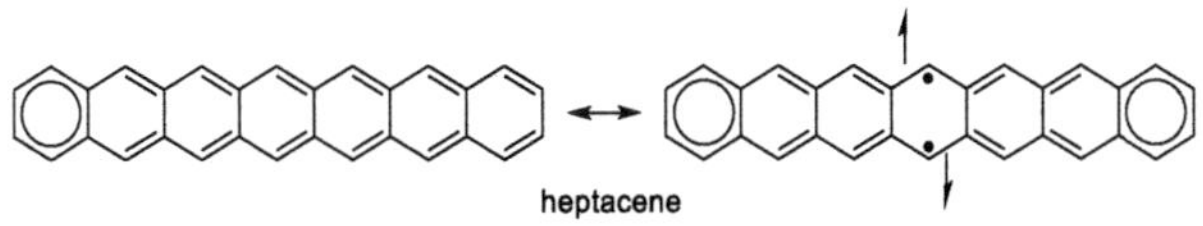

Scheme 6 Resonance formula for heptacene using Clar sextet

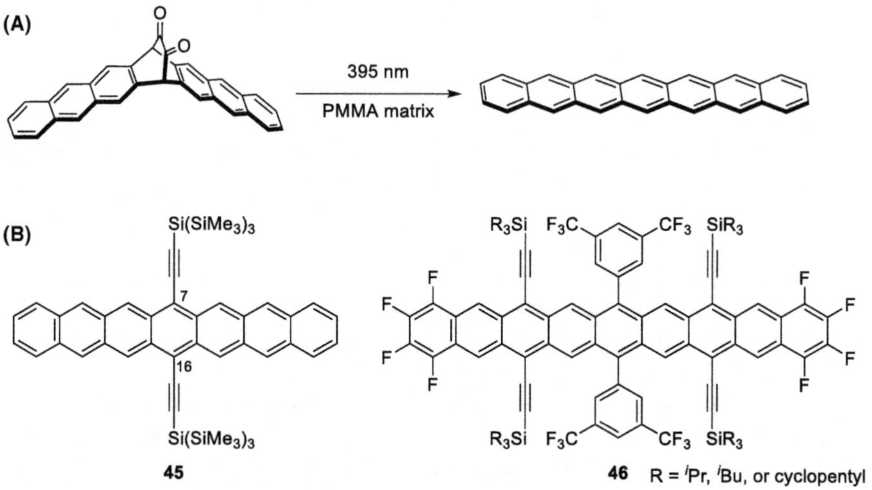

Fig. 15 a Photogeneration of non-substituted heptacene and **b** molecular structures of substituted heptacene **45** and nonacene **46**

the aromatic region due to the existence of magnetic species, probably thermally excited triplet species

On-surface chemical reactions have proved to be a versatile tool for the access to highly reactive molecules from stable and soluble precursors. Even parent triangulene **43** that is a triplet biradical were generated on metal surface at cryogenic temperature and under ultra-high vacuum conditions (see also below) [121]. In the chemistry of higher acenes, on-surface syntheses have achieved significant results. Higher acenes from heptacene [122, 123] to nonacene [124] and decacene [125] have been synthesized on the metal surface and their molecular structures and electronic properties have been observed by STM and AFM measurements.

4.2 Periacenes

4.2.1 Electronic Structure of Periacenes

Peri-condensation of acenes leads to the rectangular PAHs, which are referred to as periacenes, and can exhibit the prominent diradical character. The diradical character of acenes, which has been narrowly observed even at a larger acene than hexacene, would be enhanced by *peri*-condensation since the *o*-quinoid structures in the acene can be effectively accumulated in such a way as to stabilize the open-shell resonance forms. Moreover, because of the existence of both zigzag and armchair edges in their periphery, periacenes have been frequently taken as finite-size models of hydrogen-passivated graphene nanoribbons/nanoflakes (Fig. 16). The number of six-membered rings in each direction of the zigzag and the armchair ring-fusion is indicated in brackets as [*m.n*] in the order of the zigzag and the armchair ring-fusion.

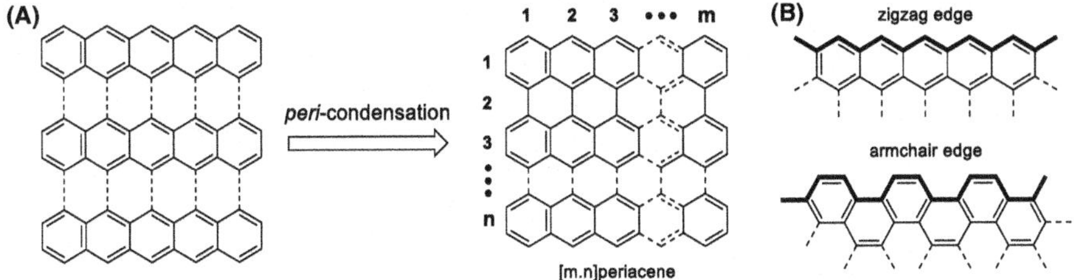

Fig. 16 **a** *Peri*-condensation of acenes toward [*m.n*]periacene and **b** two types of the edge structures of periacene

The diradical character of periacenes has been widely assessed by various methods [126–129]. With the extension of their zigzag edges or *peri*-condensation, the prominent increase in diradical character and the manifestation of multiradical character were predicted [130]. The extent of singlet diradical character (y_0) and tetraradical character (y_1) of [*m.n*]periacene was calculated by using the index defined by Yamaguchi [131–133] coupled to the symmetry-broken UBHandHLYP/6-31G* calculation [134]. Table 1 shows the y_0 values of acenes and [*m.n*]peri-acene. [3.3]periacene, namely bisanthene, has small y_0 (12%), suggesting that the molecule can be classified into a closed-shell molecule. A larger homolog, [3.5]periacene (namely teranthene), gives a large value of y_0 (59%), clearly indicating appreciable singlet diradical character. Lager systems, in the right lower region of Table 1, possess perfect singlet diradical character. A tetraradical state becomes more likely in larger systems of periacenes. (Table 2) The y_1 of [6.7]periacene is found to be large (0.44), suggesting an appreciable contribution of tetraradical structure to the ground state. The largest polyperiacene, [7.7]periacene, in this paper has very large y_0 (1.00) and y_1 (0.77), indicating the singlet tetraradical ground state.

For understanding singlet open-shell state of acenes and periacenes, the resonance formula consisted of Kekulé and diradical forms is helpful. As mentioned above, acenes and periacenes contain *o*-QDM structures. The molecular structure of *o*-QDM should be drawn as a resonance hybrid of Kekulé and diradical forms, in contrast to that most Kekulé compounds have negligible contribution of a diradical form in the ground state. In the formula of [3.*n*]periacenes, namely anthenes, two unpaired electrons appear on both *meso-* positions of anthenes when one draws the structure with a maximum number of Clar sextets, and at the same time, the bonds denoted by *a* assume double bond character (Scheme 7). Therefore, the singlet diradical contribution to the ground state would lead to shortening of the bonds

Table 1 Singlet diradical character y_0 of [*m.n*]periacenes

y_0	$m = 2$	3	4	5	6	7
$n = 1$	0.00	0.00	0.01	0.07	0.20	0.40
3	0.00	0.12	0.60	0.84	0.94	0.98
5	0.01	0.59	0.91	0.98	0.99	1.00
7	0.05	0.84	0.98	1.00	1.00	1.00

Table 2 Singlet tetraradical character y_1 of [$m.n$]periacenes

y_1	$m = 2$	3	4	5	6	7
$n = 1$	0.00	0.00	0.00	0.00	0.01	0.02
3	0.00	0.00	0.01	0.02	0.07	0.20
5	0.00	0.00	0.01	0.05	0.23	0.54
7	0.00	0.00	0.01	0.10	0.44	0.77

Scheme 7 Resonance formula of anthenes. The six-membered rings highlighted by *bold lines* represent the Clar sextets

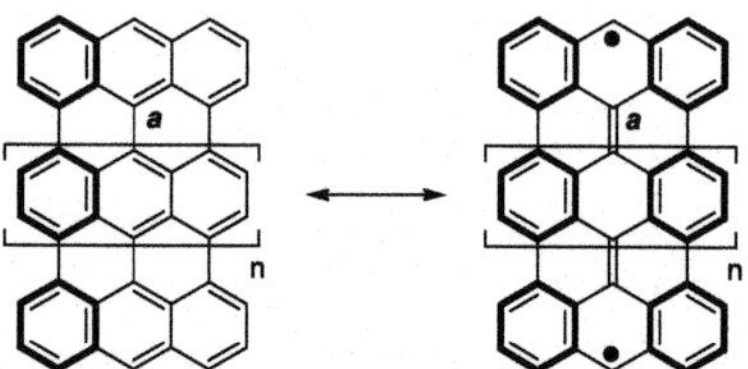

a. The aromatic stabilization energy of benzene based on the homodesmic stabilization energy is ca. 90 kJ/mol [135, 136], whereas the destabilization energy due to C–C π-bond cleavage is ca. 270 kJ/mol [137]. This energetic relationship suggests that the diradical forms would dominantly contribute to the ground state when the difference in the number of Clar sextet between the Kekulé and diradical forms is more than two, because the instability of a π-bond breaking is substantially compensated by the formation of extra Clar sextets.

The same strategy is applicable to estimate the ground state of larger periacenes, such as [4.3]periacenes (namely peritetracene) and [5.3]periacenes (namely peripentacene) (Fig. 17). For both molecules, the difference in the number of Clar sextet between the biradical and the Kekulé form is three, which would lead to the open-shell singlet ground state. Furthermore, peripentacene can be described as a

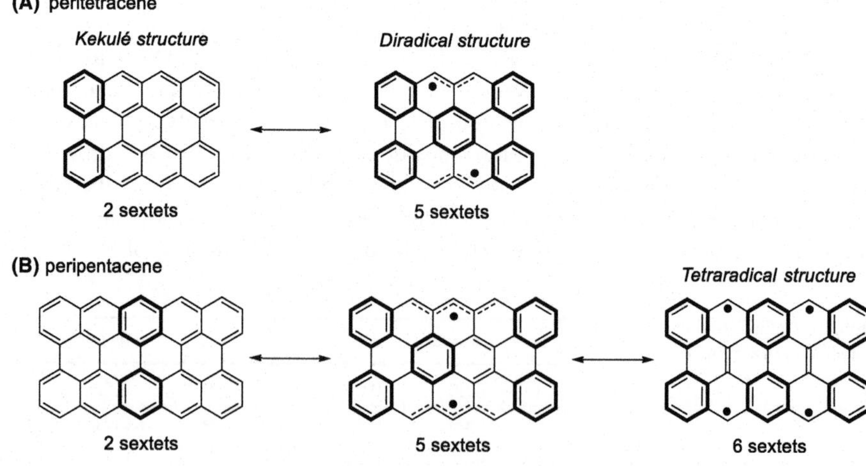

Fig. 17 Resonance formulas of **a** peritetracene and **b** peripentacene. The six-membered rings highlighted by *bold lines* represent the Clar sextets

resonance formula including a tetraradical formula when it has the maximum number of Clar sextets, which is in agreement with the theoretical prediction that the multi-spin singlet state could be the ground state [130].

4.2.2 Synthetic Attempts Toward Peritetracene and Peripentacene

Periacenes isolated in a crystalline form so far are very limited: only bisanthene ([3.3]), teranthene ([3.5]), and quateranthene ([3.7]periacene) (see below). To date, the isolation of periteracene and peripentacene is still not successful although some attempts have been done by several research groups.

Peripentacene was detected in gas phase by mass spectroscopy in 2005 [138], but there has been no structural information. Very recently, Fischer et al. reported the surface-assisted synthesis of peripentacene [139]. Thermally induced cyclodehydrogenation of bispentacene precursor on Au(111) in ultrahigh-vacuum afforded peripentacene stabilized by surface interactions with the underlying metallic substrate (Fig. 18a). They successfully observed rectangular molecular flakes that are assigned to peripentacene featuring parallel zigzag- and armchair edges using STM and atomic-resolution non-contact AFM imaging techniques. Attempts to synthesize peripentacene in solution phase, including the introduction of bulky substituents [140] and the benzo-annulations [141, 142], have been carried out. The isolation of a derivative was also attempted by Müllen et al., who synthesized a planer tetraone **47** as a possible precursor of peripentacene [143]. However, the low solubility of the tetraone hampered the use of analytical methods and further synthetic treatments (Fig. 18b).

The same group also attempted to synthesize peritetracene. Recently, they reported the synthesis and electronic properties of tetrabenzo[a,f,j,o]perylene **48**, bearing a partial structure of peritetracene (Fig. 18b) [144]. ^{1}H NMR signals of **48** showed the line sharpening upon cooling, and the intensities of ESR signals of **48** increased upon increasing the temperature, indicating the open-shell singlet ground state. The singlet–triplet energy gap (ΔE_{S-T}) of **48** given by the temperature dependency of the ESR signal intensities was -14.2 kJ/mol, which is in good agreement with the estimated value of the DFT calculations.

4.3 Anthenes

Non-substituted bisanthene was first reported as an air-sensitive blue solid in 1934 [145]. The introduction of bulky groups, such as aryl or triisopropylsilylethynyl,

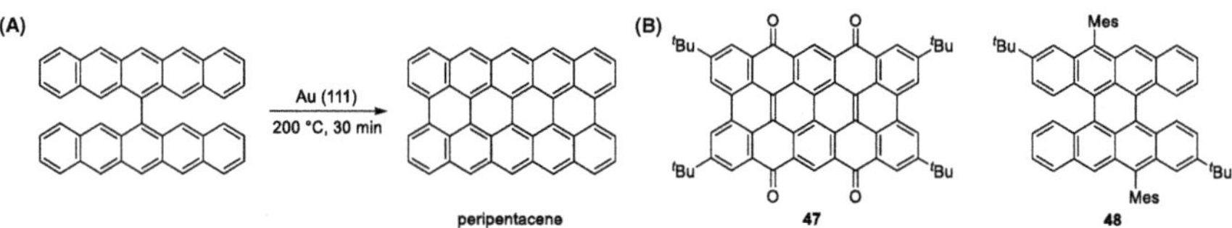

Fig. 18 **a** Metal surface-assisted synthesis of peripentacene and **b** molecular structures of planer tetraone **47** and tetrabenzo[a,f,j,o]perylene **48**

into the *meso*-position effectively allows us to isolate and characterize bisanthene derivatives [146, 147]. In contrast to the stable derivatives, a tetra-*tert*-butyl derivative of bisanthene **49** that is freely capped at the *meso*-position was synthesized for the investigation of singlet diradical character contribution to the ground state (Fig. 19) [148]. **49** was found to be moderately stable in air under room light ($\tau_{1/2} = 19$ days in a toluene solution at room temperature) and was fully characterized by X-ray crystallography, electrochemical, and spectroscopic methods. As described above, the length of the central C–C bond, indicated by *a* in Scheme 7, would serve as a good indicator of the singlet diradical character. X-ray crystallographic analysis of **49** revealed the length of 1.451(2) Å for the bond *a*, which is almost comparable to the usual $C(sp^2)$–$C(sp^2)$ single bond length (1.45 Å). This means that bisanthene has very small diradical character in the ground state. The relatively large HOMO–LUMO energy gap (1.87 and 1.68 eV determined by optical and electrochemical methods) and large $\Delta E_{S–T}$ (– 52.4 kJ/mol estimated by an UB3LYP/6-31G* method) are consistent with the small diradical character. The variable-temperature ^{1}H NMR experiment also supported the closed-shell nature of **49**, which gave no signal broadening of aromatic core even at 383 K and the assignment of all protons was accomplished.

The unique behaviors induced by prominent diradical character were experimentally detected through the characterization of a teranthene derivative. In 2010, a kinetically stabilized teranthene **50** was synthesized and isolated as a dark green solid, which showed a gradual decomposition under ambient condition ($\tau_{1/2} = 3$ days in a toluene solution) [149]. Scheme 8 shows the synthetic route to **50** and the key reaction conditions for the synthesis of **50** are a partial cyclization with KOH/quinoline and a full cyclization with DDQ/Sc(OTf)$_3$.

The X-ray crystallographic analysis of **50** showed an obvious constriction of the bond *a* (1.424(4) Å), indicating the increase in the double bond character for the bond (Fig. 20a). In addition, the side six-membered rings had large benzenoid character, which was confirmed by the harmonic oscillator model of aromaticity (HOMA) values [150, 151], the geometry-based aromaticity index. The HOMO values of the side six-membered rings were much closer to unity for **50** (0.854 and 0.832) than for **49** (0.707), as shown in Fig. 20b. These experimental findings

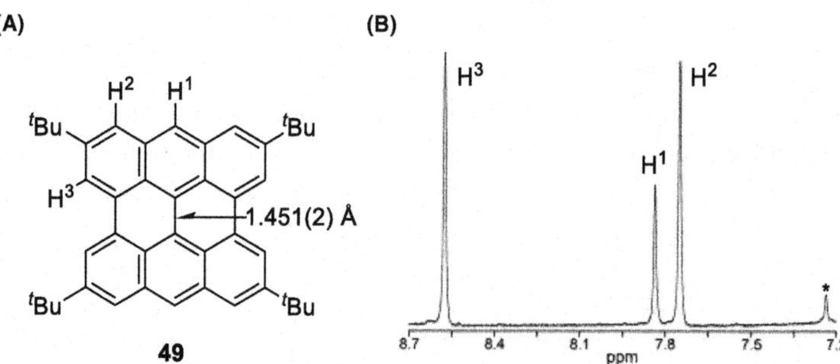

Fig. 19 **a** Tetra-*tert*-butyl bisanthene **49** and **b** ^{1}H NMR spectrum of **49** at 383 K in *p*-xylene-d_{10} with the assignment of protons

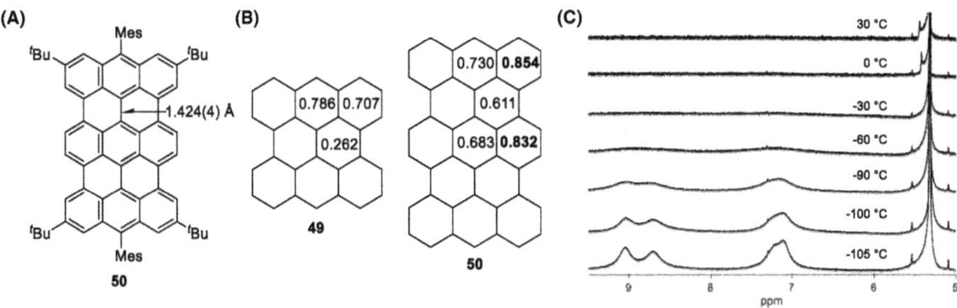

Scheme 8 Synthetic procedure for teranthene **50**. Reagents and conditions: **a** nBuLi, ether, 0 °C, then 1,5-dichloroanthraquinone, 0 °C to rt; **b** NaI, $NaH_2PO_2 \cdot H_2O$, AcOH, 150 °C, 90% (2 steps); **c** KOH, quinoline, 190 °C, 78%; **d** Mestylceriumbromide, THF/ether (4:1), − 30 °C; **e** NaI, $NaH_2PO_2 \cdot H_2O$, AcOH, 150 °C, 75% (2 steps); **f** DDQ, Sc(OTf)$_3$, toluene, reflux; **g** $N_2H_4 \cdot H_2O$, toluene, rt, 72% (2 steps)

Fig. 20 a Tetra-*tert*-butyl-dimesitylteranthene **50**, **b** HOMA values of **49** and **50**, and **c** variable temperature ^{1}H NMR spectra of **50**

suggest the large contribution of the diradical form to the ground state. The magnetic and optical properties of **50** also supported the large contribution of the diradical form. The ^{1}H NMR spectrum of **50** exhibited the severe line broadening at room temperature and upon decreasing the temperature, progressive line sharpening was observed (Fig. 20c). Thermally excited triplet species would responsible for the signal broadening and the small ΔE_{S-T} gap was experimentally determined to be − 16.0 kJ/mol by SQUID measurements. Electronic absorption spectrum of **50** showed an intense absorption band at 878 nm ($\varepsilon = 97,800$ M^{-1} cm^{-1}) accompanied by a broad and weak absorption band at 1054 nm ($\varepsilon = 8,000$ M^{-1} cm^{-1}). The former intense absorption band corresponds to a HOMO → LUMO single excitation, on the other hand, the latter weak and broad absorption band can be assigned to a HOMO, HOMO → LUMO, LUMO double excitation with the help of a strongly contracted second-order n-electron valence state perturbation theory (NEVPT2) calculation that allows multielectron excitation [152]. As Negri et al. reported that a weak low-energy band of thienoquinoid compounds with relatively large diradical character can be assigned to a double excitation [153], one actual signature of the diradical character is the presence of the low-lying excited state dominated by the doubly excited configuration.

The largest homologue of reported anthenes is quateranthene that consists of the quarterly *peri*-condensed anthracenes, which behaves as "pure" diradical molecules due to its larger diradical character ($y_0 = 84\%$ at the UBHandHLYP level of calculation). A kinetically stabilized derivative of quateranthene **51** was prepared via the same synthetic strategy for the teranthene **50** and isolated as a bluish-black solid, which showed a rapid decomposition with $\tau_{1/2} = 15$ h [154]. The molecular geometry, optical and magnetic properties and chemical reactivity of **51** ascertained large diradical character. X-ray crystallographic analysis revealed that the bond *a* of **51a** is quite short (1.416(3) Å in average), supporting the dominant contribution of the diradical resonance form to the ground state (Fig. 21). Electronic absorption spectrum of **51a** showed a low-energy band centered at 917 nm ($\varepsilon = 15,300$ $M^{-1}cm^{-1}$) with a broad and weak band at 1147 nm ($\varepsilon = 870$ $M^{-1}cm^{-1}$). As the similarity to the case of teranthene **50**, these bands can be assigned to a single and a double excitation, respectively. ^{1}H NMR spectra of **51b** gave no signals from the main core ring even at 183 K. The ΔE_{S-T} determined by SQUID measurements was only $- 2.9$ kJ/mol, from which the population of the triplet species of **51a** is estimated to be $\sim 50\%$ at room temperature on the basis of the Boltzmann distribution. Due to the high population of the triplet species, the UV–Vis–NIR spectrum of **51a** showed the temperature dependency. However, the spectral changes were small: a low-energy band at 920 nm observed at 300 K bathochromically shifted only by 50 nm at 180 K along with an isosbestic point. The similarity of the spectra may indicate similar distribution of unpaired electrons around the *meso*-positions in the singlet and triplet states. The substantially small ΔE_{S-T}, that is, very weak coupling of electrons, strongly supports the edge localization of unpaired electrons in quateranthene. In fact, **51b** easily reacts with oxygen at less-protected carbon atoms with a large spin density (see also the spin density map in Fig. 21b) around the zigzag edge, giving rise to an oxygen-adduct one-dimensional polymer **52** in a crystalline form.

The consecutive change in the electronic structure of the ground state from bisanthene, teranthene to quateranthene, varying from a closed-shell to an open-shell singlet ground state, comes from the aromatic stabilization of the additional sextets in the diradical form. The destabilization of a π-bond cleavage and stabilization of aromaticity are well balanced, resulting in the large contribution of the diradical form to the ground state.

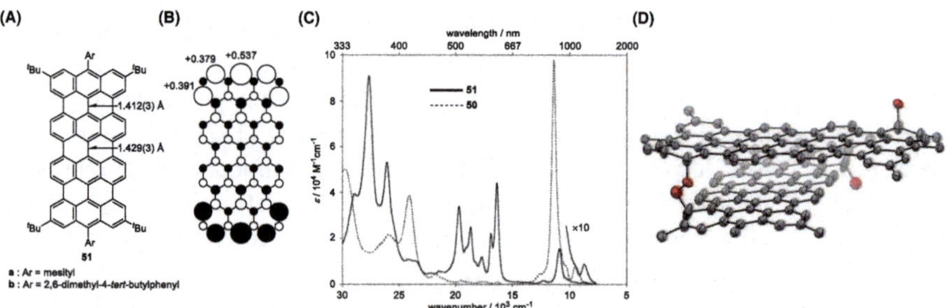

Fig. 21 **a** Tetra-*tert*-butyl-diarylquateranthene **51**, **b** spin density map of **51**, **c** electronic absorption spectra of **51** and **50**, and **d** oxygen adduct of **51**. Some carbon atoms of *tert*-butyl and aryl groups and all hydrogen atoms are omitted for clarity

4.4 Graphene Nanoribbons (GNRs)

The diradical properties of anthenes are closely related to the "edge state" in the zigzag-edged graphene nanoribbons (ZGNRs), which is a peculiar magnetic state originating from multiple electron–electron correlations in an edge-localized nonbonding π-state (Fig. 22). The edge state has been attempted to be observed by various experimental measurements [155–163], and especially, scanning tunneling microscopy and spectroscopy (STM/S) have achieved to propose the good evidence of the edge state around the zigzag edges in GNRs [155, 156, 163]. The edge localization was first understood by a physics approach, that is, by a tight binding band calculation combined with a Hubbard model [164–166]. On the other hand, unveiling the mechanism of the edge localization of unpaired electrons in anthenes is a bottom-up chemical approach. Aromatic sextet formation is a key mechanism of the manifestation of the magnetic state of anthenes, and this classical and simple idea would indeed dominate the electronic structure of nanographenes.

Recent advances in the direct synthesis of GNRs on the metal surface have allowed the preparation of ZGNRs with the precise edge structure and the observation of the electronic structure. Müllen's group, which collaborated with the group of Fasel, successfully achieved atomically precise preparation of GNR on Au(111) surfaces [167]. They established a general protocol for the on-surface fabrication of atomically precise GNRs as follows; (1) vacuum sublimation of dihalogenated monomers onto metal surface and generation of diradical species via the cleavage of carbon-halogen bonds upon thermal activation; (2) radical polymerization among diradical species on the metal surface; and (3) surface-assisted cyclodehydrogenation by annealing at a higher temperature. By the means of the protocol, they have synthesized the various types of GNRs [168–170]. They initially reported the surface-assisted syntheses of 7-armchair GNRs **52**, which are the laterally extended homologues of anthenes [167]. Diradical species generated from 10,10′-dibromo-9,9′-bianthryl **53** by heating at 473 K on an Au(111) surface formed a polymeric compound **54**, and subsequent heating at higher temperature led to anthracene-rings fusion, giving a width-controlled GNR **52** (Fig. 23a and b). The width and the edge-shape of **52** are defined by the structure of the molecular precursor **53**. The edge termination of the resulting GNR **52** by hydrogen atoms is confirmed by STM experiments and the high electron density distribution around

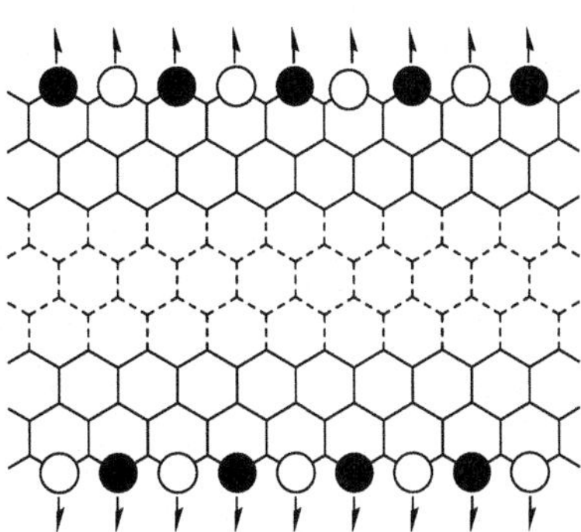

Fig. 22 Edge-localized nonbonding π-state of ZGNRs

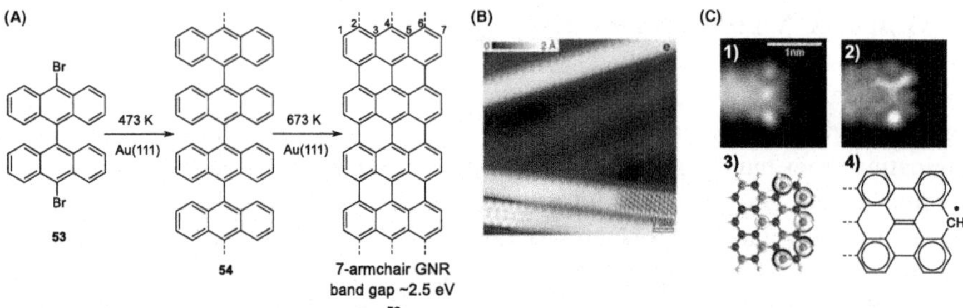

Fig. 23 **a** Reaction scheme from **53** to GNR **52** on an Au(111) surface, **b** high-resolution STM image of **52**. After Ref. [154], and **c** (*1*) STM image of 7-armchair GNR **52** termini on Au (111), (*2*) STM simulation for the termini, (*3*) electron density isosurface based on a DFT calculation of the termini, and (*4*) Clar formula with maximum number of Clar sextets. After Ref. [158]

the zigzag edged termini of the GNRs, which is almost identical to the edge state predicted theoretically, was estimated by the large-scale density functional theory calculations (Fig. 23c) [171, 172]. It is worth noting that the observed electron density map of the terminated zigzag edge of **52** is reminiscent of the spin density map of quateranthene **51**, where the unpaired electron distributes around the zigzag edges.

Very recently, the same group achieved the bottom-up synthesis of ZGNRs **55** with atomically precise zigzag edges through surface-assisted polymerization and cyclodehydrogenation of a U-shaped precursor monomer **56** (Fig. 24) [173]. After polymerization of **56** at 475 K on an Au(111) surface, further annealing the sample at 625 K resulted in a complete planarization of the linear structures. The preparation of atomically precise zigzag edges was directly observed by non-contact atomic force microscopy (nc-AFM) imaging with a CO-functionalized tip, which can confirm that the observed width and edge morphology corresponded to the expected 6-ZGNRs structure as defined by the design of the precursor **56**. The analyses of the local density of state (DOS) and the differential conductance (dI/dV) of **55** observed on NaCl plates unambiguously revealed the small band gap (1.5–1.9 eV) and the high electron density localized on the zigzag edges. These features, in conjunction with the agreement between experimentally determined and simulated DOS maps, clearly evidences the experimental observation of the unperturbed zigzag edge state.

Fig. 24 Reaction scheme from **56** to GNR **55** on an Au(111) surface

5 Zethrene and Related Compounds

Zethrene [174, 175], a Z-shaped polycyclic hydrocarbon, was first reported more than 60 years ago and has experienced a renaissance as a unique biradicaloid molecule. The interests in connection with singlet diradical character were first evoked by Nakano et al. They pointed out potential diradical character in zethrene and its larger homologues by theoretical calculations based on the spin-projection UHF theory [176]. Due to the diradical character, a large static second-order hyperpolarization (γ) [176] and a high singlet fission efficiency [177] are expected for zethrene and related compounds.

The non-substituted zethrene **56** was first synthesized by Clar et al. in 1955 through a long synthetic step and isolated as relatively photo-sensitive deep violet needles (Fig. 25) [178]. **56** gives an intense absorption peak at 544 nm ($\varepsilon = 42{,}900\ \mathrm{M}^{-1}\ \mathrm{cm}^{-1}$) in a benzene solution accompanied by fluorescent

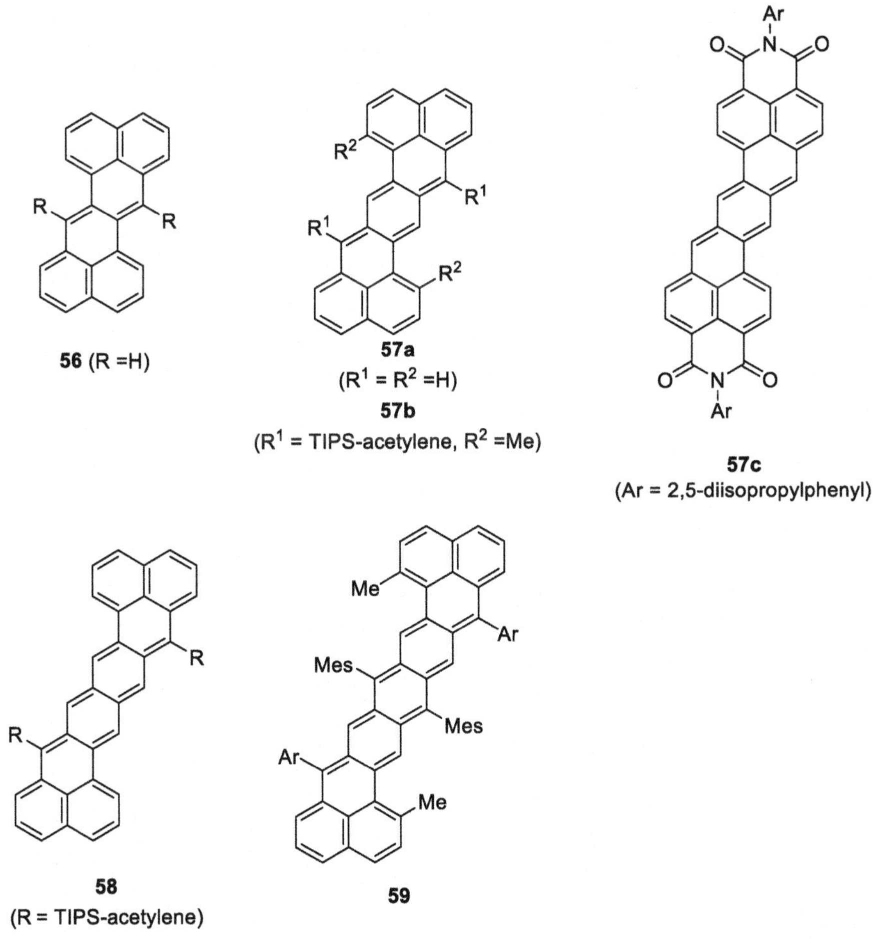

Fig. 25 Molecular structures of zethrene family

($\phi_{PL} = 0.34$) at 571 nm. Clar pointed out that zethrene would contain two formally fixed double bonds in the center six-membered rings. The developments of the more efficient synthetic methods allowed the characterization of the molecular geometry. In 2010, the X-ray crystallographic analyses were achieved by Wu [179] and Tobe [180], which disclosed the fixed double bond character and indicated that the ground state of zethrene can be best represented as the closed-shell singlet state. Sharp NMR signals well support the negligible diradical character.

A larger homologue, which is constituted by seven six-membered rings, of zethrene is named heptazethrene. The significant difference between zethrene and heptazetherene is that a p-QDM skeleton is incorporated into the heptazethrene framework, suggesting the potential for the manifestation of singlet diradical character. The non-substituted heptazethrene **57a** was first prepared as green solids by Clar and Macpherson in 1962, but its high reactivity hampered the reliable characterization [181]. Kinetically stabilized derivatives **57b** and **c** were successfully isolated by Wu. The electronic absorption of **57b** gave an intense typical p-band at 634 nm together with a weak fluorescence at 704 nm. ^{1}H NMR spectrum gave sharp signals from the main core rings, indicating the closed-shell character [182]. The optically and electrochemically determined HOMO–LUMO gaps are relatively large (1.46 and 1.82 eV, respectively). Interestingly, the attachment of diimide groups at the both end of heptazethrene led to biradicaloid behavior. This derivative **57c** showed no ^{1}H NMR signals in the aromatic region at room temperature [183].

The same group achieved the syntheses of the kinetically blocked octazethrene **58** [182] and nonazethrene **59** [184] derivatives, which contain naphthoquinoid and anthraquinoid skeletons, respectively. The octazethrene derivative **58** showed an intense absorption band at 668 nm accompanied by weak broad bands at 719 and 795 nm. The optically and electrochemically determined HOMO–LUMO gaps of **58** are smaller (1.13 and 1.50 eV, respectively) than those of the heptazethrene derivative **57b**. This derivative **58** behaved as a singlet biradicaloid molecule, which gave broad ^{1}H NMR signals in aromatic region, small ΔE_{S-T} (–16.2 kJ/mol), and an ESR signal in solid state. The nonazetherene derivative **59** has an open-shell singlet ground state with moderate diradical character ($y_0 = 0.48$ estimated by UCAM-B3LYP calculation) and a small singlet–triplet gap ($\Delta E_{S-T} = -21.7$ kJ/mol determined by SQUID measurements). The detailed theoretical calculations revealed that the zethrene family fast decrease the singlet–triplet energy gap and fast increase the diradical character with increasing the molecular size compared to acene family, which is brought about by the difference of the localization of the frontier molecular orbitals. The unpaired electrons in larger acenes are localized at the center of the molecule, while they are at the edges in extended zethrenes. The spatial localization of the frontier molecular orbitals in zethrenes would play a very important role in the nature of radical character as well as the excitation energy.

The effects of the benzene-fusions to a heptazetherene or zethrene core on diradical character are also investigated by Wu. Dibenzoheptazetherene isomers **60a** and **60b** [185], and dibenzozetherene isomers **61a** [186] and **61b** [187] were prepared (Fig. 26). As described above in 4-2-1, the extra aromatic stabilization in the diradical resonance form can play a main role to determine the electronic

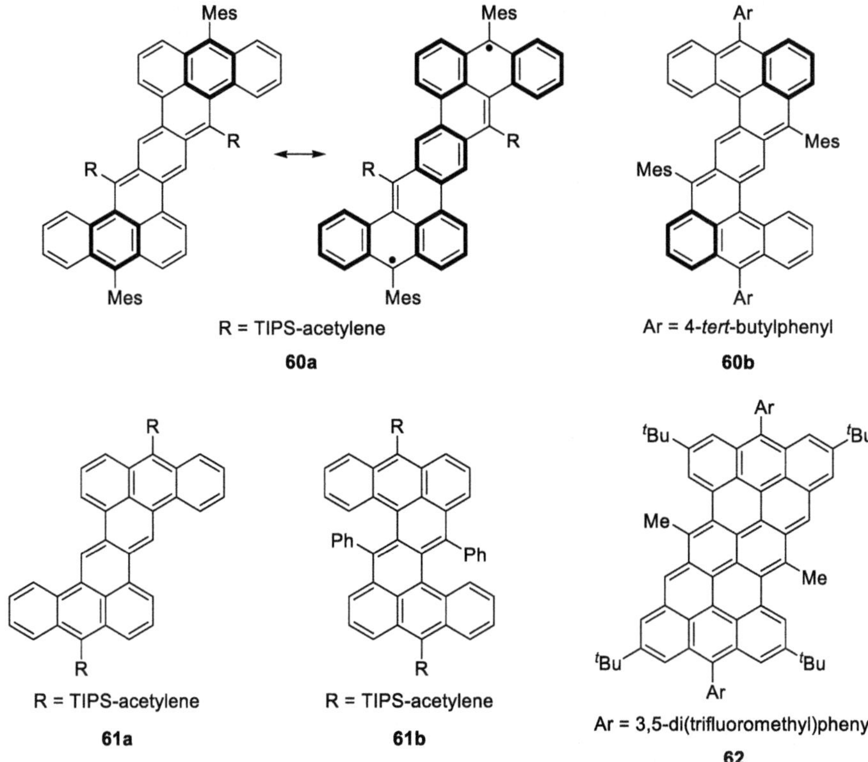

Fig. 26 Molecular structures of π-extended zethrene family. The six-membered rings highlighted by *bold lines* represent the Clar sextets

configuration of the singlet ground state. The isomers **60a** and **61a**, which can be represented as the diradical resonance form with more extra aromatic sextets, are typical singlet biradicaloid molecules characterized by the thermally excited paramagnetic activity, whereas **60b** and **61b** behave as more like closed-shell molecules. Very recently, the same group successfully synthesized a laterally extended heptazethrene molecule **62**, namely super-heptazethrene, which is calculated to have larger diradical character ($y_0 = 0.71$ at the UCAM-B3LYP/6-31G** level) than that of heptazetherene ($y_0 = 0.17$) [188]. This derivative **62** also showed biradicaloid behavior: broad ^{1}H NMR signals in aromatic region, small ΔE_{S-T} (− 3.93 kJ/mol), and an ESR signal in solid state.

6 Conclusions

The interest in QDMs stems from their unusual electronic structures that highly depend on the π-conjugated manners, which have attracted much attention of many chemists in both synthetic and theoretical research fields. The synthetic efforts of π-extension of QDMs have opened the door to thermodynamically and kinetically stabilized QDMs, which enables us to assess diradical behaviors and redox properties of QDMs precisely. The recent studies on benzenoid QDMs have shown wide progressions that contains not only diradicaloid molecules but also multi-spin polarized edge state of GNRs. The application of QDMs to spintronics, as well as novel optical or electronic materials, would be one of the next challenges of QDMs. At the same time, the fundamental understandings of QDMs, which related to the aromaticity, diradical character and the singlet–triplet energy gap, still remain to be answered. The strong collaboration of synthesis, measurements, and theoretical studies is essential for unveiling unusual but fascinating properties of QDMs.

References

1. Kubo T (2015) Recent progress in quinoidal singlet biradical molecules. Chem Lett 44:111–122
2. Abe M (2013) Diradicals. Chem Rev 113:7011–7088
3. Zeng Z, Wu J (2015) Stable π-extended *p*-quinodimethanes: synthesis and tunable ground states. Chem Rec 15:322–328
4. Ni Y, Wu J (2016) Diradical approach toward organic near infrared dyes. Tetrahedron Lett 57:5426–5434
5. Sun Z, Zeng Z, Wu J (2013) Benzenoid polycyclic hydrocarbons with an open-shell biradical ground state. Chem Asian J 8:2894–2904
6. Sun Z, Ye Q, Chi C, Wu J (2012) Low band gap polycyclic hydrocarbons: from closed-shell near infrared dyes and semiconductors to open-shell radicals. Chem Soc Rev 41:7857
7. Nishinaga T (2016) Organic redox systems. Wiley, Hoboken
8. Szwarc M (1947) The C–H bond energy in toluene and xylenes. Nature 160:403
9. Brown CJ, Farthing AC (1949) Preparation and structure of di-*p*-xylylene. Nature 164:915
10. Coulson CA, Craig DP, Maccoll A, Pullman A (1947) *p*-Quinodimethane and its diradical. Discuss Faraday Soc 2:36
11. Williams DJ, Pearson JM, Levy M (1970) Nuclear magnetic resonance spectra of quinodimethanes. J Am Chem Soc 92:1436–1438
12. Pearson JM, Six HA, Williams DJ, Levy M (1971) Spectroscopic studies of quinodimethanes. J Am Chem Soc 93:5034–5036
13. Yamakita Y, Furukawa Y, Tasumi M (1993) Observation of the infrared and Raman spectra of *p*-benzoquinodimethane in low-temperature Ar matrices. Chem Lett 22:311–314
14. Yamakita Y, Tasumi M (1995) Vibrational analyses of *p*-benzoquinodimethane and *p*-benzoquinone based on ab initio Hartree–Fock and second-order Moller–Plesset calculations. J Phys Chem 99:8524–8534
15. Koenig T, Wielesek R, Snell W, Balle T (1975) Helium(I) photoelectron spectrum of *p*-quinodimethane. J Am Chem Soc 97:3225–3226
16. Mahaffy PG, Wieser JD, Montgomery LK (1977) An electron diffraction study of *p*-xylylene. J Am Chem Soc 99:4514–4515
17. Bobrowski M, Skurski P, Freza S (2011) The electronic structure of *p*-xylylene and its reactivity with vinyl molecules. Chem Phys 382:20–26
18. Thiele J, Balhorn H (1904) Ueber einen chinoïden Kohlenwasserstoff. Ber Dtsch Chem Ges 37:1463–1470
19. Tschitschibabin AE (1907) Über einige phenylierte Derivate des p, p-ditolyls. Ber Dtsch Chem Ges 40:1810–1819
20. Cava MP, Napier DR (1957) Condensed cyclobutane aromatic systems. II. Dihalo derivatives of benzocyclobutene and benzocyclobutadiene dimer. J Am Chem Soc 79:1701–1705
21. Segura JL, Martín N (1999) *o*-Quinodimethanes: efficient intermediates in organic synthesis. Chem Rev 99:3199–3246

22. Yoshida H, Ohshita J, Kunai A (2010) Aryne, *ortho*-quinone methide, and *ortho*-quinodimethane: synthesis of multisubstituted arenes using the aromatic reactive intermediates. Bull Chem Soc Jpn 83:199–219
23. Charlton JL, Alauddin MM (1987) Orthoquinodimethanes. Tetrahedron 43:2873–2889
24. Quinkert G, Wiersdorff W-W, Finke M, Opitz K (1966) Über das tetraphenyl-*o*-xylylen. Tetrahedron Lett 7:2193–2200
25. Fishwick CWG, Jones DW (1988) *ortho*-Quinonoid Compounds. In: Patai S, Rappoport Z (eds) The chemistry of the quinonoid compounds. Wiley, New York, pp 403–453
26. Flynn CR, Michl J (1974) π, π-Biradicaloid hydrocarbons. *o*-Xylylene. Photochemical preparation from 1,4-dihydrophthalazine in rigid glass, electric spectroscopy, and calculations. J Am Chem Soc 96:3280–3288
27. Flynn CR, Michl J (1973) Photochemical preparation of *o*-xylylene from 1,3-dihydrophthalazine in rigid glass. J Am Chem Soc 95:5802–5803
28. Tseng KL, Michl J (1977) An approach to biradical-like species. Spectroscopy of *o*-xylylene in argon matrix. J Am Chem Soc 99:4840–4842
29. Miller RD, Kolc J, Michl J (1976) Photochemical generation of stable *o*-xylylene derivatives by the electrocyclic ring opening of some polycyclic benzocyclobutene derivatives. J Am Chem Soc 98:8510–8514
30. McMahon RJ, Chapman OL (1987) Direct spectroscopic observation of intramolecular hydrogen shifts in carbenes. J Am Chem Soc 109:683–692
31. Baird NC (1972) Quantum organic photochemistry. II. Resonance and aromaticity in the lowest $3\pi\pi^*$ state of cyclic hydrocarbons. J Am Chem Soc 94:4941–4948
32. Roth WR, Biermann M, Erker G, Jelich K, Gerhartz W, Görner H (1980) Isolierung und Energiemulde des 2,3-dimethylen-1,4-cyclohexadiyl-Diradikals. Chem Ber 113:586–597
33. Roth WR, Scholz BP (1982) Zur Energiedelle von Diradikalen, II. Das 2,3-dimethylen-1,4-cyclohexadiyl. Chem Ber 115:1197–1208
34. Roth WR, Biermann M, Dekker H, Jochems R, Mosselman C, Hermann H (1978) Das Energieprofil des *o*-Chinodimethan-Benzocyclobuten-Gleichgewichtes. Chem Ber 111:3892–3903
35. Roth WR, Scholz BP (1981) Das Energieprofil des *o*-Chinodimethan ⇌ Benzocyclobuten-Gleichgewichtes, II. Chem Ber 114:3741–3750
36. Kametani T, Honda T, Ebisawa Y, Ichikawa H (1985) An Mo theoretical study on the reaction of benzocyclobutene with ethylene. Concerted vs stepwise. Tetrahedron 41:3643–3653
37. Chou CH, Trahanovsky WS (1986) The [4 + 4] dimerization of 2,3-dimethylene-2,3-dihydrofuran: secondary deuterium kinetic isotope effect evidence for a two-step mechanism. J Am Chem Soc 108:4138–4144
38. Trahanovsky WS, Macias JR (1986) Direct observation of *o*-xylylene (*o*-quinodimethane) in solution. Dimerization kinetics of some *o*-quinodimethanes. J Am Chem Soc 108:6820–6821
39. Migirdicyan E, Baudet J (1975) Electron spectra of *o*- and *m*-xylylenes and their methylated derivatives. experimental and theoretical study. J Am Chem Soc 97:7400–7404
40. McCullough JJ (1980) *o*-Xylylenes and isoindenes as reaction intermediates. Acc Chem Res 13:270–276
41. Davidson ER, Borden WT (1977) Some aspects of the potential surface for singlet trimethylenemethane. J Am Chem Soc 99:2053–2060
42. Borden WT (1982) Diradicals. Wiley, New York
43. Schlenk W, Brauns M (1915) Zur Frage der Metachinoide. Ber Dtsch Chem Ges 48:661–669
44. Neuhaus P, Grote D, Sander W (2008) Matrix isolation, spectroscopic characterization, and photoisomerization of *m*-xylylene. J Am Chem Soc 130:2993–3000
45. Wright BB, Platz MS (1983) Electron spin resonance spectroscopy of the triplet state of *m*-xylylene. J Am Chem Soc 105:628–630
46. Wenthold PG, Kim JB, Lineberger WC (1997) Photoelectron spectroscopy of m-xylylene anion. J Am Chem Soc 119:1354–1359
47. Zhang G, Li S, Jiang Y (2003) Effects of substitution on the singlet–triplet energy splittings and ground-state multiplicities of m-phenylene-based diradicals: a density functional theory study. J Phys Chem A 107:5573–5582
48. Wang T, Krylov AI (2005) The effect of substituents on electronic states' ordering in *meta*-xylylene diradicals: qualitative insights from quantitative studies. J Chem Phys 123:104304
49. Wang T, Krylov AI (2006) Electronic structure of the two dehydro-*meta*-xylylene triradicals and their derivatives. Chem Phys Lett 425:196–200

50. Goodman JL, Berson JA (1984) Formation and intermolecular capture of *m*-quinodimethane. J Am Chem Soc 106:1867–1868
51. Goodman JL, Berson JA (1985) *m*-Quinodimethane, parent hydrocarbon of the m-quinonoid nonkekule series. low-temperature isolation and solution-phase chemical reactivity. J Am Chem Soc 107:5409–5424
52. Berson JA (1988) *meta*-Quinonoid Compounds. In: Patai S, Rappoport Z (eds) The chemistry of the quinonoid compounds. Wiley, New York, pp 455–536
53. Müller E, Müller-Rodloff I (1935) Magnetochemische Untersuchungen organischer Stoffe. 1. Mitteilung. Zur Frage der Existenz von Biradikalen. Justus Liebig's Ann der Chemie 517:134–151
54. Schwab G-M, Agliardi N (1940) Einwirkung von organischen Radikalen auf p-Wasserstoff. II. Mitteil.: zur Frage nach der Natur der Biradikale. Ber Dtsch Chem Ges (A B Series) 73:95–98
55. Jarrett HS, Sloan GJ, Vaughan WR (1956) Paramagnetic resonance absorption in some organic biradicals. J Chem Phys 25:697
56. Sloan GJ, Vaughan WR (1957) Stable organic biradicals. J Org Chem 22:750–761
57. Müller E, Rieker A, Scheffler K, Moosmayer A (1966) Applications and limitations of magnetic methods in free-radical chemistry. Angew Chem Int Ed Engl 5:6–15
58. Cavalieridoro P, Mangini A, Pedulli GF, Spagnolo P, Tiecco M (1970) On the nature of the radical formed from tetraphenyl-*p*-xylene dichloride and zinc dust. Mol Phys 18:861–863
59. Hutchison CA, Kowalsky A, Pastor RC, Wheland GW (1952) A note on the detection of free radicals by means of paramagnetic resonance absorption biradicals. J Chem Phys 20:1485
60. Reitz DC, Weissman SI (1960) Spin exchange in biradicals. J Chem Phys 33:700
61. McConnell HM (1960) The biradical paradox. J Chem Phys 33:1868
62. Waring RK, Sloan GJ (1964) Association in biradical solutions. J Chem Phys 40:772
63. Brauer H-D, Stieger H, Hyde JS, Kispert LD, Luckhurst GR (1969) ENDOR of biradicals. Mol Phys 17:457–471
64. Montgomery LK, Huffman JC, Jurczak EA, Grendze MP (1986) The molecular structures of Thiele's and Chichibabin's hydrocarbons. J Am Chem Soc 108:6004–6011
65. Cava MP, Schlessinger RH (1963) Reactive *o*-quinonoid aromatic hydrocarbons of the pleiadene Series. J Am Chem Soc 85:835–836
66. Cava MP, Schlessinger RH (1965) Pleiadene systems—IV. Tetrahedron 21:3073–3081
67. Kolc J, Michl J (1970) Photochemical synthesis of matrix-isolated pleiadene. J Am Chem Soc 92:4147–4148
68. Downing J, Dvořák V, Kolc J, Manzara A, Michl J (1972) Direct observation of a doubly excited state of pleiadene. Chem Phys Lett 17:70–73
69. Kolc J, Michl J (1973) π, π-Biradicaloid hydrocarbons. Pleiadene family. I. Photochemical preparation from cyclobutene precursors. J Am Chem Soc 95:7391–7401
70. Müller E, Pfanz H (1941) Über biradikaloide Terphenylderivate. Ber Dtsch Chem Ges (A B Series) 74:1051–1074
71. Schmidt R, Brauer H-D (1971) The energetic positions of the lowest singlet and triplet state of the Schlenk and of the Müller hydrocarbon. Angew Chem Int Ed Engl 10:506–507
72. Zeng Z, Sung YM, Bao N, Tan D, Lee R, Zafra JL, Lee BS, Ishida M, Ding J, Navarrete JTL, Li Y, Zeng W, Kim D, Huang KW, Webster RD, Casado J, Wu J (2012) Stable tetrabenzo-Chichibabin's hydrocarbons: tunable ground state and unusual transition between their closed-shell and open-shell resonance forms. J Am Chem Soc 134:14513–14525
73. Lim Z, Zheng B, Huang K-W, Liu Y, Wu J (2015) Quinoidal oligo(9,10-anthryl)s with chain-length-dependent ground states: a balance between aromatic stabilization and steric strain release. Chem Eur J 21:18724–18729
74. West R, Jorgenson JA, Stearley KL, Calabrese JC (1991) Synthesis, structure and semiconductivity of a *p*-terphenoquinone. J Chem Soc Chem Commun 1991:1234
75. Rebmann A, Zhou J, Schuler P, Stegmann HB, Rieker A (1996) Synthesis, EPR spectroscopy and voltammetry of a *p*-quaterphenyl biradical/quinone. J Chem Res Synop 7:318–319
76. Rebmann A, Zhou J, Schuler P, Stegmann HB, Rieker A (1996) Synthesis, EPR, spectroscopy and voltammetry of a *p*-quaterphenyl biradical/quinone. J Chem Res Miniprint 7:1765–1783
77. Zhou J, Rieker A (1997) Electrochemical and spectroscopic properties of a series of *tert*-butyl-substituted *para*-extended quinones. J Chem Soc Perkin Trans 2:931–938
78. Canesi EV, Fazzi D, Colella L, Bertarelli C, Castiglioni C (2012) Tuning the quinoid versus biradicaloid character of thiophene-based heteroquaterphenoquinones by means of functional groups. J Am Chem Soc 134:19070–19083

79. Zhang K, Huang K, Li J, Luo J, Chi C, Wu J (2009) A soluble and stable quinoidal bisanthene with NIR absorption and amphoteric redox behavior. Org Lett 11:4854–4857
80. Ueda A, Nishida S, Fukui K, Ise T, Shiomi D, Sato K, Takui T, Nakasuji K, Morita Y (2010) Three-dimensional intramolecular exchange interaction in a curved and nonalternant π-conjugated system: corannulene with two phenoxyl radicals. Angew Chem Int Ed 49:1678–1682
81. Schmidt D, Son M, Lim JM, Lin M-J, Krummenacher I, Braunschweig H, Kim D, Würthner F (2015) Perylene bisimide radicals and biradicals: synthesis and molecular properties. Angew Chem Int Ed 54:13980–13984
82. Takahashi T, Matsuoka K, Takimiya K, Otsubo T, Aso Y (2005) Extensive quinoidal oligothiophenes with dicyanomethylene groups at terminal positions as highly amphoteric redox molecules. J Am Chem Soc 127:8928–8929
83. Ponce Ortiz R, Casado J, Hernández V, López Navarrete JT, Viruela PM, Ortí E, Takimiya K, Otsubo T (2007) On the biradicaloid nature of long quinoidal oligothiophenes: experimental evidence guided by theoretical studies. Angew Chem Int Ed 46:9057–9061
84. Zeng Z, Ishida M, Zafra JL, Zhu X, Sung YM, Bao N, Webster RD, Lee BS, Li R-W, Zeng W, Li Y, Chi C, Navarrete JTL, Ding J, Casado J, Kim D, Wu J (2013) Pushing extended p-quinodimethanes to the limit: stable tetracyano-oligo(N-annulated perylene)quinodimethanes with tunable ground states. J Am Chem Soc 135:6363–6371
85. Zeng Z, Lee S, Son M, Fukuda K, Burrezo PM, Zhu X, Qi Q, Li R-W, Navarrete JTL, Ding J, Casado J, Nakano M, Kim D, Wu J (2015) Push-Pull type oligo(N-annulated perylene)quinodimethanes: chain length and solvent-dependent ground states and physical properties. J Am Chem Soc 137:8572–8583
86. Zeng Z, Lee S, Zafra JL, Ishida M, Zhu X, Sun Z, Ni Y, Webster RD, Li R-W, López Navarrete JT, Chi C, Ding J, Casado J, Kim D, Wu J (2013) Tetracyanoquaterrylene and tetracyanohexarylenequinodimethanes with tunable ground states and strong near-infrared absorption. Angew Chem Int Ed 52:8561–8565
87. Zhu X, Tsuji H, Nakabayashi K, Ohkoshi S, Nakamura E (2011) Air- and heat-stable planar tri-p-quinodimethane with distinct biradical characteristics. J Am Chem Soc 133:16342–16345
88. Quinkert G, Wiersdorff W-W, Finke M, Opitz K, van de Haar F-G (1968) Lichtinduzierte Reaktionen, V. Darstellung und elektrocyclische Isomerisierungen des Tetraphenyl-o-chinodimethans. Chem Ber 101:2302–2325
89. Iwashita S, Ohta E, Higuchi H, Kawai H, Fujiwara K, Ono K, Takenaka M, Suzuki T (2004) First stable 7,7,8,8-tetraaryl-o-quinodimethane: isolation, X-ray structure, electrochromic response of 9,10-bis(dianisylmethylene)-9,10-dihydrophenanthrene. Chem Commun 2076
90. Suzuki T, Sakano Y, Iwai T, Iwashita S, Miura Y, Katoono R, Kawai H, Fujiwara K, Tsuji Y, Fukushima T (2013) 7,7,8,8-Tetraaryl-o-quinodimethane stabilized by dibenzo annulation: a helical π-electron system that exhibits electrochromic and unique chiroptical properties. Chem Eur J 19:117–123
91. Ghereg D, El Kettani SE-C, Lazraq M, Ranaivonjatovo H, Schoeller WW, Escudié J, Gornitzka H (2009) An isolable o-quinodimethane and its fixation of molecular oxygen to give an endoperoxide. Chem Commun 5:4821
92. Shimizu A, Tobe Y (2011) Indeno[2,1-a]fluorene: an air-stable ortho-quinodimethane derivative. Angew Chem Int Ed 50:6906–6910
93. Sato C, Suzuki S, Kozaki M, Okada K (2016) 2,11-Dibromo-13,14-dimesityl-5,8-dioxapentaphene: a stable and twisted polycyclic system containing the o-quinodimethane skeleton. Org Lett 18:1052–1055
94. Banerjee M, Lindeman SV, Rathore R (2007) Structural characterization of quaterphenyl cation radical: X-ray crystallographic evidence of quinoidal charge delocalization in poly-p-phenylene cation radicals. J Am Chem Soc 129:8070–8071
95. Banerjee M, Shukla R, Rathore R (2009) Synthesis, optical, and electronic properties of soluble poly-p-phenylene oligomers as models for molecular wires. J Am Chem Soc 131:1780–1786
96. Kayahara E, Kouyama T, Kato T, Yamago S (2016) Synthesis and characterization of [n]CPP (n = 5, 6, 8, 10, and 12) radical cation and dications: size-dependent absorption, spin, and charge delocalization. J Am Chem Soc 138:338–344
97. Kayahara E, Kouyama T, Kato T, Takaya H, Yasuda N, Yamago S (2013) Isolation and characterization of the cycloparaphenylene radical cation and dication. Angew Chem Int Ed 52:13722–13726

98. Toriumi N, Muranaka A, Kayahara E, Yamago S, Uchiyama M (2015) In-plane aromaticity in cycloparaphenylene dications: a magnetic circular dichroism and theoretical study. J Am Chem Soc 137:82–85

99. Kothe G, Denkel K, Sümmermann W (1970) Schlenk's biradical—a molecule in the triplet ground state. Angew Chem Int Ed Engl 9:906–907

100. Rajca A, Utamapanya S, Xu J (1991) Control of magnetic interactions in polyarylmethyl triplet diradicals using steric hindrance. J Am Chem Soc 113:9235–9241

101. Rajca A, Utamapanya S (1992) π-Conjugated systems with unique electronic structure: a case of "planarized" 1,3-connected polyarylmethyl carbodianion and stable triplet hydrocarbon diradical. J Org Chem 57:1760–1767

102. Rajca A (1990) A polyarylmethyl quintet tetraradical. J Am Chem Soc 112:5890–5892

103. Rajca A (1990) A polyarylmethyl carbotetraanion. J Am Chem Soc 112:5889–5890

104. Utamapanya S, Rajca A (1991) Topological control of electron localization in π-conjugated polyarylmethyl carbopolyanions and radical anions. J Am Chem Soc 113:9242–9251

105. Rajca A, Utamapanya S, Thayumanavan S (1992) Poly(arylmethyl) Octet ($S = 7/2$) heptaradical and undecet ($s = 5$) decaradical. J Am Chem Soc 114:1884–1885

106. Rajca A, Utamapanya S (1993) Poly(arylmethyl) quartet triradicals and quintet tetraradicals. J Am Chem Soc 115:2396–2401

107. Latif IA, Hansda S, Datta SN (2012) High magnetic exchange coupling constants: a density functional theory based study of substituted Schlenk diradicals. J Phys Chem A 116:8599–8607

108. Clar E (1972) Aromatic sextet. Wiley, London

109. Ovchinnikov AA (1978) Multiplicity of the ground state of large alternant organic molecules with conjugated bonds—(Do Organic Ferromagnetics Exist?). Theor Chim Acta 47:297–304

110. Borden WT, Davidson ER (1977) Effects of electron repulsion in conjugated hydrocarbon diradicals. J Am Chem Soc 99:4587–4594

111. Morita Y, Suzuki S, Sato K, Takui T (2011) Synthetic organic spin chemistry for structurally well-defined open-shell graphene fragments. Nat Chem 3:197–204

112. Inoue J, Fukui K, Kubo T, Nakazawa S, Sato K, Shiomi D, Morita Y, Yamamoto K, Takui T, Nakasuji K (2001) The first detection of a Clar's hydrocarbon, 2,6,10-tri-*tert*-butyltriangulene: a ground-state triplet of non-Kekulé polynuclear benzenoid hydrocarbon. J Am Chem Soc 123:12702–12703

113. Li Y, Huang K-W, Sun Z, Webster RD, Zeng Z, Zeng W, Chi C, Furukawa K, Wu J (2014) A kinetically blocked 1,14:11,12-dibenzopentacene: a persistent triplet diradical of a non-Kekulé polycyclic benzenoid hydrocarbon. Chem Sci 5:1908

114. Clar E (1964) Polycyclic hydrocarbons: v.1, and v.2. Academic Press Inc., London

115. Bendikov M, Duong HM, Starkey K, Houk KN, Carter EA, Wudl F (2004) Oligoacenes: theoretical prediction of open-shell singlet diradical ground states. J Am Chem Soc 126:7416–7417

116. Mondai R, Shah BK, Neckers DC (2006) Photogeneration of heptacene in a polymer matrix. J Am Chem Soc 128:9612–9613

117. Tönshoff C, Bettinger HF (2010) Photogeneration of octacene and nonacene. Angew Chem Int Ed 49:4125–4128

118. Einholz R, Fang T, Berger R, Grüninger P, Früh A, Chassé T, Fink RF, Bettinger HF (2017) Heptacene: characterization in solution, in the solid state, and in films. J Am Chem Soc 139:4435–4442

119. Payne MM, Parkin SR, Anthony JE (2005) Functionalized higher acenes: hexacene and heptacene. J Am Chem Soc 127:8028–8029

120. Purushothaman B, Bruzek M, Parkin SR, Miller A-F, Anthony JE (2011) Synthesis and structural characterization of crystalline nonacenes. Angew Chem Int Ed 50:7013–7017

121. Pavliček N, Mistry A, Majzik Z, Moll N, Meyer G, Fox DJ, Gross L (2017) Synthesis and characterization of triangulene. Nat Nanotechnol 12:308–311

122. Zugermeier M, Gruber M, Schmid M, Klein BP, Ruppenthal L, Müller P, Einholz R, Hieringer W, Berndt R, Bettinger HF, Gottfried JM (2017) On-surface synthesis of heptacene and its interaction with a metal surface. Nanoscale. doi:10.1039/C7NR04157H

123. Urgel JI, Hayashi H, Di Giovannantonio M, Pignedoli CA, Mishra S, Deniz O, Yamashita M, Dienel T, Ruffieux P, Yamada H, Fasel R (2017) On-surface synthesis of heptacene organometallic complexes. J Am Chem Soc. doi:10.1021/jacs.7b05192

124. Zuzak R, Dorel R, Krawiec M, Such B, Kolmer M, Szymonski M, Echavarren AM, Godlewski S (2017) Nonacene generated by on-surface dehydrogenation. ACS Nano. doi:10.1021/acsnano.7b04728

125. Krüger J, García F, Eisenhut F, Skidin D, Alonso JM, Guitián E, Pérez D, Cuniberti G, Moresco F, Peña D (2017) Decacene: on-surface generation. Angew Chem Int Ed. doi:10.1002/anie.201706156

126. Jiang D, Sumpter BG, Dai S (2007) First principles study of magnetism in nanographenes. J Chem Phys 127:124703

127. Hod O, Barone V, Scuseria GE (2008) Half-metallic graphene nanodots: a comprehensive first-principles theoretical study. Phys Rev B 77:1–6

128. Moscardó F, San-Fabián E (2009) On the existence of a spin-polarized state in the n-periacene molecules. Chem Phys Lett 480:26–30

129. Jiang DE, Dai S (2008) Circumacenes versus periacenes: HOMO–LUMO gap and transition from nonmagnetic to magnetic ground state with size. Chem Phys Lett 466:72–75

130. Plasser F, Pašalić H, Gerzabek MH, Libisch F, Reiter R, Burgdörfer J, Müller T, Shepard R, Lischka H (2013) The multiradical character of one- and two-dimensional graphene nanoribbons. Angew Chem Int Ed 52:2581–2584

131. Yamaguchi K (1990) Self-Consistent Field: Theory and Applications. In: Carbo R, Klobukowski M (eds) Instability chemical bonding. Elsevier, Amsterdam, pp 727–823

132. Yamanaka S, Okumura M, Nakano M, Yamaguchi K (1994) EHF theory of chemical reactions Part 4. UNO CASSCF, UNO CASPT2 and R(U)HF coupled-cluster (CC) wavefunctions. J Mol Struct (Theochem) 310:205–218

133. Yamaguchi K, Kawakami T, Takano Y, Kitagawa Y, Yamashita Y, Fujita H (2002) Analytical and ab initio studies of effective exchange interactions, polyradical character, unpaired electron density, and information entropy in radical clusters $(R)_N$: allyl radical cluster ($N = 2–10$) and hydrogen radical cluster ($N = 50$). Int J Quantum Chem 90:370–385

134. Shimizu A, Hirao Y, Kubo T, Nakano M, Botek E, Champagne B (2012) Theoretical consideration of singlet open-shell character of polyperiacenes using Clar's aromatic sextet valence bond model and quantum chemical calculations. In: AIP conference proceedings American Institute of Physics, p 399–405

135. Glukhovtsev MN, Bach RD, Laiter S (1997) Isodesmic and homodesmotic stabilization energies of [n]annulenes and their relevance to aromaticity and antiaromaticity: is absolute antiaromaticity possible? J Mol Struct (Theochem) 417:123–129

136. Slayden SW, Liebman JF (2001) The energetics of aromatic hydrocarbons: an experimental thermochemical perspective. Chem Rev 101:1541–1566

137. Douglas J (1955) Kinetics of the thermal *Cis-Trans* isomerization of dideuteroethylene. J Chem Phys 23:315

138. Roberson LB, Kowalik J, Tolbert LM, Kloc C, Zeis R, Chi X, Fleming R, Wilkins C (2005) Pentacene disproportionation during sublimation for field-effect transistors. J Am Chem Soc 127:3069–3075

139. Rogers C, Chen C, Pedramrazi Z, Omrani AA, Tsai H, Jung HS, Lin S, Crommie MF, Fischer FR (2015) Closing the nanographene gap: surface-assisted synthesis of peripentacene from 6,6′-bipentacene precursors. Angew Chem Int Ed 54:15143–15146

140. Zhang X, Li J, Qu H, Chi C, Wu J (2010) Fused bispentacenequinone and its unexpected michael addition. Org Lett 12:3946–3949

141. Matsumoto A, Suzuki M, Kuzuhara D, Hayashi H, Aratani N, Yamada H (2015) Tetrabenzoperipentacene: stable five-electron donating ability and a discrete triple-layered β-graphite form in the solid State. Angew Chem Int Ed 54:8175–8178

142. Dorel R, Manzano C, Grisolia M, Soe W-H, Joachim C, Echavarren AM (2015) Tetrabenzocircumpyrene: a nanographene fragment with an embedded peripentacene core. Chem Commun 51:6932–6935

143. Zöphel L, Berger R, Gao P, Enkelmann V, Baumgarten M, Wagner M, Müllen K (2013) Toward the *peri*-pentacene framework. Chem Eur J 19:17821–17826

144. Liu J, Ravat P, Wagner M, Baumgarten M, Feng X, Müllen K (2015) Tetrabenzo[a, f, j, o]perylene: a polycyclic aromatic hydrocarbon with an open-shell singlet biradical ground state. Angew Chem Int Ed 54:12442–12446

145. Scholl R, Meyer K (1934) Der blaue aromatische Grundkohlenwasserstoff des *meso*-Naphthodianthrons und seine Überführung durch Maleinsäure-anhydrid in Anthro-dianthren. Ber Dtsch Chem Ges (A B Series) 67:1236–1238

146. Li J, Zhang K, Zhang X, Huang KW, Chi C, Wu J (2010) *meso*-Substituted bisanthenes as soluble and stable near-infrared dyes. J Org Chem 75:856–863

147. Fort EH, Donovan PM, Scott LT (2009) Diels–Alder reactivity of polycyclic aromatic hydrocarbon bay regions: implications for metal-free growth of single-chirality carbon nanotubes. J Am Chem Soc 131:16006–16007

148. Hirao Y, Konishi A, Matsumoto K, Kurata H, Kubo T, Simos TE, Maroulis G (2012) Synthesis and electronic structure of bisanthene: a small molecular-sized graphene with zigzag edges. In: AIP conference proceedings American Institute of Physics, pp. 863–866

149. Konishi A, Hirao Y, Nakano M, Shimizu A, Botek E, Champagne B, Shiomi D, Sato K, Takui T, Matsumoto K, Kurata H, Kubo T (2010) Synthesis and characterization of teranthene: a singlet biradical polycyclic aromatic hydrocarbon having Kekulé structures. J Am Chem Soc 132:11021–11023

150. Kruszewski J, Krygowski TM (1972) Definition of aromaticity basing on the harmonic oscillator model. Tetrahedron Lett 13:3839–3842

151. Krygowski TM (1993) Crystallographic studies of inter- and intramolecular interactions reflected in aromatic character of π-electron systems. J Chem Inf Model 33:70–78

152. Angeli C, Pastore M, Cimiraglia R (2007) New perspectives in multireference perturbation theory: the *n*-electron valence state approach. Theor Chem Acc 117:743–754

153. Di Motta S, Negri F, Fazzi D, Castiglioni C, Canesi EV (2010) Biradicaloid and polyenic character of quinoidal oligothiophenes revealed by the presence of a low-lying double-exciton state. J Phys Chem Lett 1:3334–3339

154. Konishi A, Hirao Y, Matsumoto K, Kurata H, Kishi R, Shigeta Y, Nakano M, Tokunaga K, Kamada K, Kubo T (2013) Synthesis and characterization of quarteranthene: elucidating the characteristics of the edge state of graphene nanoribbons at the molecular level. J Am Chem Soc 135:1430–1437

155. Pan M, Girão EC, Jia X, Bhaviripudi S, Li Q, Kong J, Meunier V, Dresselhaus MS (2012) Topographic and spectroscopic characterization of electronic edge states in CVD grown graphene nanoribbons. Nano Lett 12:1928–1933

156. Tao C, Jiao L, Yazyev OV, Chen Y-C, Feng J, Zhang X, Capaz RB, Tour JM, Zettl A, Louie SG, Dai H, Crommie MF (2011) Spatially resolving edge states of chiral graphene nanoribbons. Nat Phys 7:616–620

157. Hou Z, Wang X, Ikeda T, Huang S-F, Terakura K, Boero M, Oshima M, Kakimoto M, Miyata S (2011) Effect of hydrogen termination on carbon *K*-edge X-ray absorption spectra of nanographene. J Phys Chem C 115:5392–5403

158. Suenaga K, Koshino M (2010) Atom-by-atom spectroscopy at graphene edge. Nature 468:1088–1090

159. Joly VLJ, Kiguchi M, Hao S-J, Takai K, Enoki T, Sumii R, Amemiya K, Muramatsu H, Hayashi T, Kim YA, Endo M, Campos-Delgado J, López-Urías F, Botello-Méndez A, Terrones H, Terrones M, Dresselhaus MS (2010) Observation of magnetic edge state in graphene nanoribbons. Phys Rev B 81:245428

160. Sugawara K, Sato T, Souma S, Takahashi T, Suematsu H (2006) Fermi surface and edge-localized states in graphite studied by high-resolution angle-resolved photoemission spectroscopy. Phys Rev B 73:45124

161. Kobayashi Y, Fukui K, Enoki T, Kusakabe K (2006) Edge state on hydrogen-terminated graphite edges investigated by scanning tunneling microscopy. Phys Rev B 73:125415

162. Niimi Y, Matsui T, Kambara H, Tagami K, Tsukada M, Fukuyama H (2005) Scanning tunneling microscopy and spectroscopy studies of graphite edges. Appl Surf Sci 241:43–48

163. Kobayashi Y, Fukui K, Enoki T, Kusakabe K, Kaburagi Y (2005) Observation of zigzag and armchair edges of graphite using scanning tunneling microscopy and spectroscopy. Phys Rev B 71:193406

164. Nakada K, Fujita M, Dresselhaus G, Dresselhaus MS (1996) Edge state in graphene ribbons: nanometer size effect and edge shape dependence. Phys Rev B 54:17954–17961

165. Fujita M, Wakabayashi K, Nakada K, Kusakabe K (1996) Peculiar localized state at zigzag graphite edge. J Phys Soc Jpn 65:1920–1923

166. Tanaka K, Yamashita S, Yamabe H, Yamabe T (1987) Electronic properties of one-dimensional graphite family. Synth Met 17:143–148

167. Cai J, Ruffieux P, Jaafar R, Bieri M, Braun T, Blankenburg S, Muoth M, Seitsonen AP, Saleh M, Feng X, Müllen K, Fasel R (2010) Atomically precise bottom-up fabrication of graphene nanoribbons. Nature 466:470–473

168. Bieri M, Treier M, Cai J, Aït-Mansour K, Ruffieux P, Gröning O, Gröning P, Kastler M, Rieger R, Feng X, Müllen K, Fasel R (2009) Porous graphenes: two-dimensional polymer synthesis with atomic precision. Chem Commun 6919

169. Bieri M, Nguyen M-T, Gröning O, Cai J, Treier M, Aït-Mansour K, Ruffieux P, Pignedoli CA, Passerone D, Kastler M, Müllen K, Fasel R (2010) Two-dimensional polymer formation on surfaces: insight into the roles of precursor mobility and reactivity. J Am Chem Soc 132:16669–16676

170. Narita A, Feng X, Müllen K (2015) Bottom-up synthesis of chemically precise graphene nanoribbons. Chem Rec 15:295–309

171. Talirz L, Söde H, Cai J, Ruffieux P, Blankenburg S, Jafaar R, Berger R, Feng X, Müllen K, Passerone D, Fasel R, Pignedoli CA (2013) Termini of bottom-up fabricated graphene nanoribbons. J Am Chem Soc 135:2060–2063

172. Wang S, Talirz L, Pignedoli CA, Feng X, Müllen K, Fasel R, Ruffieux P (2016) Giant edge state splitting at atomically precise graphene zigzag edges. Nat Commun 7:11507

173. Ruffieux P, Wang S, Yang B, Sánchez-Sánchez C, Liu J, Dienel T, Talirz L, Shinde P, Pignedoli CA, Passerone D, Dumslaff T, Feng X, Müllen K, Fasel R (2016) On-surface synthesis of graphene nanoribbons with zigzag edge topology. Nature 531:489–492

174. Hu P, Wu J (2017) Modern zethrene chemistry. Can J Chem 95:223–233

175. Sun Z, Zeng Z, Wu J (2014) Zethrenes, extended *p*-quinodimethanes, and periacenes with a singlet biradical ground state. Acc Chem Res 47:2582–2591

176. Nakano M, Kishi R, Takebe A, Nate M, Takahashi H, Kubo T, Kamada K, Ohta K, Champagne B, Botek E (2007) Second hyperpolarizability of zethrenes. Comput Lett 3:333–338

177. Minami T, Nakano M (2012) Diradical character view of singlet fission. J Phys Chem Lett 3:145–150

178. Clar E, Lang KF, Schulz-Kiesow H (1955) Aromatische Kohlenwasserstoffe, LXX. Mitteil. 1): zethren (1.12; 6.7-dibenztetracen). Chem Ber 88:1520–1527

179. Wu T-C, Chen C-H, Hibi D, Shimizu A, Tobe Y, Wu Y-T (2010) Synthesis, structure, and photophysical properties of dibenzo[de, mn]naphthacenes. Angew Chem Int Ed 49:7059–7062

180. Umeda R, Hibi D, Miki K, Tobe Y (2009) Tetradehydrodinaphtho[10]annulene: a hitherto unknown dehydroannulene and a viable precursor to stable zethrene derivatives. Org Lett 11:4104–4106

181. Clar E, Macpherson IA (1962) The significance of Kekulé structures for the stability of aromatic systems—II. Tetrahedron 18:1411–1416

182. Li Y, Heng W-K, Lee BS, Aratani N, Zafra JL, Bao N, Lee R, Sung YM, Sun Z, Huang K-W, Webster RD, López Navarrete JT, Kim D, Osuka A, Casado J, Ding J, Wu J (2012) Kinetically blocked stable heptazethrene and octazethrene: closed-shell or open-shell in the ground state? J Am Chem Soc 134:14913–14922

183. Sun Z, Huang K-W, Wu J (2011) Soluble and stable heptazethrenebis(dicarboximide) with a singlet open-shell ground state. J Am Chem Soc 133:11896–11899

184. Huang R, Phan H, Herng TS, Hu P, Zeng W, Dong S, Das S, Shen Y, Ding J, Casanova D, Wu J (2016) Higher-order π-conjugated polycyclic hydrocarbons with open-shell singlet ground state: nonazethrene versus nonacene. J Am Chem Soc 138:10323–10330

185. Sun Z, Lee S, Park KH, Zhu X, Zhang W, Zheng B, Hu P, Zeng Z, Das S, Li Y, Chi C, Li R-W, Huang K-W, Ding J, Kim D, Wu J (2013) Dibenzoheptazethrene isomers with different biradical characters: an exercise of Clar's aromatic sextet rule in singlet biradicaloids. J Am Chem Soc 135:18229–18236

186. Yadav P, Das S, Phan H, Herng TS, Ding J, Wu J (2016) Kinetically blocked stable 5,6:12,13-dibenzozethrene: a laterally π-extended zethrene with enhanced diradical character. Org Lett 18:2886–2889

187. Sun Z, Zheng B, Hu P, Huang K-W, Wu J (2014) Highly twisted 1,2:8,9-dibenzozethrenes: synthesis, ground state, and physical properties. ChemPlusChem 79:1549–1553

188. Zeng W, Sun Z, Herng TS, Gonçalves TP, Gopalakrishna TY, Huang K-W, Ding J, Wu J (2016) Super-heptazethrene. Angew Chem Int Ed 55:8615–8619

Top Curr Chem (Z) (2018) 376:12
https://doi.org/10.1007/s41061-018-0189-0

REVIEW

Quinodimethanes Incorporated in Non-Benzenoid Aromatic or Antiaromatic Frameworks

Yoshito Tobe[1]

Received: 4 September 2017 / Accepted: 12 February 2018 / Published online: 20 March 2018
© Springer International Publishing AG, part of Springer Nature 2018

Abstract Three isomers of quinodimethanes (QDMs) adopt different electronic configurations and geometries, generating their own characteristic physical properties. Incorporation of QDMs into non-benzenoid aromatic or antiaromatic frameworks not only planarizes the whole π system optimizing conjugation, but also changes the electronic properties inherent to QDMs, sometimes drastically, due to the topology of the π system, through interaction with the remaining part of the molecules. In non-benzenoid systems, molecular orbital levels and orbital distribution are uneven compared to benzenoid systems, thereby polarizing the ground state and leading to unique behavior in excited states. In antiaromatic systems, open-shell, diradical character, which is inherent to QDMs, may be enhanced due to small HOMO–LUMO energy gap. In this chapter, effects of incorporating QDMs into non-benzenoid aromatic or antiaromatic frameworks are discussed focusing on the open-shell, diradical character with respect to their molecular structures, antiaromaticity, and physical properties related to the open-shell character and molecular orbital levels and materials applications, as well as covering historical works to the current state-of-the-art achievements.

Keywords Quinodimethanes · Open-shell character · Kekulé structure · Aromaticity · Antiaromaticity

Chapter 3 was originally published as Tobe, Y. Top Curr Chem (Z) (2018) 376: 12. https://doi.org/10.1007/s41061-018-0189-0.

✉ Yoshito Tobe
tobe@chem.es.osaka-u.ac.jp

[1] The Institute of Scientific and Industrial Research, Osaka University, 8-1 Mihogaoka, Ibaraki, Osaka 567-0047, Japan

1 Introduction

Three structural isomers of quinodimethanes (QDMs), *o*-quinodimethane (*o*-QDM), *m*-quinodimethane (*m*-QDM), and *p*-quinodimethane (*p*-QDM) have different ground state electronic configurations, closed-shell *o*-QDM and *p*-QDM and open-shell *m*-QDM, which can be easily understood by chemical structure drawing of Kelulé structures for the *o*- and *p*-isomers and non-Kelulé structure for the *m*-isomer (Fig. 1). The structural difference is explained more accurately in terms of Hückel molecular orbital (HMO) calculations [1] and on the basis of spin polarization [2, 3]. No unsubstituted QDMs are stable enough for isolation. Derivatives of *p*- and *m*-QDM have been known since the dawn of physical organic chemistry, whereas *o*-QDMs have been investigated as versatile intermediates for construction of the naphthalene skeleton in modern synthetic organic chemistry [4]. The prototypical *p*-QDM, Thiele's hydrocarbon (**1**), was isolated in 1904 [5], and its extended homologue **2**, called Chichibabin's hydrocarbon, was reported just 2 years later [6]. However, until the detailed ESR study of **2** was undertaken in 1964 [7], its electronic structure was not fully understood due to irregular results observed by different researchers under different conditions. It was further clarified by the X-ray crystallographic structure analysis of **1** and **2** reportd in 1986 [8], by which different behavior of **1** and **2** are explained in terms of different open-shell character, in other words, a different magnitude of diradical character [1, 9]. This notion was further confirmed by the ESR study [10] for further extended molecule **3**, already known in 1941 [11], which exhibits enhanced diradical character. Similarly, *m*-QDM derivative **4**, known as Schlenk's hydrocarbon, was reported as early as 1915 [12]. However, chemistry of *m*-QDM has advanced significantly as reactive intermediates [13, 14] and high-spin molecules [15–17] because of its triplet ground state [1, 9]. In contrast, stable derivatives of *o*-QDM have not been synthesized until relatively recent years.

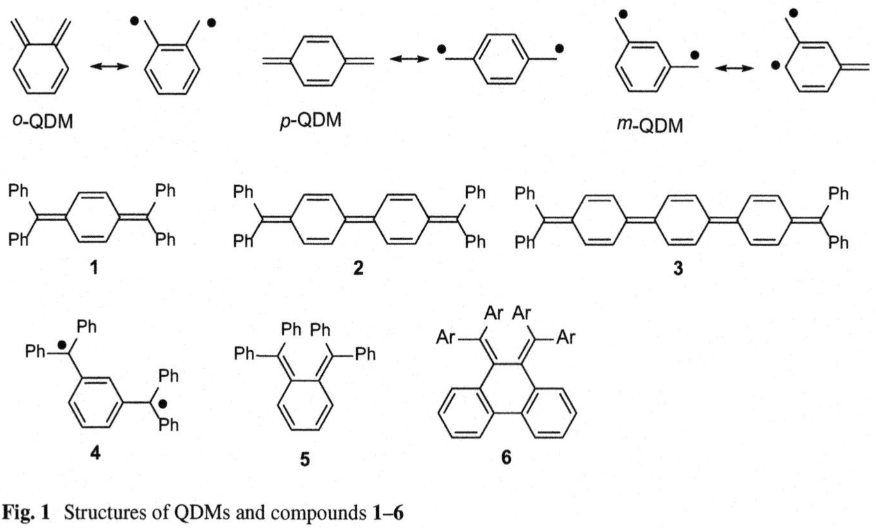

Fig. 1 Structures of QDMs and compounds **1–6**

Whereas tetraphenyl *o*-QDM **5** can only detected spectroscopically in rigid glass matrices at low temperature [18, 19], tetraaryl derivatives **6** incorporated in a phenanthrene framework have enough stability for isolation [20, 21]. The properties of QDMs are often explained in terms of the relative weight of closed-shell and open-shell diradical resonance canonical structures as described in Fig. 1, in other words "pro-aromaticity", which means the driving force for the formation of diradicals is the aromatic stabilization energy gained by breaking double bonds in a non-aromatic system, such as QDMs, that can surpass the energy required to break the double bonds [22].

Because of the recent progress and advances in theoretical, synthetic, physical, and materials aspects of singlet diradicaloids, QDMs have attracted increasing interest [23–27]. The common strategy in the recent studies is to incorporate QDM unit(s) into an aromatic or antiaromatic framework, which (i) planarizes the π system optimizing conjugation and, more importantly, (ii) changes electronic properties sometimes drastically due to the topology of the π system, through interaction with the remaining part of the molecules. For example, incorporation into a non-benzenoid framework will change the total number of electrons with that of benzenoid aromatic molecules, often making the system antiaromatic. Because antiaromatic molecules such as cyclobutadiene and planar cyclooctatetraene possess open-shell character due to small HOMO–LUMO energy gap [22, 28, 29], open-shell, diradical character inherent to QDMs may be enhanced. The open-shell character is defined theoretically, in most cases by Yamaguchi scheme [Eq. (1-1)], by twice the weight of the doubly excited configuration in the ground state [30], though simply the occupation numbers are used in some instances, e.g. Equation (2) shown later [31].

$$y_i = 1 - \frac{2T_i}{1 + T_i^2} \tag{1-1}$$

where T_i (i = 0 for HOMO/LUMO, 1 for HOMO $-$ 1/LUMO $+$ 1, and so on) is the orbital overlap between the HOMO $-$ i and LUMO $+$ i. It is defined by occupation numbers (n_i) of UHF or UDFT natural orbitals by Eq. (1-2).

$$T_i = \frac{n_{\text{HOMO}-i} - n_{\text{LUMO}+i}}{2} \tag{1-2}$$

Moreover, in non-benzenoid systems, molecular orbital levels and orbital distribution are uneven compared to benzenoid systems, thereby polarizing the ground state or leading to unique behavior in excited states [32, 33]. In this respect, QDMs incorporated into a non-benzenoid aromatic or antiaromatic framework are expected to exhibit unconventional characteristics.

On the basis of this background, in this chapter, effects of incorporation of QDMs into non-benzenoid aromatic or antiaromatic frameworks are discussed focusing on an open-shell, diradical character with respect to their molecular structures, antiaromaticity, and physical properties related to the open-shell character and molecular orbital levels and materials applications, as well as covering historical works to the current state-of-the-art achievements. Representative structures are shown Fig. 2. It starts with pleiadene (**7a**), in which an *o*-QDM unit is conjoined with a cyclohepta[*de*]naphthalene core, intensively studied spectroscopically decades ago to reveal the existence of doubly exited state in the ground state electronic configuration. The next section gives a detailed overview of *s*-indacene (**8a**) and its congeners, in which *p*-QDM is incorporated in a 5–6–5 conjugated ring system, because of their relevance as constituent units at the central portion of the QDM molecules that will appear in the latter sections. In the subsequent sections, a class of compounds called bisphanelenyl (**9a**), in which **8a** is flanked by two phenalenyls, followed by indenofluorene isomers, such as **10a** and **11a**, in which **8a** is flanked by two benzene rings, will be described. In connection with indenofluorenes, tetraradicaloids like **12a** and **13a** are described followed by miscellaneous systems such as the most recent molecule **14a**.

2 Pleiadene System

Benzo[5,6]cyclohepta[1,2,3-*de*]naphthalene (**7a**), coined pleiadene by Fieser [34], is an *o*-QDM derivative conjoined into a cyclohepta[*de*]naphthalene (**15**) framework (Fig. 3). It has attracted significant interest for several decades because of its *o*-QDM substructure [35], elusiveness for isolation, and relevance to **15** which was isolated as a stable entity [36, 37].

In a series of papers, Cava et al. [38–40] reported generation of **7a** and its ace derivatives, acepleiadene (**16a**) and acepleiadylene (**16b**), in solution. The most persistent **16b** was prepared by acid-catalyzed double dehydration of diol **17b** and was characterized the blue color of the solution with the longest wavelength absorption

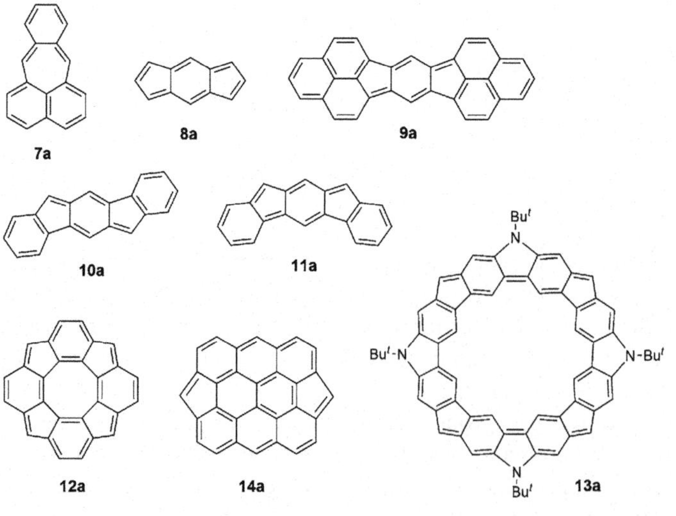

Fig. 2 Structures of compounds **7a–14a**

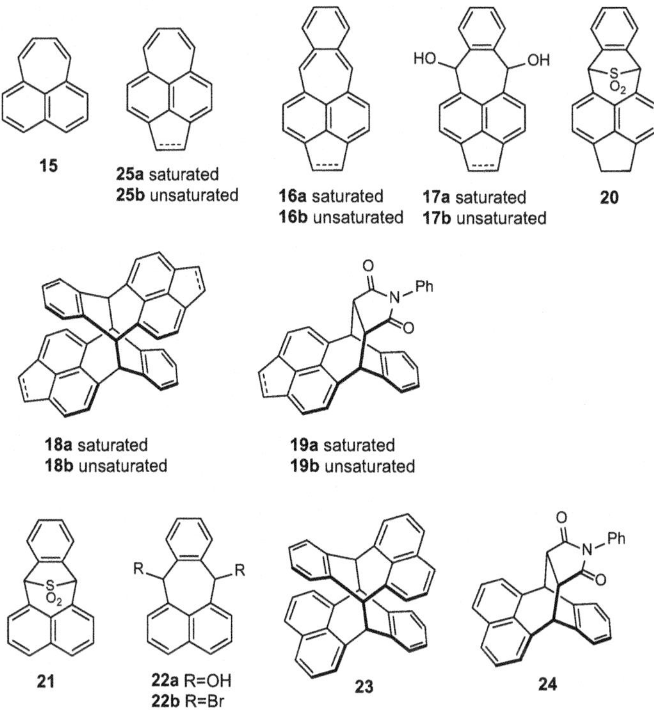

Fig. 3 Structures of compounds **15–25**

at 674 nm. Later, an absorption spectrum of **16b**, generated by treatment of **17b** with HCl gas at − 40 °C, was measured in 2-methyltetrahydrofuran (2-MTHF) at − 196 °C [41]. However, all attempts to isolate **16b** in the solid state failed to yield anti dimer **18b** in 30–50% yield together with polymeric materials. Treatment of a solution of **16b** with *N*-phenylmaleimide (NPM) at room temperature afforded adduct **19b**. The anti-stereochemistry of dimer **18b** and [4 + 2] adduct **19b** was assumed based on the similarity with those of the corresponding products obtained from **7a**, whose stereochemistry was determined on the basis of spectroscopic considerations. These results indicate that **16b** is kinetically stable enough to survive for some moment at temperatures lower than room temperature [38]. The dihydo derivative, acepleiadene (**16a**), was generated by heating sulfone **20** at 240 °C, which was prepared from diol **17a** by treatment with phosphorous pentasulfide, followed by oxidation with peracetic acid, yielding dimer **18a** in 75% yield. Photoirradiation of a benzene solution of **20** at 10 °C also generated **16a**, although no visual evidence of a colored intermediate was observed. Instead, dimer **18a** was obtained in 54% yield. Heating **20** at 210 °C in the presence of NPM gave adduct **19a** in 74% yield. Because hydrogenation of the corresponding dimer **18b** and NPM adduct **19b** gave **18a** and **19a**, respectively, the same anti stereochemistry of the latter products was confirmed [39]. The parent pleiadene (**7a**) was also generated from sulfone **21**, prepared from diol **22a**, by pyrolysis in the molten state at 250 °C or in diphenyl phthalate at 200 °C, yielding dimer **23** in ca. 50% yields. Pyrolysis in the presence of NPM gave adduct **24** in 68% yield. UV irradiation at room temperature gave dimer **23** too. Dimer **23** was also obtained by reduction of dibromide **22b** with zinc or copper-bronze at temperatures ranging from 25 to 150 °C [40]. The anti-stereochemistry of dimer **23** and [4 + 2] adduct **24** was determined on the basis of comparison of the ^{1}H NMR chemical shifts with appropriate model compounds [40]. These results indicate that even though **7a** and **16a** are generated as short-lived intermediates; they are too reactive to observe spectroscopically at near room temperature. Clearly shortening the distance between the peri positions of the naphthalene rings on the order of **7a** (no bridge), **16a** (an ethylene bridge), and **16b** (an ethenylene bridge) reduced their reactivity by shortening the distance between the most reactive carbons of the *o*-QDM substructure. It is worth mentioning that the stability order of pleiadene homologues parallels that of **15** and its ace derivatives, cyclohepta[*fg*]acenaphthylene (**25a**) and 1,2-dihydrocyclohepta[*fg*]acenaphthylene (**25b**), all of which were isolated [36, 37]. In boiling benzene, [4 + 2] addition of **15** with maleic anhydride completed in 10 h, whereas 2.5 days were required for the same reaction with **25a**. Moreover, with **25b** no reaction took place under the same conditions after 8.5 days. The remarkably different reactivity must be steric, but not electronic origin (i.e., number of π electrons). Similar steric (distance-controlling) effect of naphthalene-acenaphthene-acenaphthylene units has been reported for peri-substituted acridinium dications [42].

Spectroscopic characterization of pleiadene (**7a**) in solvent glass matrices was achieved by Michl et al. [41, 43–48]. In addition to **7a**, characterization of 2,3-naphthoquinodimethne homologue **26** and its 1,2-anologue **27,** as well as methyl-substituted derivatives **7b** and **7c** was achieved (Fig. 4). More importantly, they also pointed out theoretically the contribution of doubly excited state in ground state

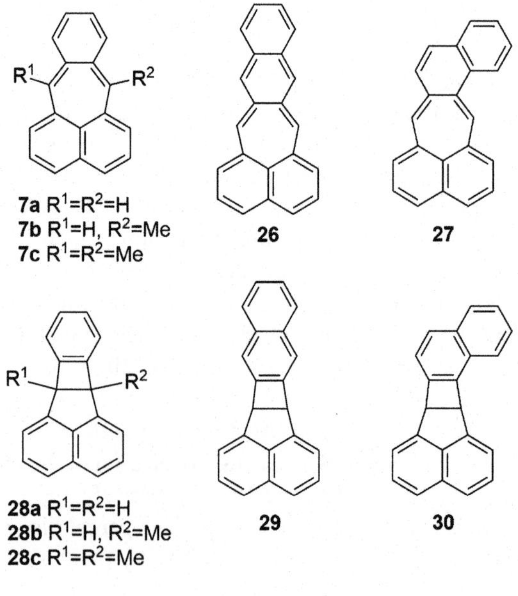

Fig. 4 Structures of pleiadenes **7a–c**, **26**, and **27** and their precursors **28a–c**, **29**, and **30**

7a R^1=R^2=H
7b R^1=H, R^2=Me
7c R^1=R^2=Me

26

27

28a R^1=R^2=H
28b R^1=H, R^2=Me
28c R^1=R^2=Me

29

30

electronic configuration of **7a** and demonstrated experimentally by the observation of absorption due to this electronic transition involving a doubly excited state.

All pleiadene congeners are generated by ring opening of the corresponding benzo-(or naphtho-)fused 6b,8a-dihydrocyclobuta[a]acenaphthylenes **28a–c**, **29**, and **30** [41, 43, 47]. The photochemical process involves higher excited states as described below for the conversion of **28a–7a** as a representative example [44, 48]. While precursor **28a** is thermally inert, because its conrotatory ring opening is symmetry forbidden, it produced pleiadene (**7a**) upon photo-irradiation only in frozen solvent matrices at low temperature (typically 77 K), conditions under which both fluorescence and phosphorescence of the precursors **28a** were observed. When the solvent matrix was melted, dimer **23** was formed, as characterized by the absorption spectrum and TLC. The reaction did not take place—no dimer **23** was formed—in fluid solution at room temperature or in melted solvent glass of 3-methylpentane (3-MP) at − 100 °C. In addition to the reaction medium, there are a few uncommon features in the photochemical ring opening reaction of **28a**. Upon broadband irradiation at wavelengths of > 300 nm or with monochromatic radiation at 303 nm, the initial rate of product formation is proportional to the second power of light intensity. Moreover, by irradiation of **28a** in 3-MP glass (77 K) using two directly opposed beams impinging on the sample, one with a weak monochromatic beam of 303 nm light and the other an intense broadband beam filtered by a band-pass filter (330–410 nm), **7a** was formed at least an order of magnitude faster than with the first beam alone. The wavelength region of the second beam was not absorbed by the precursor **28a** and it corresponds naphthalene triplet–triplet absorption (390 and 415 nm), the ring opening reaction is assumed to take place from high triplet excited states of **28a** generated via (i) first excitation to S_1 by the 303 nm light, (ii) intersystem crossing to T_1, (iii) further excitation to an upper triplet state by second (long wavelength) light absorption, (iv) at which ring opening takes place. Moreover, by 229 and 214 nm radiation the rate of the formation of **7a** was found to be proportional to the beam intensity, and the quantum yields of the reaction depended on the wavelength: $\phi\,(214) = 0.009$, $\phi\,(229) = 0.006$, $\phi\,(254) < 0.0003$. These observations indicate the existence of one-photon process from upper and/or hot excited singlet states in addition to the above-mentioned two-photon processes from a triplet state.

Naphtho derivatives **26** and **27** behave similarly; they could be characterized only in the solvent glass matrix [41]. Upon melting the glass by warming, the yellow and green color disappear to give insoluble materials mainly containing dimers. The dilute solution (10^{-4} M) of **26** is least stable and **7a** is only a little more. In contrast, the dilute solution of **27** is considerably more stable; the color persists after warming up to room temperature for a while (4–5 min). After warming to room temperature, both **26** and **27** gave products presumably by dimerization, though the structures of the products were not characterized. It is worth noting that there are a number of possible dimer structures in addition to those corresponding to **23**, not only due to orientation of dimerization (for **27**), but also due to the naphthoquinodimethane substructure, which may provide additional reactive centers. In contrast, dimethyl derivative **7b**, generated by photo irradiation of **28b**, is more stable than **7a** and its yellowish-green color persists for minutes at 0 °C, reverting quantitatively back to the precursor **28b** with an activation energy of 21.3 kcal/mol (log A = 15.6 s^{-1}) [47]. No dimer corresponding to **23** was formed. Monomethyl derivative **7c** behaves between **7a** and **7b**. Its yellowish-green color persists up to about − 25 °C, giving a mixture of **7c** and poorly soluble materials of a dimeric nature. Although the conrotatory ring closure like the conversion of **7b–28b** is symmetry forbidden, the facile ring closure which takes place at temperatures below 0 °C is ascribed to the involvement of doubly excited state with a diradicaloid (open-shell) character in the ground state of **7b**.

Certainly, the most important insight in a series of work on pleiadene done by Michl et al. is the contribution of doubly excited state, in other words, diradicaloid character, in its ground state electronic configuration [41, 45, 46]. As shown in Table 1 for electronic configurations of **7a**, **26**, and **27** calculated by semiempirical π-electron PPP (Pariser-Parr-Pople) SCF-CI method, which includes some double excited configurations, the occupation number of the highest occupied natural orbital (HONO) of **7a** is 1.90 and that of the lowest unoccupied natural orbital (LUNO) is 0.09, indicating diradical nature of its ground state. This tendency is more remarkable in **26,** with an occupation number of 1.85 and 0.14 for HONO and LUNO, respectively. In contrast, the diradical character of **27** is significantly smaller than **7a** and **26**, consistent with the observed longer persistency of **27** than the other two in solution. Calculations also indicate that the singlet → triplet excitation energy of **7a** is very small and nearly zero in the case of **26**, suggesting the potential occupation of thermally excited triplet states.

On the basis of the PPP SCF-CI calculations, in addition to the low energy HOMO → LUMO transition due to small HOMO–LUMO gap, low-lying singlet states of **7a** and **26** are calculated to contain large amounts of a two-electron excitation (HOMO,HOMO → LUMO,LUMO), whereas in **27** this configuration contributes almost equally to several low-lying states [41]. Moreover, due to the small energy of excitation, the two-electron excitation bands of **7a** and **26** are predicted not to overlap with other transitions. Indeed, the two-electron excitation bands of **7a** and **26** were experimentally observed as shown in Fig. 5 for absorption spectrum of **7a** in 3-MP at 77 K in which the absorption bands are numbered from 1 to 11 [46]. The assignments of the electronic transitions were done based on the vibrational analysis and peak intensities, and the measured polarization direction for the

Table 1 Electronic configurations of **7a**, **26**, and **27** calculated by PPP SCF-CI method [41]

Compd	n_H^a	n_L^b	^1E (eV)^c	$E_{H,H \to L,L}$ (eV)^d	^3E (eV)^e
7a	1.90	0.09	1.75 (0.87)	3.62	0.39
26	1.85	0.14	1.54 (0.84)	2.87	0.02
27	1.96	0.03	2.00 (0.88)	4.40	0.76

^{a}Occupation number of HONO

^{b}Occupation number of LUNO

^{c}Excitation energy of the lowest singlet state with the weight of HOMO → LUMO configuration in this state in parentheses

^{d}Energy of the HOMO,HOMO → LUMO,LUMO configuration

^{e}Excitation energy of triplet state

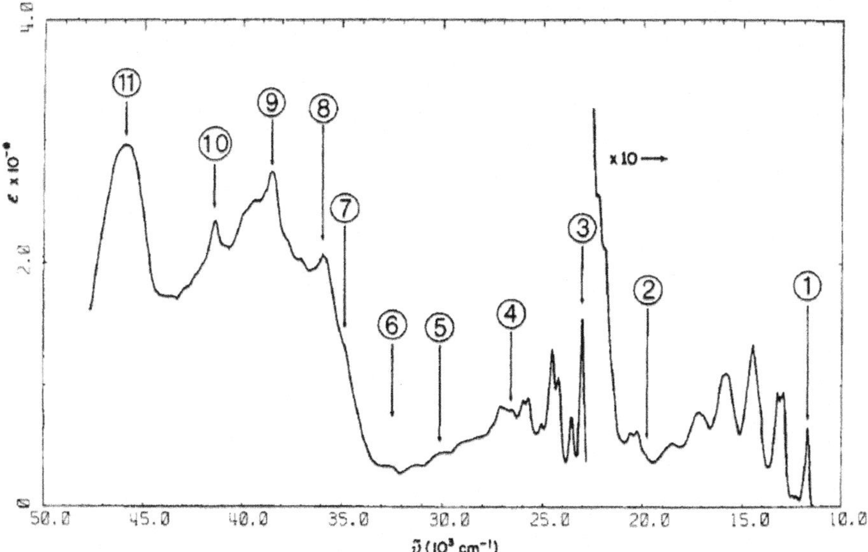

Fig. 5 Absorption spectrum of **7a** in 3-MP at 77 K. Reproduced with permission from Ref. [46]. Copyright 1976 American Chemical Society

lowest five transitions and analysis of the small spectral shifts due to methyl substitution of several methyl derivatives of **7a**, which bears a methyl group attached on the naphthalene or the six-membered ring. On the basis of these theoretical and experimental studies, the second lowest energy band of **7a** marked by a circled 2 in Fig. 5, with its longest wavelength feature (19840 cm^{-1}) observed as a weak shoulder located on top of a tail of transition 1, is attributed to the two-photon transition. The experimental observation of the two-electron excitation confirms the contribution of doubly excited state, i.e., diradicaloid character, in the ground state electronic configuration of **7a**.

Facile thermal [4 + 4] dimerization of **7a** to give **23**, which takes place below room temperature, is also related to the diradicaloid character of **7a** [41]. The mechanism of the reaction, whether it is a stepwise one via a diradical intermediate or a concerted one, is not proven. In the first case, the activation energy of the reaction must be reduced, because the ground state has already diradicaloid character. In the latter case, though [4 $_s$ + 4 $_s$] cycloaddition is typically symmetry-forbidden, symmetry-allowed or symmetry-forbidden reaction paths lose much of their significance, since the two electrons in HOMO of **7a** are considerably non-bonding. Moreover, in either of the mechanisms, secondary orbital interaction must favor a transition state leading to head-to-tail dimer rather than head-to-head dimer as was indeed observed, although steric hindrance would also favor the observed product.

3 Indacene System

There are two structural isomers for indacenes, *s*-indacene (**8a**) with linearly fused 5–6–5 membered rings and *as*-indacene (**31**) with angularly fused 5–6–5 rings (Fig. 6) [49, 50]. Whereas **8a** contains a *p*-QDM unit in the tricyclic framework, **31** has two Kekulé structures containing a *p*-QDM (**31-1**) or *o*-QDM (**31-2**) unit. Compared to *as*-indacene (**31**) for which only a few papers have been reported, intense theoretical and synthetic works have been done for *s*-indacene (**8a**), because of its high-symmetric framework which is included in many π-conjugated systems related to *p*-QDM.

3.1 *s*-Indacene

Namely, **8a** can adopt either bond-localized, low symmetry C_{2h} structure or bond-delocalized high symmetry D_{2h} structure. In the C_{2h} structure, a *p*-QDM unit enclosed in the π-conjugated system is apparent. The high symmetry D_{2h} structure locates a transition state of degenerate C_{2h} structures which are interconverting (Scheme 1).

The parent hydrocarbon **8a** was generated by the electrocyclic reaction of an amino-bisfulvene intermediate **32** and subsequent elimination of dimethylamine (Scheme 2) [51]. Despite attempts it eluded isolation even at temperatures below − 30 °C. However, it was characterized by ^{1}H NMR (signals between 7.6–6.3 ppm in an intensity ratio of 1:2:1) and chemical transformation, for example, by catalytic hydrogenation to tetrahydroindacene and reduction with sodium in liquid anion giving the corresponding dianion salt of 14 π-electrons [51, 52]. Although no attempts seem to have been made to investigate further **8a** experimentally, a few derivatives substituted by amino groups have been synthesized, principally based on the same strategy as that used for the generation of **8a** [53, 54]. All of them have one or two dimethylamino group(s) at the central position(s) (C4, C8) and frequently one or two *t*-butyl group(s) at C2, C6 of the five-membered ring(s). Because mono-dimethyl-amino derivative is not very stable, substitution by the electron-donating groups is

Fig. 6 Structures of *s*-indacenes **8a,b** and *as*-indacene (**31**)

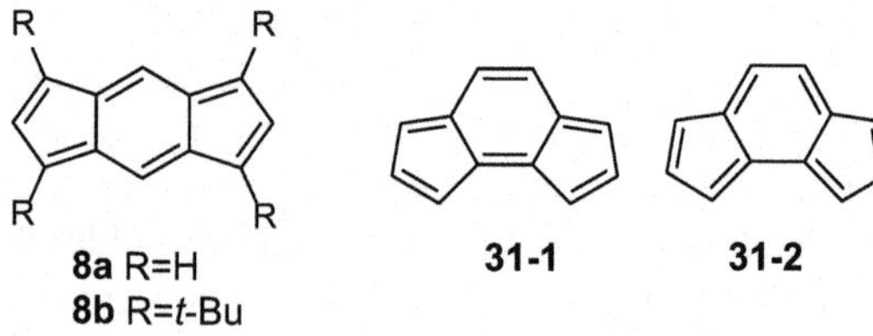

8a R=H
8b R=*t*-Bu

31-1   **31-2**

Scheme 1 Bond-localized, low symmetry (C_{2h}) structure and bond-delocalized, high symmetry (D_{2h}) structure of *s*-indacene (**8a**) with carbon numbering

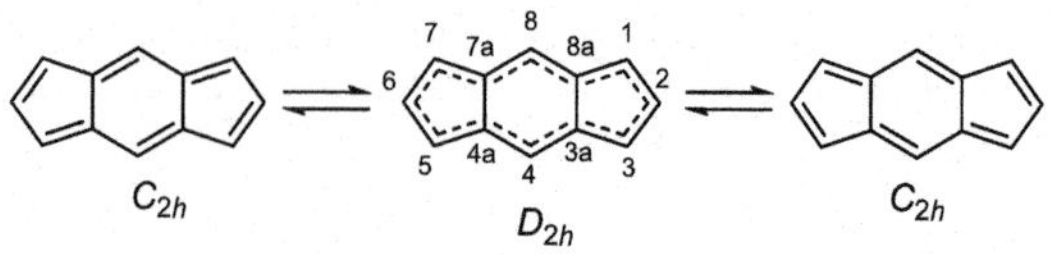

C_{2h} D_{2h} C_{2h}

Scheme 2 Synthetic scheme for the generation of *s*-indacene (**8a**)

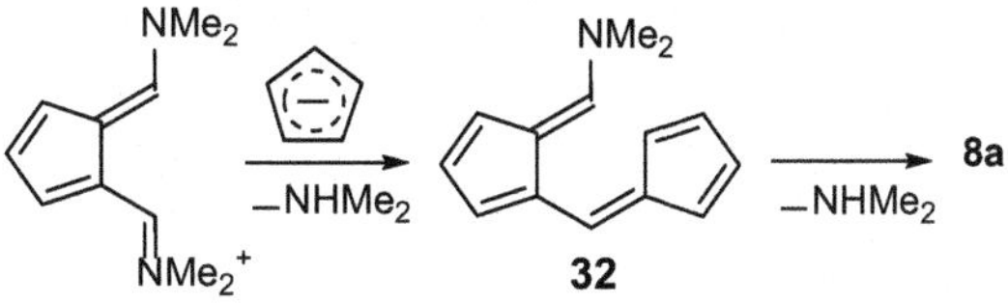

necessary to stabilize the molecules, presumably by perturbing the intrinsic electronic structure of **8a**.

1,3,5,7-Tetrta-*tert*-butyl-*s*-indacene (**8b**), synthesized via an analogous method by Hafner in 1986, is the only known hydrocarbon derivative of **8a** [55]. While it is stable in the solid state, it is extremely sensitive towards oxygen and traces of acid in solution. Although the longest wavelength absorption of **8b** reported in [55] is at 545 nm (log ε 4.64), it appears to exhibit a low intensity absorption band at significantly longer wavelength region as described below. In ^{1}H NMR spectrum, the vinylic protons of **8b** appear at 5.26 (H2, H6) and 6.90 (H4, H8) ppm, being consistent with paratropicity in the 12 π-electron circuit. Although preliminary X-ray structural analysis at room temperature was conducted to reveal a highly symmetric structure, more precise analyses were done afterwards. Since even by lowering the temperature to $- 130$ °C, the shape of the ^{13}C NMR signals did not change, it was not possible to conclude whether (i) **8b** adopts interconverting low-symmetry C_{2h} structures with minimal energy barrier or (ii) the π-system is delocalized adopting a D_{2h} structure (Scheme 1).

Crystallographic structural analysis of **8b** at low temperature (100 K) was conducted [56]. As shown in Table 2, the bond lengths of **8b** indicate that it adopts an approximate D_{2h} structure. Although there exists a possibility of disorder between two sets of aromatic positions for the skeletal atoms, it is excluded by detailed analysis of the anisotropic displacement parameter of the carbon atoms. Reexamination of the X-ray analysis of **8b** was conducted to confirm the D_{2h}-like structure as seen from the bond lengths listed in Table 2, as well as to reveal additional information [57]. The topological analysis on experimental and theoretical electron densities indicates the existence of bond critical points and the corresponding bond paths linking a hydrogen atom attached to C4 (C8) with four methyl hydrogen atoms of the *t*-butyl groups attached to C3 and C5 (C1 and C7) as shown in Fig. 7. These results indicate the presence of weak intramolecular dihydrogen interaction, $C(sp^3)$–H $\cdots$ H–C(sp^2) with the H $\cdots$ H distances of 2.17(2) – 2.19(2) Å, which may contribute to stabilize the D_{2h}-like structure.

In addition to the fundamental physical and structural properties related to antiaromatic character, **8b** exhibits interesting photophysical properties. First, it exhibits a weak symmetry-forbidden $S_0 \rightarrow S_1$ absorption in the wavelength ranging from 9000 to 16,000 cm^{-1} (600–1100 nm) in cyclohexane with a maximum extinction coefficient of ca. 70 M^{-1} cm^{-1}, which is typical in antiaromatic molecules, in addition to intense narrow band in the visible region at 19,000 cm^{-1} (ε 50,000 M^{-1} cm^{-1}) and in the UV at 33,000 cm^{-1} [58]. More interestingly, though emission from the S_1 state was not detected, suggesting effective radiationless deactivation of the S_1 level, a weak fluorescence (quantum yield: 9×10^{-5}) was observed between 15,000 and 19,000 cm^{-1} by excitation at frequencies lower than 19,000 cm^{-1} with a small

Table 2 Bond lengths (Å) determined by X-ray crystallographic analysis of **8b**

Bond	Ref. [55][a]	Ref. [56][b]	Ref. [57][b]
1–2	1.409	1.406	1.411
1–8a	1.435	1.438	1.441
2–3	1.397	1.408	1.410
3–3a	1.446	1.434	1.443
3a–4	1.392	1.395	1.400
4–4a	1.408	1.394	1.400
3a–8a	1.440	1.442	1.446

Because of the presence of a center of symmetry, only a half of the bond length data are listed. For carbon numbering, see Scheme 1

[a] At room temperature

[b] At 100 K

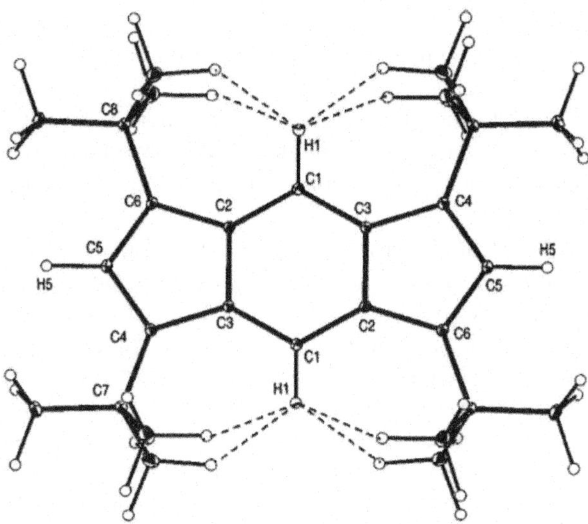

Fig. 7 Molecular structure of **8b** showing the intramolecular dihydrogen bond interaction. Reproduced with permission [57] of the International Union of Crystallography

Stokes shift of 600 cm^{-1}. The lifetimes of S_1 and S_2 states were determined to be 18 and 2.5 ps, respectively, by time-resolved measurements. CNDO/SDCI calculations for **8a** suggest that radiationless $S_2 \rightarrow S_0$ decay is more effective than $S_2 \rightarrow S_1$ decay pathway on the basis of the molecular geometries of the S_0, S_1 and S_2 states; the former is optimized to adopt a C_{2h} structure, while the latter two have D_{2h} structures of similar geometry. More detailed analysis of the fluorescence of **8b** from the second excited state was done by the measurements of $S_0 \rightarrow S_2$ absorption, fluorescence excitation, and fluorescence in rigid glass solution of isopentane-ether at 77 K [59]. Namely, several emissions were observed, which were assigned to different conformers of **8b** due to the torsion of the *t*-butyl group with respect to the molecular plane. The observed fluorescence spectra are explained assuming that the indacene core of **8b** in the S_2 states adopts a D_{2h} structure, whereas that in the S_0 state has less symmetric C_{2h} structure. The structural change by excitation to S_2 from S_0 of **8b** is also supported by the analysis of absorption and resonance Raman spectroscopy in glass solution at 77 K [60]. For the observed $S_0 \to S_2$, $S_0 \to S_3$, and $S_0 \to S_6$ transitions, major vibration modes are assigned on the basis of semiempirical QCFF-PI calculations for **8a**, which give a C_{2h} structure of better agreement with the experimental D_{2h}-like structure of **8b** than SCF methods (see Table 3). It is worth noting that the occurrence of $S_2 \to S_0$ emission violating Kasha's rule, a common unusual feature of non-alternant hydrocarbons **8b** and azulene [61–63], is closely related to (i) the large energy gap between the S_2 and S_1 states and (ii) the high oscillator strength of the $S_0 \to S_2$ transition. The short S_1 lifetime and absence of fluorescence from S_1 are explained by theoretically locating the conical intersection between the S_0 and S_1 states of **8a** by molecular mechanics with valence bond (MMVB) and complete active space SCF (CASSCF) calculations [64]. Namely, because the intersection minimum is close in energy and geometry to that on the S_1 state and can be readily reached along the coordinate from the Frank–Condon geometry to the S_1 minimum, non-radiative decay crossing the S_1/S_0 surface can occur easily.

The molecular structure of **8a**, with regard to the C_{2h} or D_{2h} structure, is favored in its ground state (Scheme 1), has been a topic of significant theoretical interest in connection with the distortion from symmetric structure, a phenomenon well-documented both by theory and experiments in antiaromatic annulenes cyclobutadiene and cycooctatetraene [28, 29, 65], as well as the mechanism of heavy atom tunneling between low-symmetry isomers of antiaromatic molecules [65–69]. Additionally, because the experimentally determined structure of tetra-*t*-butyl derivative **8b** is D_{2h}-like, both **8a** and **8b** continue to be subjects of further theoretical study. Early HMO [70] and semiempirical calculations [71] show that the low-symmetry C_{2h} structure is more favored over the delocalized D_{2h}. Hatree-Fock (HF) MO

Table 3 Theoretical bond lengths (Å) of **8a** and **8b**

Bond	HF/3-21 G[a] C_{2h} **8a** (D_{2h} **8b**)	MP2/6-31G*[b] D_{2h} **8a**	LDA + BP/TZP[b] D_{2h} **8a**	B3LYP/6-31G*[c] C_{2h} **8a**/D_{2h} **8a**	B3LYP/6-31G[d] (D_{2h} **8b**)
1–2	1.341 (1.398)	1.404	1.408	1.383/1.406	(1.414)
1–8a	1.473 (1.411)	1.429	1.427	1.455/1.423	(1.443)
2–3	1.479 (1.398)	1.404	1.408	1.432/1.406	(1.414)
3–3a	1.347 (1.411)	1.429	1.427	1.400/1.423	(1.443)
3a–4	1.450 (1.387)	1.397	1.401	1.419/1.399	(1.402)
4–4a	1.335 (1.387)	1.397	1.401	1.380/1.399	(1.402)
3a–8a	1.476 (1.477)	1.437	1.444	1.452/1.488	(1.454)

For carbon numbering, see Scheme 1

[a]Bond lengths for **8b** are given in parentheses. Ref. [72]

[b]Reference [75]

[c]Reference [76]

[d]Bond lengths for **8b**. Ref. [78]

calculations at medium sized basis sets also result in favor of C_{2h} structures by 18.2 (STO-3G) to 5.6 (double-ζ valence: DZV) kcal/mol than the D_{2h} structure [72]. As an example, an optimized geometry using 3-21G basis set, which is 7.9 kcal/mol more stable than the D_{2h} structure, is shown in Table 3. In view of the D_{2h}-like structure observed by X-ray crystallography of **8b** as described above and the earlier suggestion based on quantitative consideration on bond–bond polarizability that substitution at the 1, 3, 5, and 7 positions decreases the double bond localization in the indacene core [73], Salvi et al. [72] studied the structure of **8b** theoretically. As a result, though C_{2h} structure is still favored, the energy difference becomes much smaller (0.9 kcal/mol), suggesting almost degenerate C_{2h} and D_{2h} structures of **8b**. The bond lengths of the D_{2h} structure optimized using 3-21G basis set agree well with the experimental structure as listed in Table 3. Moreover, among the various conformers arising from rotation of the $C(sp^3)$- $C(sp^2)$ bond of the *t*-butyl groups with respect to the indacene plane, the one in which the one of the methyl carbons of a *t*-butyl group is placed in the molecular plane and orient to the opposite direction to the central six-membered ring, is calculated to be most stable. This conformation is the same one as that observed by X-ray crystallography (Fig. 2) and can be assumed to be the only one predominates at room temperature.

On the other hand, higher level quantum chemical calculations reveal the lowest energy singlet state of **8a** adopts D_{2h} structures [74, 75]. Ab initio calculations based on Møller–Plesset perturbation theory (MP2/6-31G*) including electron correlations favors the C_{2h} structure compared to D_{2h} by 0.7 kcal/mol. However, single-determinant-based methods, such as MP2, are not suited to describe properly these molecules as is indicated by large spin contaminations. Complete active space SCF (CASSCF) calculations including 12 π-electrons in 12 MOs for the geometry optimized by the MP2/6-31G* are more stable than D_{2h} by 5.7 kcal/mol. Moreover, excitation to the LUMO is prominent, as it is occupied by 0.45 electrons in D_{2h} and 0.29 electrons in the C_{2h} structures, respectively, indicating substantial singlet diradical character. Further refinement of the level of theory using the CASPT2 method, a multireference version of second-order, many-body perturbation theory with full CASSCF wavefunction was carried out to the D_{2h} structure is more stable than the localized C_{2h} structure by 3.1 kcal/mol. Bond lengths of the D_{2h} structure is included in Table 3. As another approach, density functional theory employing the local density approximation (LDA) with nonlocal gradient corrections for exchange and correlation (BP) was employed to optimize the structure. As a result, whereas the D_{2h} structure was optimized as shown in Table 3, C_{2h}-symmetric minimum was not found. The structural parameters derived by both MP2 and LDA + BP methods agree well with the experimental structure of **8b** (Table 2) with the root mean square deviations for the C–C bonds of only 0.05 Å and 0.06 Å, respectively. It was also pointed out that even though the π-electrons are delocalized throughout the whole molecule, singlet *s*-indacene is probably better described as built from two allyl radicals connected to a benzene ring, because four pairs of the bonds have a length around 1.4 Å, similar to those of benzene (1.39 Å) and allyl radical (1.387 Å) and three pairs with a longer distance of about 1.43 Å or slightly longer [75]. Triplet states have also been computed to have D_{2h}-symmetric minima, with a large singlet–triplet energy gap of 17.4 kcal/mol (CASPT2) or 18.5 kcal/mol (BP/TZP).

DFT calculations for **8a** using B3LYP/6-31G* functional and basis set, which successfully reproduce structures of a variety of π-conjugated systems, were carried out [76, 77]. As a result, the C_{2h} structure (Table 3) appears a minimum while the D_{2h}-structure located 0.1 kcal/mol above the C_{2h} minimum is transition state for the double bond shifting with an imaginary frequency and a transition vector connecting C_{2h} to D_{2h} structures. However, if the zero point energy is included, the D_{2h} structure becomes 0.6 kcal/mol more stable than C_{2h}. The calculated D_{2h} structure (Table 3) matches well the experimental geometry of **8b**. Therefore, the authors call *s*-indacene a "quasi-delocalized" molecule. In view of the effect of 1, 3, 5, 7-tetraalkyl substitution [73], the **8b** may well favor the delocalized D_{2h} structure. However, another complication is pointed out by Salvi et al. who investigated the theoretical structure of tetra-*t*-butyl derivative **8b** by B3LYP method [78]. Though the most stable conformer (see Fig. 7) due to the rotation around the $C(sp^3)$–$C(sp^2)$ bond adopts D_{2h} geometry (Table 2) in accord with the HF calculations described above, there exist a few thermally attainable conformers (within 3 kcal/mol from the minimum with barriers of smaller than 5 kcal/mol) which adopt either D_{2h} or C_{2h} structure depending on the direction of the methyl groups. In conclusion, it is deduced that though the most stable structure of unsubstituted **8a** must have a D_{2h} geometry, since the energy difference between the C_{2h} and D_{2h} structures are small, the structure of isolable derivative should depend on the electronic properties of substituents, as well as their substitution patterns.

In addition to **8a**, the structural and energetic aspects of its higher homologues with additional benzene ring(s) in the center of the framework are also theoretically investigated, though none of the higher homologues including their derivatives have been synthesized yet. For example, Aihara pointed out a unique behavior of **8a** by comparing the energetic and magnetic criteria of aromaticity, topological resonance energy (TRE) and magnetic resonance energy (MRE: aromatic stabilization energy derived from the magnetic response of a cyclic π-system) developed by himself, for pentalene (**33**) and the higher homologues of **8a** (Fig. 8) with increasing number (N) of six-membered ring(s), **34 (4)**–**34 (8)** (the number in parentheses = N) [79]. %TRE (defined as 100 times TRE derided by the π-electron energy of the polyene reference) of antiaromatic **33** is negative (− 2.02) and that of **8a** is very small (0.34), indicating its non-aromatic character based on the energetic criterion. Similarly, whereas MRE of **33** is also negative (− 0.154) in accord with paratropicity, that of **8a** (0.17) is positive in sign. Therefore, **8a** can be regarded as a prototype of paratropic, but non-antiaromatic system. Higher homologues of **8a**, however, have increasing positive %TRE values (from **34 (4)** to **34 (8)** 0.89, 1.07, 1.16, 1.20, 1.24, respectively), as well as positive MRE (from **34 (4)** to **34 (8)** 0.259, 0.318, 0.373, 0.430, 0.488, respectively) indicative of aromaticity, in spite of their $4n$ π-electrons. These results suggest that the conjugated circuit models are not always applicable to non-alternant hydrocarbons.

Aromaticity of **8a** was evaluated from geometrical, energetic, and magnetic criteria [76]. First, the degree of bond length alternation was evaluated using the maximum deviation of the C–C bond length from the mean value (Δr_m) and the Julg parameter ($A = 1 - (255/n)\Sigma[1(r_i/r)]^2$, where n is the number of delocalized C–C bonds, r_i the length of an individual C–C bond, and r is the mean C–C bond length:

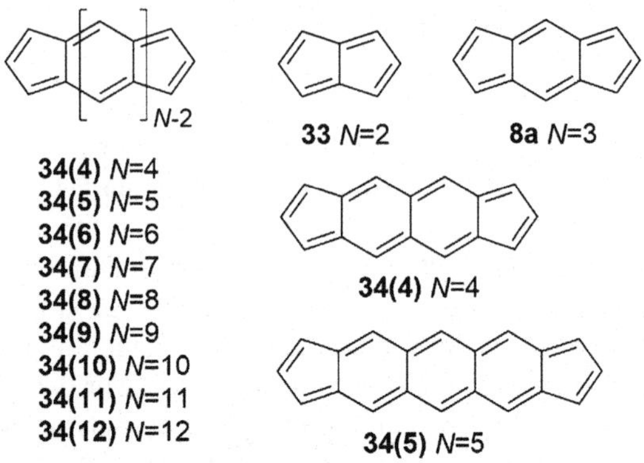

Fig. 8 Structures of pentalene (**33**) and higher homologues **34** (**4**)–**34** (**12**) of *s*-indacene (**8a**)

34(4) N=4
34(5) N=5
34(6) N=6
34(7) N=7
34(8) N=8
34(9) N=9
34(10) N=10
34(11) N=11
34(12) N=12

33 N=2   **8a** N=3

34(4) N=4

34(5) N=5

$A = 1$ for benzene.) [80]. For theoretically optimized (B3LYP/6-31G*) C_{2h} and D_{2h} structures of **8a**, Δr_m are 0.036 and 0.033 Å, respectively. The Julg parameters for theoretical C_{2h} and D_{2h} structures and experimental structure of **8b** are 0.920, 0.970, and 0.961, respectively. Comparison of these parameters with those of anthracene, $\Delta r_m = 0.039$ Å and $A = 0.919$, indicates that **8a** and **8b** are more delocalized than anthracene, in spite of $4n$ π-electrons. Next, aromatic stabilization energy (ASE) was estimated by homodesmotic reactions for **8a** and anthracene, yelding a much smaller value for **8a** than that of anthracene and even smaller than that of benzene. Therefore, **8a** is designated as non-aromatic. Finally, as the magnetic criterion the NICS(0) values are calculated NICS(0) for the six- and five-membered rings of C_{2h} structure (B3LYP/6-31G*) of **8a** are $+$ 17.6 and $+$ 21.6, and those for the D_{2h} structure are $+$ 20.8 and $+$ 25.8, respectively. These NICS values clearly indicate antiaromaticity of the *s*-indacene system. On the basis of these considerations, it is therefore reasonable to describe **8a** as a fully delocalized non-aromatic system with a paramagnetic ring current.

This view was confirmed by a theoretical study on ring currents and magnetic properties of **8a** by Lazzeretti et al. [81]. The structure of **8a** was optimized by the B3LYP level of theory using the 6-31G** and 6-311G** basis sets. Although the optimized geometries are similar to that obtained by using the 6-31G* basis set (Table 3), the optimum geometry corresponds the C_{2h} symmetry, and the D_{2h} geometry has one imaginary frequency value. Figure 9 shows the total (both π and σ components) integral trajectories of current density flux at 0.8 bohr above the molecular plane of **8a**. The paramagnetic π ring current is partly submerged by intense σ diamagnetic flow, though the paramagnetic flow at the center of the molecule is maintained. It shows a characteristic feature of magnetic properties of *s*-indacene in which interplay of σ and π currents determines the unique response. Current densities are also investigated for a series of formally antiaromatic $4n$ π-electron compounds, pentalene (**33**), **8a**, and its homologues **34** (**4**)–**34** (**7**) [82]. Structure optimization at the B3LYP/6-31G** level of theory results in the C_{2h} goemetries for **33** and **8a**, and D_{2h} geometries for **34** (**4**)–**34** (**7**). The current density map (Fig. 10) shows that the ring current changes from paratropic in **33** and **8a**, through quenched in **34** (**4**) and **34** (**5**), to diatropic in **34** (**6**) and **34** (**7**). Valence bond calculations for

Fig. 9 Total integral trajectories of current density flux at 0.8 bohr above the molecular plane of **8a**. Anticlockwise trajectory is paratropic and clockwise one is diatropic. Reproduced with permission from [81]. Copyright 2002 American Chemical Society

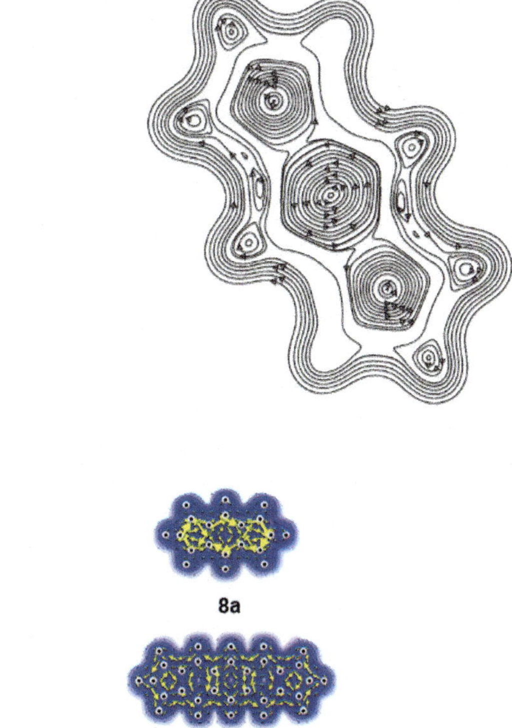

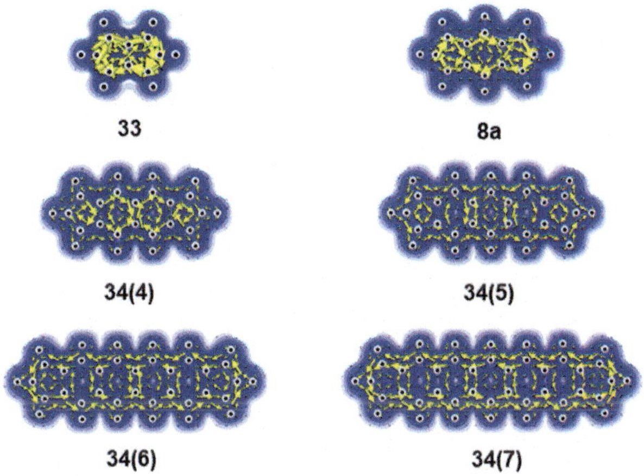

Fig. 10 Total current density map for **33**, **8a**, and **34** (**4**)–**34** (**7**) plotted 1 a_0 above the molecular. Anticlockwise circulation is diatropic and clockwise one is paratropic. Reproduced with permission from [82]. Copyright 2004 The Royal Society of Chemistry

the π system composed of atomic p-orbitals show that the electronic structures of the small members of the series (**33** and **8a**) are described in terms of two closed-shell Kekulé resonance structures, while in the larger molecules such as **34** (**6**) and **34** (**7**), allyl-polyacene-allyl biradical structures prevail, due to the large resonance energies gained from the polyacene portion of the molecules.

Nakano et al. undertook theoretical study on the relationship between the electronic, magnetic, and nonlinear optical properties for **33**, **8a**, and its homologues **34** (**4**)–**34** (**12**) using long-range corrected spin-unrestricted density functional theory (LC-UBLYP/6-31G*) to evaluate open-shell characters of these system which are critically related to the above properties [83]. Multiple diradical characters (y_i: i = 0 and 1) defined as Eq. (2) by the occupation numbers (n_k) of the natural orbitals

(NOs), [31] together with magnetic properties [nucleus independent chemical shift (NICS) 1 Å above the molecular plane, out-of-plane diagonal element of the σ tensor ($-\sigma_{zz}$) 1 Å above the molecular plane, and anisotropy of magnetic susceptibility ($\Delta\chi$)], and hyperpolarizabilities are listed in Table 4. Spin density distributions calculated by the LC-(U)BLYP/6-31G* method for **34 (5)**–**34 (9)** are shown in Fig. 11. As shown in Table 4, the diradical character y_0 increases rapidly with increasing the number of fused rings from $N = 2$ to 6 and reaches constant at $N = 10$. In contrast, y_1 starts to increase at $N = 8$, which is consistent with the fact that the spin densities appear at the acene portion of the molecules from $N = 8$ (Fig. 11). Namely, Fig. 11 shows that whereas the spin polarization for **34 (5)**–**34 (7)** is characterized by longitudinal components, that of **34 (8)** and **34 (9)** exhibits both longitudinal and transversal components, indicating that y_0 and y_1 correspond the increase of spin polarization in longitudinal and transversal directions, respectively. Moreover, comparison of the aromaticity indexes, NICS(1) and $-\sigma_{zz}$, with the diradical characters indicates that the longitudinal spin polarization (correlated to y_0) is associated with the antiaromatic ($N = 2, 3$) and non-aromatic ($N = 4, 5$) character at the terminal five-membered rings, whereas the transversal spin-polarization (correlated to y_1) is inversely related to the aromaticity in the acene part of the molecule. The longitudinal second hyperpolarizability is significantly enhanced with an increasing number of fused rings, which is associated with an increase in the longitudinal spin polarization between the terminal five-membered rings. The unique behaviors of relatively small members of compounds ($N < 9$) are ascribed to the increase in diradical

Table 4 Diradical characters (y_0 and y_1), magnetic properties (NICS(1), $-\sigma_{zz}$, and $\Delta\chi/N$), and longitudinal hyperpolarizability for **33**, **8a**, and **34 (4)**–**34 (12)** [83]

Compd	y_0^a	y_1^a	NICS(1) (central)[b]	NICS(1) (terminal)[b]	$-\sigma_{zz}$ (1Å) (central)[c]	$-\sigma_{zz}$ (1 Å) (terminal)[c]	$\Delta\chi/N^d$	γ/N^e
33	0.212	0.002	–	109.63	–	335.25	248.47	5.5
8a	0.393	0.007	4.40	9.85	19.57	36.86	− 0.15	21.7
34 (4)	0.602	0.012	− 4.57	0.59	− 6.71	9.27	− 5.88	65.0
34 (5)	0.744	0.015	− 9.62	− 1.82	− 21.60	2.16	− 7.96	131
34 (6)	0.836	0.018	− 10.98	− 2.70	− 25.61	− 0.43	− 9.07	208
34 (7)	0.895	0.021	− 12.94	− 3.09	− 31.43	− 1.57	− 9.80	294
34 (8)	0.929	0.146	− 11.22	− 2.76	− 26.20	− 0.55	− 9.56	328
34 (9)	0.947	0.305	− 9.51	− 2.52	− 20.96	− 0.16	− 9.23	310
34 (10)	0.956	0.485	− 7.71	− 2.25	− 15.45	0.97	− 8.98	363
34 (11)	0.961	0.609	− 6.46	− 2.16	− 11.65	1.25	− 8.89	447
34 (12)	0.965	0.690	− 6.13	− 2.16	− 10.63	1.24	− 8.88	538

[a]Defined by Eq. (2)

[b]Nucleus independent chemical shift 1 Å above the molecular plane

[c]Out-of-plane diagonal element of σ tensor 1 Å above the molecular plane

[d]Anisotropy of magnetic susceptibility per fused ring

[e]Longitudinal hyperpolarizability per fused ring in 10^5 a.u

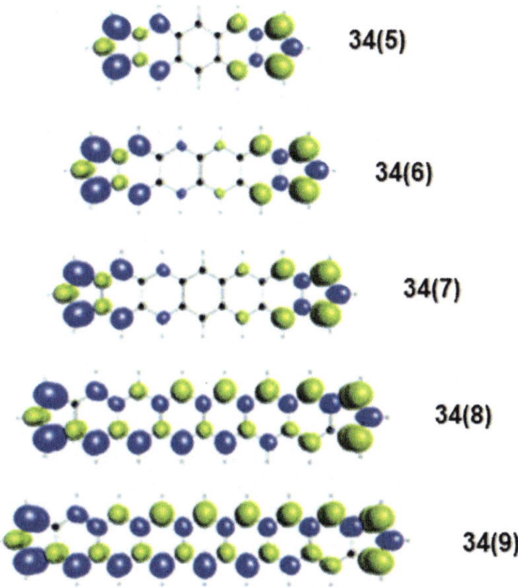

Fig. 11 Spin density distributions of **34 (5)**–**34 (9)**. The yellow and blue surfaces represent α and β densities with iso-surfaces of 0.005 a.u., respectively. Reproduced with permission from [83]. Copyright 2011 The Royal Society of Chemistry

character y_0 correlated with the antiaromaticity or non-aromaticity of terminal rings and the corresponding emergence of a global aromatic character.

$$y_i = n_{\text{LUNO}+i} = 2 - n_{\text{HONO}-i} \qquad (2)$$

As described in [81], **8a** provides a remarkable model for a peculiar magnetic response and represents a prototype of a "paratropic, but non-antiaromatic system." Since **8a** and its higher homologue **34 (4)** and **34 (5)**, which are yet unknown, but should be accessible in view of their medium open-shell character, are theoretically predicted to occupy a position at the borderline between the peculiar π-conjugated system and polyacene-like system, there remains much more to be clarified in the physical organic chemistry of *s*-indacene and its homologues.

3.2 *as*-Indacene

In contrast to *s*-indacene (**8a**), there are two possible bond-localized isomers **31–1** and **31–2** of *as*-indacene (**31**) in addition to bond-delocalized isomer **31–3** (Scheme 3). Notably, all of these belong to the same C_{2v} point group symmetry,

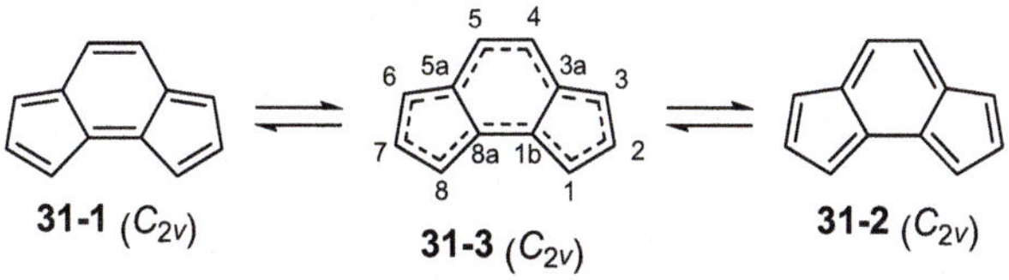

Scheme 3 Two bond-localized structures **31–1** and **31–2** and bond-delocalized structure **31–3** (all C_{2v} symmetric) of *as*-indacene (**31**) with carbon numbering

Springer

Table 5 Theoretical bond lengths (Å) and relative energies (kcal/mol) of **31–1**, **31–2**, and **31–3** [75] [a]

Bond	MP2/6-31G*			LDA + BP/TZP		
	31–1	**31–2**	**31–3**	**31–1**	**31–2**	**31–3**
2–3	1.469	1.364	1.390	1.474	1.362	1.409
3–3a	1.374	1.462	1.443	1.374	1.471	1.430
1–2	1.363	1.466	1.386	1.360	1.469	1.406
1b–8a	1.363	1.439	1.399	1.360	1.441	1.397
4–5	1.366	1.449	1.394	1.366	1.453	1.394
1b–3a	1.440	1.367	1.399	1.441	1.365	1.397
Relative energy	0.0	2.5	22.2	0.0	2.4	1.4

[a]For carbon numbering, see Scheme 3

making the situation more complicated. Theoretical calculations by at HF/6-31G*, MP2/6-31G* CASCSF/6-31G* CASPT/6-31G* and LDA + BT/TZP levels of theory were undertaken [75]. While **31–1** and **31–2** were localized on the singlet potential energy surfaces, no minimum structure corresponding to the delocalized geometry was found with all theoretical methods employed. Geometries are summarized in Table 5. The calculated bond lengths show the expected alternation for **31–1** and **31–2** and equalization in **31–3** as well. At most, in theory **31–1** is more stable than **31–2** by 2–3 kcal/mol. However, **31–1** is estimated to be less stable than **31–3** by 5.0 (MP2), 6.3 (CASPT2), and 8.7 (BP) kcal/mol, whereas CASSCF gives reversed result of − 7.6 kcal/mol. Although the transition state of bond shifting between **31–1** and **31–2** was not localized, it was estimated using the geometry of the triplet state to be 5.1 (CASSCF) kcal/mol or 3.1 (CASPT2) kcal/mol, suggesting the isomerization would take place with a low energy barrier.

In spite of the attempts to prepare *as*-indacene (**31**) [84], as well as an indication as a reactive intermediate generated under high temperature conditions [84], it was not detected. None of its stabilized derivatives are known either, thus are waiting for experimental verification of the structure and properties of a theoretically challenging system.

3.3 Indacene-Related Systems

Homologues of indacenes in which (a) five-membered ring(s) are replaced by (a) seven-membered ring(s) are conceivable [86–91], and a few of them were indeed synthesized. For the 5–6–7 (14 π-electron) ring system, whereas the linearly fused cyclohepta[*f*]indene (**35**) is unknown, angularly fused cyclohepta[*e*]indene (called "phenazulene" **36a**) was detected as a deep blue solution (λ_{max} 849, 770, 642, 585, and 479 nm in dichloromethane) at low temperature [92, 93]. Although it survived at − 60 °C only for several hours, it rapidly decomposed at room temperature. In contrast, dicyano derivatives **36b,c** were isolated [94].The longest wavelength absorptions of **36b** and **36c** are observed at 680 (sh) and 700 (sh) nm, respectively, in dichloromethane. Strong negative solvatochromism is also reported for **36b** (λ_{max} 660 (sh) nm in acetonitrile, 772 (sh) nm in tetachoromethane), indicating

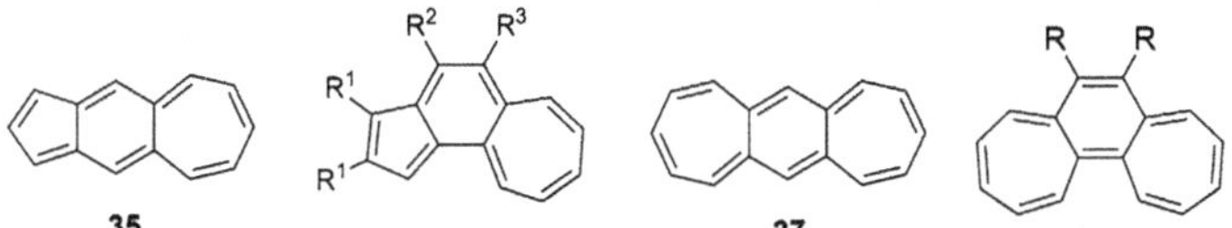

Fig. 12 Structures of 5–6–7 and 7–6–7 fused ring systems **35–38**

significantly polarized electronic structure. In the ^{1}H NMR spectra, the signals of the seven-membered ring protons appear at 9.2–9.5 ppm and 8.7–9.0 ppm, in contrast to the five-membered ring proton at 7.88 (for **36b**) and 8.06 (for **36c**). These results indicate the existence of the diamagnetic ring current, as well as the polarized nature in these compounds. [12 + 2] cycloaddition of **36b** and **36c** with both electron-rich and electron-deficient dienophiles such as 1-(1-cyclohexenyl)pyridine and dimethyl acetylenedicarboxylate took place to give the corresponding cycloadducts, which were converted to derivatives of cyclohepta[*fg*]acenaphthylene (**25a**) (see Sect. 2) by elimination reaction which took place during cycloaddition or oxidation [95]. For the 7–6–7 fused ring systems **37** and **38a**, no attempts seem to have been made to prepare stable derivatives of these 16 π-electron systems, except for attempts to synthesize the parent hydrocarbon **38a** and a dicycano derivative **38b** of the angular system [96, 97] (Fig. 12).

4 Biphenalenyl System

A series of compounds called biphenalenyls **9a–d**, **39a–b**, and **40a–b**, which consist of an *s*-indacene core or its higher congeners **34** (**4**) and **34** (**5**), respectively, flanked by two phenalenyl units (Fig. 13), were prepared by the group of Kubo et al., and their structure, physical properties, and applications were investigated by the same group collaborating with the groups of theory, spectroscopy, and applied physics, thus constructing recent foundation in singlet diradicaloids [26]. Although these are structurally more elaborate than indenofluorenes described in the next section, because of historical relevance to open-shell, singlet diradicaloids, they are described prior to indenofluorenes in this section.

As described in Sect. 3.1, because *s*-indacene can be regarded as a benzene flanked by two allyl radicals fused to the opposite edges, any radical-stabilizing units or functions at the allyl radical moiety will enhance open-shell character of the resulting π system. Phenalenyl is a well-known stable radical in which an unpaired electron is fully delocalized over the ring with spin density distributed at six equivalent positions and is utilized as a spin-delocalizing unit [98–100]. Through the studies on the homologous series of biphenalenyls, several interesting features have been elucidated [101–105]. Compounds **9b**, **39b**, and **40b** are stable in the solid state in air at least for a couple of days. However, quantum chemical calculations for the

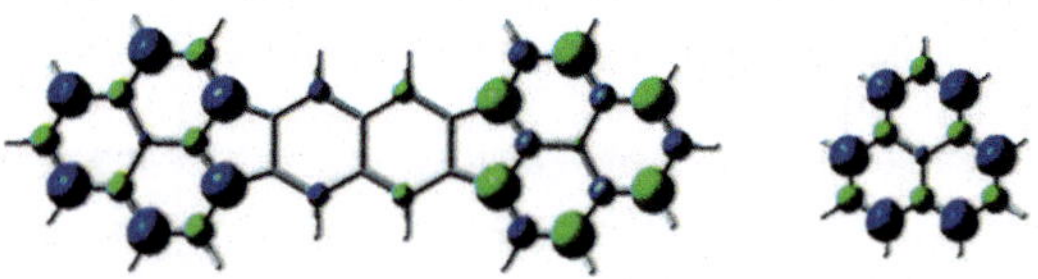

9a R^1=R^2=R^3=H
9b R^1=Ph, R^2=R^3=H
9c R^1=Ph, R^2=Me, R^3=H
9d R^1=R^2=H, R^3=t-Bu

39a R^1=R^2=R^3=H
39b R^1=Ph, R^2=H
39c R^1=Ph, R^2=t-Bu

40a R^1=R^2=R^3=H
40b R^1=Ph, R^2=H
40c R^1=Ph, R^2=t-Bu

Fig. 13 Structures of biphenalenyl derivatives **9a–d**, **39a–c**, and **40a–c**

Fig. 14 Spin densities of **39a** and phenalenyl radical (UB3LYP/6-31G**). Reproduced with permission from [105]. Copyright 2010 The American Chemical Society

basic molecule **9a** using the CASSCF(2,2)/6-31G method for the geometry optimized at the RB3LYP/6-31G** level indicate the occupation number of LUNO of 0.3, corresponding to the diradical factor y_0 [Eq. (2)] of 30%. The y_0 value increases with increasing size of the central aromatic part as 0.50 for **39a** and 0.68 for **40a**, similar to the case of homologues of s-indacene. Spin density distributions shown in Fig. 14 for **39a** clearly shows the unpaired electron is distributed to the phenalenyl units as in the phenalenyl radical itself. In accord with the y_0 values, the "*exo*-cyclic" bond in the p-QDM moieties determined by X-ray crystallographic analysis of the corresponding derivatives **9c** (1.457 Å), **39c** (1.465 Å), and **40c** (1.467 Å) becomes longer in this order. Singlet–triplet energy gaps ($\Delta E_{\text{S-T}}$) for the parent **9a**, **39a**, and **40a** are calculated to decrease in this order, namely $- 7.2$, $- 4.5$ and $- 2.9$ kcal/mol, respectively, by the UB3LYP/6-31G** method. Although information is limited due to low solubility, ^{1}H NMR spectra of **9b** and **39c** (with additional t-butyl group) exhibit severe line broadening at room temperature. The signals become sharp at low temperatures, indicating a singlet ground state admixed with a thermally excited triplet state. Magnetic measurements for **9b** show a gradual increase of the χT values

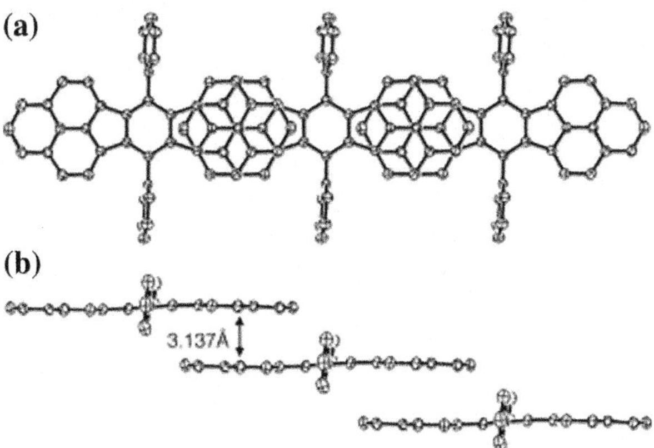

Fig. 15 Top and side views of crystal packing structure of **9b**. Reproduced with permission from [101]. Copyright 2005 Wiley–VCH

above 200 K, from which $\Delta E_{S\text{-}T}$ is estimated to be -4.3 kcal/mol [98]. Not surprisingly, the HOMO–LUMO gaps determined experimentally by cyclic voltammetry (CV) are 1.20, 1.04, 0.98 eV for **9c**, **39c**, and **40c**, respectively, and decrease in this order too. In solution, they exhibit lowest energy transitions at 746, 865, 986 nm, respectively. As a consequence of the narrow HOMO–LUMO gap, **9b** exhibits ambipolar carrier transport character, with the hole and electron mobilities of 2.6×10^{-3} and 3.2×10^{-3} $cm^2V^{-1}s^{-1}$, respectively [106].

The most interesting feature of a series of biphenalenyls is the strong *intermolecular* interaction observed in the solid state, which competes with *intramolecular* interaction between the two phenalenyl radicals located at both ends of the molecules. For example, the molecules of **9b** form a one-dimensional stack in crystals with an interlayer distance of 3.137 Å at 200 K as shown in Fig. 15. The distance is substantially shorter than the van der Waals contact of carbon atoms (3.4 Å). The phenalenyl units of **9b** overlap in the same geometry as that found in dimers of phenalenyl derivatives [98, 99, 107, 108], indicating SOMO–SOMO interaction serving as main driving force of stacking. The competing feature of the *inter-* and *intramolecular* interactions can be seen in the crystal structures of **9c** containing chlorobenzene molecules as a solvate, which show temperature-dependent variation of the interlayer distance and the bond length of the "*exo*-cyclic" bond of the *p*-QDM unit as listed in Table 6. For comparison, the data for crystals containing benzene and toluene, in which **9c** molecules stack more tightly or are isolated without interaction, are included. The π-π stacking distance decreases, whereas the "*exo*-cyclic" bond length increases with decreasing temperature. Taking the bond lengths of 1.457 and 1.476 Å as references of free and the most strongly stacked **9c** molecules, there exists a negative correlation between the distances of π-π stacking and "*exo*-cyclic" bond, indicating that enhancement of intermolecular interaction makes the unpaired electrons more localized on the phenalenyl units, thereby weakening the intramolecular bonding.

Table 6 Interlayer distances (Å) between stacked phenalenyl rings and "*exo*-cyclic" bond lengths (Å) of *p*-QDM unit of **9c** [104]

Solvate	Chlorobenzene			Benzene	Toluene
Temperature (K)	100	200	300	200	200
Distance	3.208	3.225	3.279	3.160	–
Bond Length	1.472	1.469	1.465	1.476	1.457

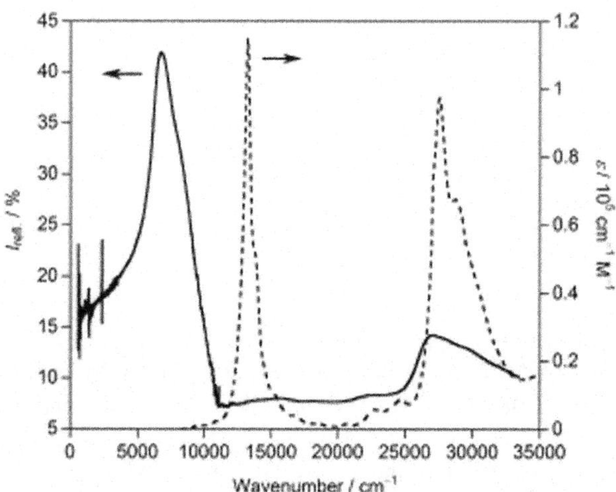

Fig. 16 Optical spectra of **9b**. Absorption spectrum in dichloromethane (dashed line) and reflection spectra of a single crystal obtained with light polarized along the *a* axis (solid line) and *b* axis (dotted line). Reproduced with permission from [101] Copyright 2005 Wiley–VCH

More evidence for the intermolecular interaction is gained from the optical conductivity spectra [101, 104, 105]. Figure 16 shows the spectra of **9b** as an example [101]. Reflection spectrum of a single crystal with a polarized light along the stacking direction (*b* axis) shows an intense peak at 6804 cm^{-1}, which is significantly shifted to lower energy compared to that in solution (13,400 cm^{-1}). In contrast, no conductivity was detected below 13,000 cm^{-1} along the *a* axis. This indicates that the electronic structure of **9b** in the solid state is affected by strong intermolecular spin–spin interaction.

It was theoretically predicted that molecules with singlet diradical character of intermediate amplitude exhibit enhanced third-order non-linear optical response, and consequently two-photon absorption (TPA) [109–111]. The static hyperpolarizability (γ) of **9a** is theoretically estimated by the UBHandHLYP/6-31G* method to be 2383×10^3 a.u., which is more than one order of magnitude larger than that of closed-shell hydrocarbon **41** (194×10^3 a.u.) of a similar π conjugation size (Fig. 17) [111]. Therefore, derivatives of **9a** and **39a** are expected to show large TPA. Indeed, TPA cross sections of **9b** and **9d** measured at their TPA peaks in chloroform are 330 ± 40 (1300 nm), 424 ± 64 (1425 nm) GM (1 GM = 10^{-50} cm^4 s/photon molecule), respectively [112]. While **39b** does not show a TPA peak, the maximum value

Fig. 17 Structures of reference compounds **41** and **42**

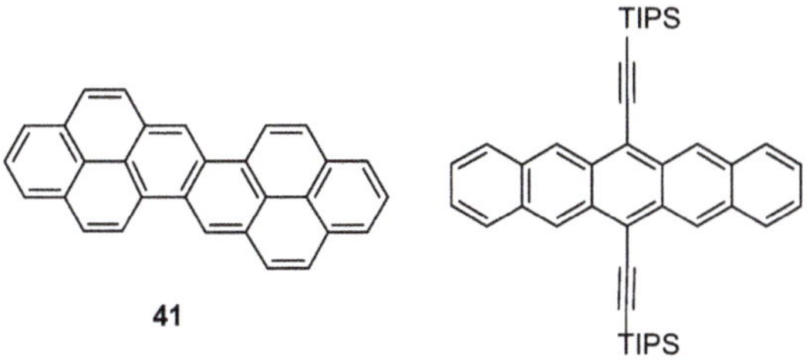

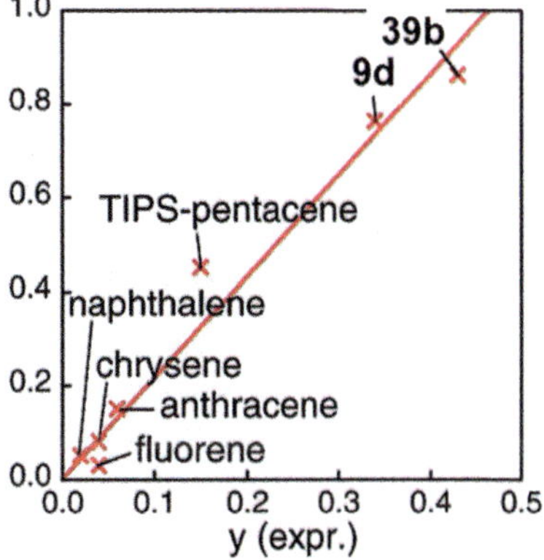

Fig. 18 Experimentally deduced versus theoretical y_0 values. Reproduced with permission from [113]. Copyright 2010 The American Chemical Society

within the wavelength limit of the measurement (< 1500 nm) is 890 ± 150 GM at 1500 nm. Though these values are comparable to those of TPA chromophores with strong donor/acceptor substituents, as nonpolar hydrocarbons, they are remarkably large. For example, a pentacene derivative **42** exhibits TPA cross section of 27 GM (875 nm) under the same conditions. These results not only support the theoretical predictions, but also allow an experimental elucidation of singlet diradical character based on theoretically derived Eq. (3) consisting of parameters experimentally accessible [113]. Namely, $E_{S1u,S1g}$ and $E_{S2g,S1g}$ correspond to the lowest-energy peaks of the one- and two-photon absorption, respectively, while $E_{T1u,S1g}$ can be obtained from phosphorescence and ESR measurement. Figure 18 shows a plot showing a linear relation between the y_0 values determined according to Eq. (2) versus the theoretical values, thus confirming the validity of this method.

$$y_0 = 1 - \sqrt{1 - \left(\frac{E_{S1u,S1g} - E_{T1u,S1g}}{E_{S2g,S1g}} \right)} \qquad (3)$$

5 Indenofluorene System

Indenofluorenes are regarded as dibenzo-fused indacenes. There are five possible constitutional isomers of indenofluorene, indeno[1,2-*b*]fluorene (**10a**),

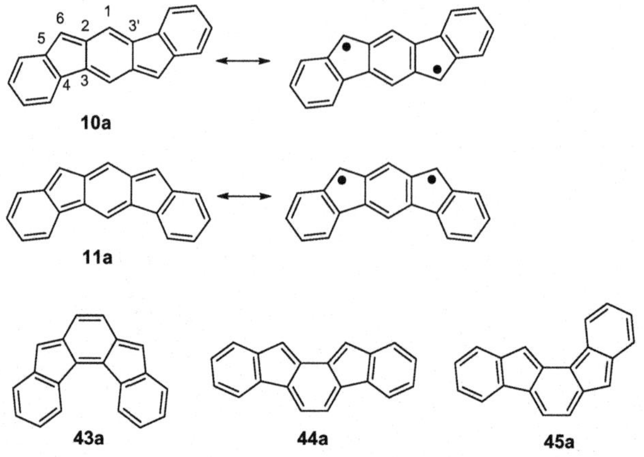

Fig. 19 Structures of five isomers of indenofluorene **10a**, **11a**, **43a**, **44a**, and **45a** with canonical resonance structures for **10a** and **11a**

[2,1-*a*] isomer **43a**, [2,1-*c*] isomer **44a**, [2,1-*b*] isomer **11a**, and [1,2-*a*] isomer **45a**, as shown in Fig. 19. Of these, although isomer **45a** and its relatives have not been fully characterized yet, because it is predicted to adopt triplet ground state configuration, derivatives of four other isomers substituted by aryl or ethynyl groups, which protect the otherwise reactive hydrocarbons from undesirable oxidative degradations or self-oligomerizations, have been synthesized in recent years mainly by the groups of Haley and Tobe, advancing the field significantly [27, 114]. Isomers **10a** and **11a** are dibenzo-fused *s*-indacenes and are regarded as *p*-QDM congeners. On the other hand, **43a** and **44a** are regarded as *as*-indacene homologues. The canonical resonance forms for **10a** and **11a** in Fig. 19 suggest the possible contribution of open-shell diradical character in the ground state electronic configurations of these systems. Namely, in the Kelulé (closed-shell) and diradical (open-shell) structures of isomers **10a** (and **43a** and **44a** as well), the number of benzenoid sextets are two and three, respectively. On the other hand, the sextets count one and three in **11a** and **45a**, suggesting increased open-shell character in **11a** and **45a** due to larger aromatic stabilization. In other words, **11a** and **45a** would enjoy more pro-aromaticity than the other isomer.

Since synthetic methods common in most of the indenofluorenes and related compounds are detailed by Haley et al. in a book review [115], synthesis is not described here. In short, they are based on either (i) Friedel–Crafts reaction of pre-cyclization precursors bearing carboxy groups to prepare diketones and subsequent nucleophilic addition of appropriate aryl or ethynyl groups followed by reductive elimination of hydroxy groups or (ii) nucleophilic addition of aryl groups to pre-cyclization precursors bearing formyl groups and subsequent acid-catalyzed cyclization accompanying dehydration. The pre-cyclization compounds are assembled by cross-coupling reactions of readily available building blocks. In this section, structures and characteristic properties including electronic applications of individual isomers are described, which are followed by those of structurally related compounds such as those containing seven membered ring(s) and more extended π-systems.

5.1 Indeno[1,2-*b*]fluorene

Because a large number of (anti)aromatic compounds related to indeno[1,2-*b*]fluorene have been reported, they are divided into two classes; compounds containing *s*-indacene core in the center of framework with peripheral aromatic rings fused at each end, in other words, homologues of **10a** called *basic* indeno[1,2-*b*]fluorenes here, and those called *modified* indeno[1,2-*b*]fluorenes, in which the *s*-indacene core is modified by extending conjugation or replacing the five- and six-membered rings to different size rings.

5.1.1 Basic Indeno[1,2-b]fluorenes

Before the intensive study by the group of Haley on derivatives of indeno[1,2-*b*] fluorene (**10a**) and their congeners was initiated in 2009 [116], only two reports had been known, one on the synthesis of tetraiodo derivative **10b** as an unstable compound generated by transannular C–C bond formation between diacetylene units of a macrocycle by Swager et al. [117] and the other by Scherf on diphenyl derivative **10d** with only a small piece of spectroscopic information [118]. The first stable derivative of **10a** is tetraethynyl derivative **10c**, which was synthesized based on the transannular cyclization method reported by Swager [119]. Although successful, this method has limitations in terms of the substituent groups that can be introduced and the substitution pattern. Then a vast variety of derivatives of **10a** bearing various substituents, as well as congeners with different frameworks were synthesized utilizing the aforementioned general synthetic methods, allowing the studies on their structural and physical properties and optoelectronic applications [115]. These include derivatives bearing aryl groups of electron-donating or -withdrawing nature on the apical positions of the five-membered rings **10d,e,i** [120–122], those with (trialkylsilyl)ethynyl groups **10j,k**, together with various substituents at the terminal benzene rings **10m,n** [122] (Fig. 20). It should be pointed out that for (trialkulsilyl) ethynyl derivative, the choice of alkyl group, i.e., their bulkiness, is important to ensure enough stability for isolation. Whereas triisopropylsilyl (TIPS) and tricyclohexylsilyl (TCHS) derivatives **10j** and **10 k** were isolated [121, 122], an attempt to synthesize trimethylsilyl (TMS) derivative **10 l** resulted in the formation of dimer, trimer, and higher oligomers arising from C–C bond formation from its allenyl form of the diradical canonical resonance structure due to partial location of spin density to the triple bonds [123]. Aryl-substituted derivatives **10f–h** are also reported by the group of Yamashita almost at the same time [124]. Indenofluorene congeners with different terminal aromatic rings such as naphthalene **46b,c**, **47b,c**, and **48b,c** [125], phenanthrene **49b,c** [125], thiophene **50b** and **51b** [126], and benzothiophne **52b–d** and **53b–d** [126, 127] fused at both ends of the *s*-indacene framework are also shown in Fig. 20. In addition, unsymmetrical analogues **54** and **55** in which only one of the benzene rings of **10a** was replaced by thiophene were added to the

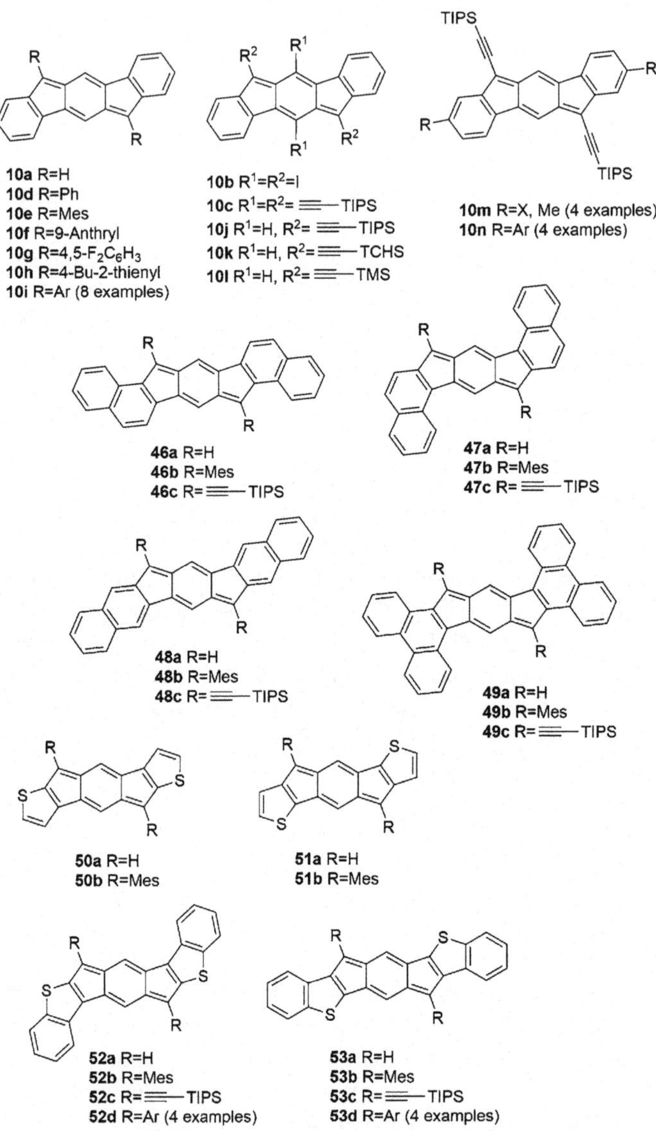

Fig. 20 Structures of derivatives **10b–10n** of indeno[1,2-*b*]fluorene (**10a**) and its congeners **46–55**

inventory [128]. The same compounds **52b,c** are also reported by Zhu later [129]. Note that for naphthalene-fused congeners, there are two isomers **46a** and **47a** of 1,2-fusion depending on the orientation of the naphthalene ring with respect to the indacene core and one 2,3-fusion isomer **48a**. For thiophene analogues **50a**, **51a**, **54**, and **55** and benzothiophene analogues **52a** and **53a**, there are also two isomers due to the orientation of fusion of the terminal rings.

Table 7 Bond lengths (Å) determined by X-ray crystallographic analysis of indeno[1,2-b]fluorenes and their congeners [a]

Bond [ref]	10c[120][b]	10j [122]	10e [120]	46c [125]	47b [125]	48c [125]	50b [126]	51b [126]	52a [126]	53a [126]
	C ≡ CTIPS	C ≡ CTIPS	Mes	C ≡ CTIPS	Mes	C ≡ CTIPS	Mes	Mes	Mes	Mes
1–2	1.438 (1.444)	1.424	1.433	1.420	1.431	1.430	1.431	1.418	1.412	1.421
1–3′	1.374 (1.379)	1.354	1.356	1.368	1.366	1.366	1.360	1.363	1.377	1.371
2–3	1.457 (1.457)	1.456	1.467	1.443	1.462	1.471	1.469	1.456	1.457	1.454
2–6	1.390 (1.396)	1.382	1.380	1.401	1.378	1.399	1.388	1.398	1.409	1.407
3–4	1.470 (1.466)	1.481	1.469	1.458	1.470	1.470	1.452	1.461	1.437	1.457
4–5	1.412 (1.463)	1.410	1.413	1.403	1.401	1.444	1.389	1.384	1.393	1.391
5–6	1.470 (1.417)	1.472	1.471	1.465	1.461	1.472	1.460	1.447	1.435	1.441

[a]Because of the presence of a center of symmetry, only a half of the bond length data are listed. For carbon numbering, see Fig. 19. [b]Theoretical bond lengths by DFT calculations (B3LYP/6-311 + G**) for desilylated compound [H instead of TIPS] are given in parentheses

Many of the molecular structures of the above compounds were determined by X-ray crystallographic analysis as listed in Table 7. Basically, all bond length data indicate characteristic p-QDM-like feature with bond length alternation; shorter C1–C3′ bonds and longer C1–C2 and C2–C3 bonds of the six-membered ring than the bond length of benzene (1.39 Å) and sorter C2–C6 bonds of the exocyclic bond than the $C(sp^3)$–$C(sp^2)$ bond of fluorene length (1.468 Å) yet longer than typical $C(sp^2)$–$C(sp^2)$ bond length (1.34 Å). For substituted indeno[1,2-b]fluorene derivatives, the bond lengths of the terminal six-membered rings are nearly homogeneous and are about 1.38–1.39 Å. DFT calculations at the B3LYP level of theory gave reasonably good agreement with the observed structure as included for the data of **10c** for example. Closer look into the effect of substituents on the basic indenofluorene backbone indicates that attachment of ethynyl groups at the central six-membered ring (**10c**) elongate the bond length of the six-membered ring (C2–C2 and C1–C3′) probably due to the extension of conjugation to the ethynyl group. Moreover, comparison of the C2–C6 bond lengths of **10c** and **10j** with that of mesityl (Mes) derivative **10e**, in which the substituent π-system cannot conjugate with the indacene core because of the nearly orthogonal geometry of the mesityl group with respect to the main backbone, suggests that substitution with silylethynyl groups at the apical position (C6) of the five-membered ring elongates C2–C6 bond also due to extension of conjugation to the ethynyl groups. Therefore, the mesityl derivatives are taken for further comparison as far as the crystallographic data are available, because the effect of substituents should be minimal. As seen from the data for **46c** and **47b**, the orientation of the 1,2-fusion of naphthalene rings does not affect the structure of the indacene core. Despite the longer C4–C5 bond due to the 2,3-fusion of naphthalene units in **48c**, the bond lengths of its indacene core are almost identical with those of **10e**, **46c**, and **47b**. On the other hand, in thiophene-fused congeners **50b** and **51b**, the corresponding C4-C5 bonds become sorter than that of **10e** because of larger double bond character due to smaller aromaticity of thiophene than benzene. This leads to longer C2–C6 bond indicating larger open-shell character. Although the C4–C5 bonds in benzothiophene congeners **52b** and **53b** are longer than those of **50b** and **51b**, their C2–C6 bonds are even longer. As a result, **52b** and **53b** may have the largest open-shell character among the indenofluorenes and their analogues shown in Fig. 20. The longer C2–C6 bonds in **52b** and **53b** parallels larger antiaromatic (paratropic) character as described in the following paragraphs.

Since theoretical HOMO and LUMO levels of **10a** are $-$ 5.53 and $-$ 3.03 eV with a relatively small gap of 2.50 eV by DFT calculations (B3LYP/6-311 + G**), its stable derivatives may be used as organic semi-conductors whose properties can be modified by introducing appropriate substituents. Introduction of ethyl groups at the apical position (6 and 12 positions) of the five-membered ring lowers the LUMO level ($-$ 3.46 eV) by 0.43 eV while maintaining the HOMO level ($-$ 5.51 eV), thus reducing the gap significantly (2.05 eV) [122]. Electrochemically determined HOMO/LUMO levels by cyclic voltammetry (CV) of C $\equiv$ CTIPS-derivative **10j**, $-$ 5.88 and $-$ 4.00 eV (gap = 1.89 eV), respectively, agree reasonably. Similarly, the experimental (by CV) HOMO/LUMO levels of derivatives **10d,e,i** with electron-donating and -withdrawing groups at the same positions vary substantially within 0.7 eV and 0.4 eV, respectively, and HOMO–LUMO gaps vary from 1.82 to 2.22 eV

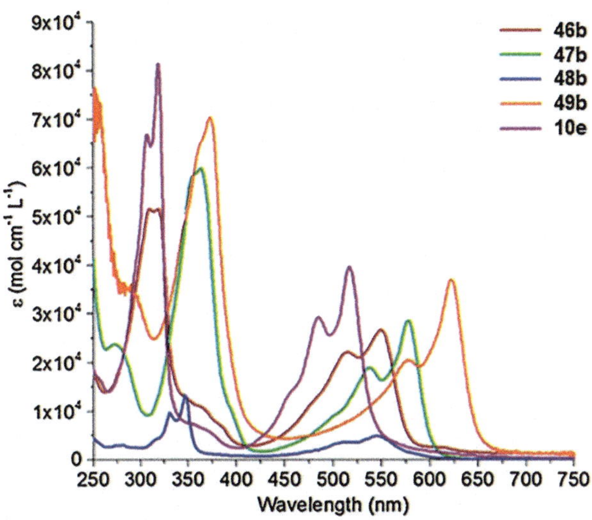

Fig. 21 Electronic absorption spectra of **46b–49b** and **10e**. Reproduced from [125]. Copyright 2016 The American Chemical Society

[120]. However, the HOMO–LUMO gaps of derivatives bearing substituents at the terminal benzene rings (2 and 8 positions) **10m,n** change within a small range of 1.8–1.94 eV because of the negligible distribution of HOMO and LUMO at these positions [122]. Extension of π-conjugation reduces the HOMO–LUMO gap significantly as expected; for a series of extended systems experimentally determined the smallest HOMO–LUMO gap by CV is that of **46c** (1.57 eV) bearing C ≡ CTIPS groups at 6,12-positions [127]. Thus, the relatively high electron affinity (i.e., low LUMO levels) and small HOMO–LUMO gaps are prominent characteristics of indenofluorenes. In electronic absorption spectra, the longest wavelengths absorptions in the visible regions of **10d,e,i** vary more remarkably than those of **10m,n** similar to the redox potentials [121, 123]. Similarly, the electronic absorption maxima of mesityl derivatives **46b–49b** are significantly red-shifted compared to **10e** as shown in Fig. 21 [125].

Being a benzo-fused *s*-indacene of 20 π-electron system, indenofluorene **10a** is expected to exhibit antiaromatic character. NICS(1) calculations for the desilylated derivative (H instead of TIPS) of the first fully characterized indenofluorene derivative **10c** showed 0.02, 1.94 and − 7.12 for central six-membered ring, five-membered ring, and terminal six-membered ring, respectively, indicating nearly atropic (non-aromatic) character at the *s*-indacene core, in contrast to the paratropicity of *s*-indacene itself (Sect. 3), and the reduced diatropicity in the outer rings. Experimentally, ¹H NMR chemical shifts of the proton attached to the central six-membered ring (H5/H11) appear at 6.6–7.0 ppm in accord with the atropicity, except for thiophene-fused **50b** and **51b** and benzothiophne-fused congeners **52b–d** and **53b–d**, which exhibit the signals at 5.9–6.3 ppm, indicating larger paratropicity. In order to compare the tropicity within a series of indenofluorene-based molecules,

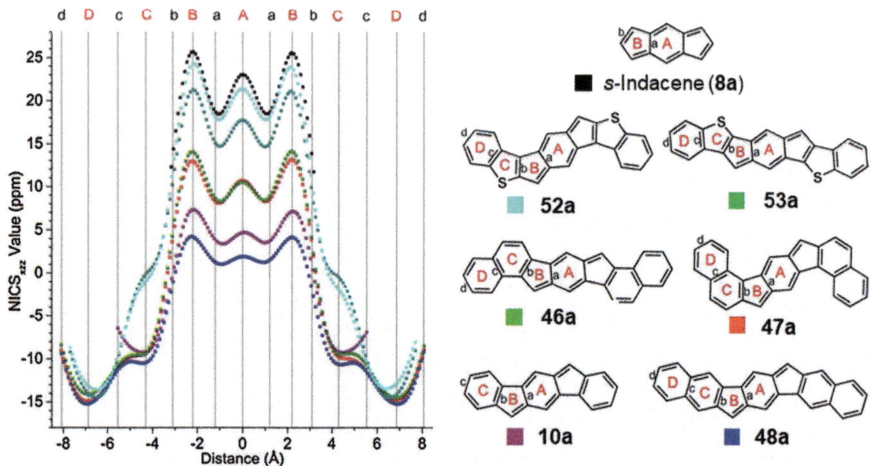

Fig. 22 NICS(1.7)$_{\pi ZZ}$-XY scan for **1a** and unsubstituted compounds for **46a**, **47a**, **48a**, **52a**, and **53a**, and *s*-indacene (**8a**) at the B3LYP/6-311 + G** level. Reproduced with permission from [114]. Copyright 2017 The American Chemical Society

Haley et al. employed NICS-XY scan using π-only models, in which the effect of σ-electrons are removed, displaying contribution to the ring current from π-electrons only [130]. As a representative example, NICS(1.7)$_{\pi ZZ}$-XY scans taken at 1.7 Å above the molecular planes for **10a** and unsubstituted compounds for **46a**, **47a**, **48a**, **52a**, and **53a** together with *s*-indacene (**8a**) are shown in Fig. 22. First, the NICS$_{\pi ZZ}$ values at rings A and B of *s*-indacene (black trace) are 23 and 26 ppm, respectively, in accord with the results of other theoretical studies (Sect. 3). Fusion of benzene rings (**10a**, purple trace) reduces the paratropicity significantly to 4 and 7 ppm at A and B rings, respectively. In contrast, 1,2-naphtho-fusion (**46a** and **47a**, green and orange traces, respectively) results in larger NICS$_{\pi ZZ}$ values of 10 and 13–14 ppm than those of **10a** due to the relatively large bond order of the naphthalene 1,2-bond (1.66). This may indicate that the *s*-indacene and naphthalene circuits in **46a** and **47a** are more independently resonated than in **10a**. This trend is enhanced in benzothiophene-fused congeners **52a** and **53a** (light blue and teal traces) with higher bond order of the fusion bonds (> 1.8), which exhibit NICS$_{\pi ZZ}$ values for the A/B rings of 18/21 and 21/24 ppm, respectively. Because the paratropicity inherent in *s*-indacene are nearly recovered in these molecules, they can be regarded as substituted *s*-indacenes with the appended phenyl groups planarized by thioether linkages. On the other hand, 2,3-naphtho-fusion (**48a**, blue trace) shows NICS$_{\pi ZZ}$ values for the A/B rings of 2/4 ppm due to smaller bond order of the naphthalene 2,3-bond (1.33), indicating the indacene circuit is merged with the naphthalene conjugation to a greater extent. To visualize the relationship between the bond order of the fusion bonds and the NICS$_{\pi ZZ}$ values, the bond lengths and orders of the fusion bonds are plotted against the maximum NICS$_{\pi ZZ}$ values in the indacene core (Fig. 23), showing a clear relationship. NICS$_{\pi ZZ}$ values of the external rings, however, are nearly

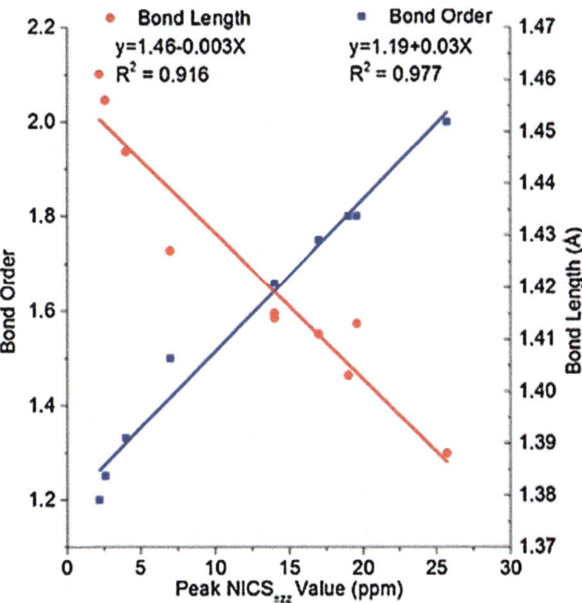

Fig. 23 Relation between the maximum $\text{NICS}_{\pi ZZ}$ value in the indacene core and the bond order (blue) and bond length (red) of the fusion bond of indenofluorenes. Reproduced from [125]. Copyright 2016 The American Chemical Society

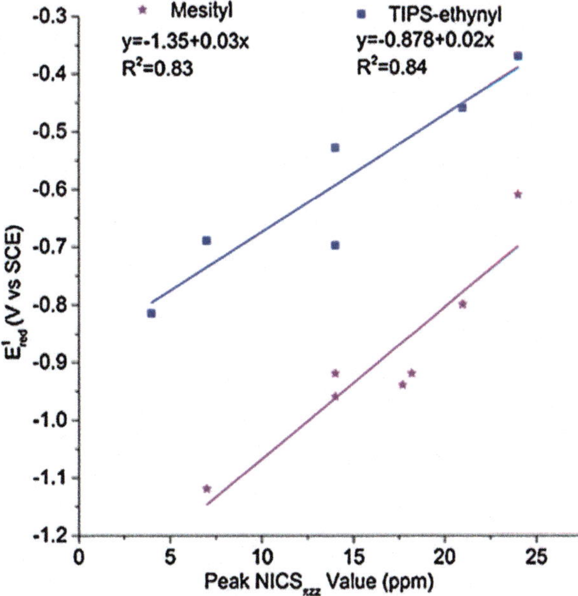

Fig. 24 Relation between the maximum $\text{NICS}_{\pi ZZ}$ value in the indacene core and first reduction potential measured by CV of indenofluorenes bearing $C \equiv CTIPS$ (blue) or Mes (red) groups. Reproduced from [125]. Copyright 2016 The American Chemical Society

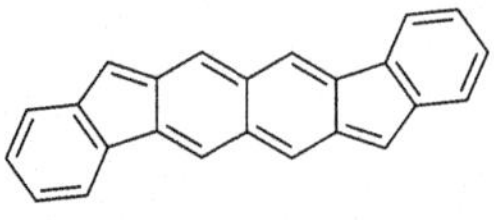

56

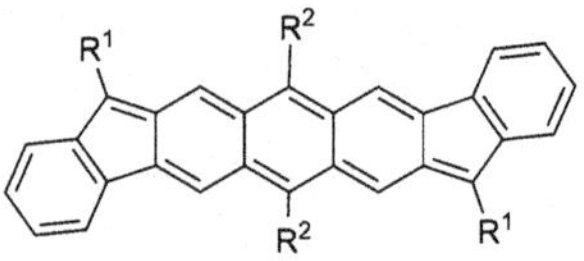

57a $R^1=R^2=H$
57b R^1=Mes, R^2= ≡—TIPS
57c R^1=Mes, R^2= ≡—TMS

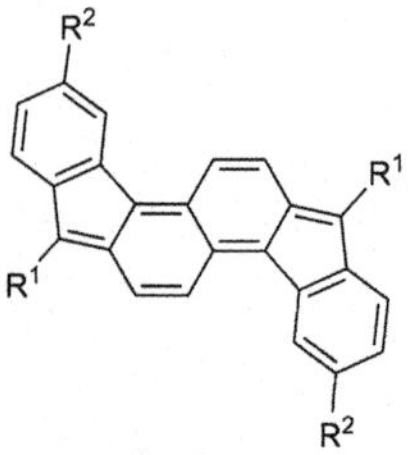

58a $R^1=R^2=H$
58b R^1=Mes, R^2=*t*-Bu

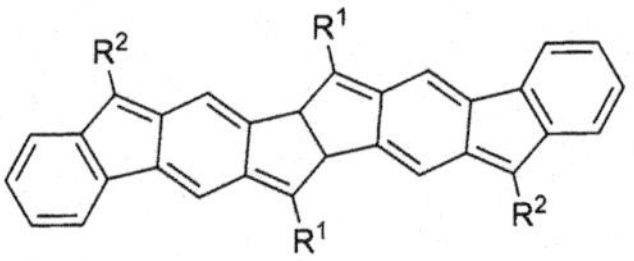

59a $R^1=R^2=H$
59b $R^1=R^2$=Mes
59c R^1=Mes, R^2=9-anthryl

Fig. 25 Structures of modified indeno[1,2-*b*]fluorenes **56a,b**, **57a–c**, **58a–c**, and **59a–c**

constant irrespective of the different aromatic rings fused to the indacene core and their fusion mode [125].

Moreover, the maximum $NICS_{\pi ZZ}$ values are plotted against the first reduction potentials in CV for the series of compounds with C ≡ CTIPS or Mes groups at the five-membered rings (Fig. 24), which shows a trend of less negative reduction potentials with increasing $NICS_{\pi ZZ}$ values. The results can be interpreted in terms of hypothesis that the electron-accepting ability of indenofluorenes is related to reduction of antiaromatic destabilization at the indacene core. Therefore, a compound with the less negative the potential (i.e., more easily reduced) would exhibit greater paratropicity.

In view of the small HOMO–LUMO gap and the high electron-accepting properties of indenofluorenes, the groups of Haley and Yamashita used indenofluorene derivatives as an active layer in organic field-effect transistors (OFETs) with particular interest in ambipolar characteristics [121, 124]. In addition, some of the indenofluorene derivative adopts two-dimensional stacking geometry of "brickwork" type, which has been shown to essential to exhibit high charge carrier mobility for pentacene derivatives with silylethynyl substituents [131–133]. However, though OFET based on **10i** bearing electron-withdrawing group (R = C_6F_5) exhibited ambipolar behavior, its hole and electron mobilities are low ($\mu_h = 7 \times 10^{-4}$ $cm^2V^{-1}s^{-1}$ and $\mu_e = 3 \times 10^{-3}$ $cm^2V^{-1}s^{-1}$) [120]. Hole mobilities of **10f**, **10g**, and **10h** (1.6×10^{-5}, 1.9×10^{-5}, and 1.1×10^{-5} $cm^2V^{-1}s^{-1}$, respectively) as well as electron mobilities of **10f** and **10g** (8.2×10^{-6} and 1.6×10^{-6} $cm^2V^{-1}s^{-1}$, respectively) are also poor [124]. The best results are obtained by using benzothiophene-fused compound **53c**, which upon a spin-coated device processing gave an average hole mobility of 0.14 $cm^2V^{-1}s^{-1}$ over 10 devices (maximum mobility 0.44 $cm^2V^{-1}s^{-1}$). Thereafter, a single crystal device using the same compound **53c** was reported to exhibit hole mobility of 0.64 $cm^2V^{-1}s^{-1}$ and electron mobility of 0.34 $cm^2V^{-1}s^{-1}$, which are the highest mobilities for OFETs so far reported using indenofluorene derivatives.

A radical anion of $C \equiv CTIPS$ derivative **10j** generated by reduction with 1 equiv. of potassium was characterized by ESR [134]. The location of the spin densities was experimentally determined by simulation and confirmed theoretically by DFT calculations (for TMS derivative instead of TIPS) using the B3PW91 method. The theoretical study reveals that the spin densities are located to a significant extent on the apical position of the five-membered ring as expected, as well as at the triple bond carbon bearing the silyl group. By reduction of **10j** with 2 equivalents of potassium, some of the ^1H NMR signals shifted downfield by ca. 1 ppm, indicating diatropic ring current due to a 22 π-electron system generated by addition of two electrons. Two crystal structures of Rb_2-18-cown-8 complexes of dianions **10 k** with $C \equiv CTCHS$ groups were solved. In one complex, two Rb^+ ions coordinate to the top and bottom faces of the central six-membered ring in η^6-corrdination mode, whereas in the other two Rb^+ ions coordinate to the bond shared by the five- and six-membered rings in η^6-corrdination mode, revealing the flexibility of coordination.

5.1.2 Modified Indeno[1,2-b]fluorenes

In view of the ready accessibility, high stability, and interesting properties of basic indenfluorenes, a number of modified congeners have been reported. Although there are a number of possible modifications on the basic framework of **10a**, they are classified into two categories; (i) one is based on lateral extension of the indacene core. i.e., replacing the formal benzene ring of indacene to naphthalene or anthracene ring and (ii) the other is replacement of the five-membered ring(s) of indacene by different size ring(s) such as seven-membered ring.

Lateral extension of the central six-membered ring of **10a** leads to **56a** and **57a** with a formal naphthalene or anthracene ring in the center. Sterically protected derivatives **56b** and **57b** of fluoreno[3,2-b]fluorene (**56a**) and diindeno[b,i]anthracene (**57a**), respectively, have been synthesized [135, 136], as well as a structural isomer of **56b**, a fluoreno[4,3-c]fluorene derivaties **58b,c** (Fig. 25) [135, 137]. Like s-indacene and its homologues (Sect. 3) lateral extension of the central portion would enhance open-shell character due to increasing aromatic stabilization in the acene-like, central part with increasing conjugation size. Indeed, whereas indeno[1,2-b]fluorene (**10a**) is closed-shell, the diradical character indices (y_0) of **56a**, **57a**, and **58a** are 0.434, 0.732, and 0.377, respectively [135]. DFT calculations for **57a** at the B3LYP/6-31G(d) level also indicate that it has an open-shell singlet ground state with a closed-shell singlet state 1.07 kcal/mol higher in energy. Experimentally, **57b** exhibited temperature-dependent ^1H NMR spectra; the signals that are slightly broadened at room temperature eventually disappear at 150 °C, while they become sharpened at $-$ 25 °C, indicating the presence of thermally excited triplet state at high temperatures. A weak Q-band ESR absorption was observed at a g value of $\sim$ 2.012, the intensity of which decreases on cooling. The magnetic property was confirmed by superconducting quantum interference device (SQUID) magnetometry in a temperature range of 4–400 K, from which singlet–triplet energy gap ΔE_{ST} of $-$ 4.18 kcal/mol was elucidated, which agrees the theoretically estimated ΔE_{S-T} of

− 4.9 kcal/mol using the TD-PBE50 spin-flip method. In accordance with the large open-shell character, the bond lengths of the peripheral bonds of the anthracene core are more or less averaged (1.359–1.446 Å), suggesting delocalized partial double bond character. The most critical feature is the long $C(sp^2)$–$C(sp^2)$ bond (1.406 Å) connecting the apical carbon to the anthracene core, due to contribution of diradical canonical structure. $NICS(1.7)_{\pi ZZ}$-XY scan and anisotropy of the induced current density (ACID) for **57c** agree with this view showing strong and weak diatropic ring currents at the external benzene ring and the central anthracene core, respectively.

In CV, **57b** showed reversible reduction peaks at − 1.28 and − 1.67 V and oxidation peaks at 0.17 and 0.70 V, from which HOMO/LUMO levels and their gap are estimated to be − 5.27/− 3.82 and 1.45 eV, respectively. The reactive intermediates, radical anion, dianion, radical cation, and dication generated from **57b** were spectroscopically characterized by UV–vis–NIR absorption and Raman spectra and their structures and electronic properties are assessed by theoretical calculations [138]. X-ray crystallographic analysis was done for three structures of dianion-K_2, revealing that K^+ ions locate not only at the center of the anthracene ring in η^6-corrdination mode but also laterally shifted positions by 1.2 Å. With this amphoteric redox character with narrow HOMO–LUMO gap, indicative of ambipolar charge transport, together with the kinetic stability of **57b** with a half-life of which in dilute solution is more than 2 months, OFET was fabricated by vapor deposition method. However, although a well-balanced ambipolar character was observed, the mobilities were not high ($\mu_h = 2 \times 10^{-3}$ $cm^2V^{-1}s^{-1}$ and $\mu_e = 4 \times 10^{-3}$ $cm^2V^{-1}s^{-1}$).

Compared to the anthracene-centered **57b**, naphthalene analogue **56b** has smaller open-shell character [135]. Although crystallographic data are not available, theoretical and experimental results for **56a** and **56b** are consistent with their quinoidal, closed-shell structure. For example, NICS-XY scan for **56a** exhibits a large negative $NICS_{\pi ZZ}$ value (− 10 ppm) for the outer six-membered rings indicating these rings are fully aromatic, whereas the inner six-membered rings are regarded essentially atropic (~ 0 ppm).

Mesityl derivatives **58b,c** of naphthalene-containing system, fluoreno[4,3-c]fluorene (**58a**), was reported by Haley et al. [135, 137]. In ^1H NMR spectrum, **58c** exhibits sharp signals without temperature-dependent line broadening up to 160 °C. It showed reversible reduction peaks in CV at − 0.87 and − 1.29 V and oxidation peaks at 0.82 and 1.27 V, from which HOMO–LUMO gap is estimated to be 1.69 eV. X-ray crystallographic analysis of **58c** revealed significant bond length alternation in the naphthalene core with short and long bond lengths of 1.355 and 1.479 Å, respectively. The critical bond length linking the apical carbon and the naphthalene core is 1.386 Å, which is comparable to those of basic indenofluorene derivatives (Table 7). All these experimental results are consistent with the closed-shell character of **58c**. Compound **58b** without *t*-butyl groups synthesized more recently exhibits structural and physical properties virtually identical to those of **58c** [135].

Pentaleno[1,2-*b*:4,5-*b'*]difluorene **59a** may also be regarded as a molecule of this class. Two derivatives **59b** and **59c** were synthesized by palladium-catalyzed dimerization of ethynylfluorenone derivatives to construct the central dibenzopentalene part of the molecule [139]. Theoretical calculations at the UB3LYP/6-311 + G(d,p) level of theory show that an open-shell singlet state of **59a** is most stable with a closed-shell singlet and triplet states by 2.0 and 1.4 kcal/mol higher in energy, respectively. The diradical factor y_0 is estimated to be 0.55. In the X-ray crystallographic structures of **59b** and **59c**, the observed bond lengths of the "indacene part" of the molecule are between those theoretically calculated for closed-shell and open-shell structures. Because of the moderate diradical character, only three singlets assigned to the mesityl groups can be observed in the ^{1}H NMR spectra of **59b** and **59c** even at the temperature as low as − 90 °C. Temperature-dependent ESR signal confirms the contribution of thermally excited triplet state as does SQUID measurements from which ΔE_{ST} of − 3.4 kcal/mol is elucidated. Both **59b** and **59c** showed reversible redox peaks in CV from which HOMO–LUMO gasp are estimated to be as small as 0.75 and 0.73 eV, respectively. Because of the small HOMO–LUMO gaps, and most likely due to local antiaromatic character of the pentalene framework, the longest wavelength electronic absorptions of **59b** and **59c** are observed at 1652 and 1702 nm, respectively. In stark contrast to indeno[1,2-*b*]fluoernes with closed-shell ground state configurations, **59a** exhibits enhanced open-shell character. This must be due to the aromaticity of the benzene rings in the dibenzopentalene unit, which is reflected in the long bond distances observed in the pentalene unit.

A hybrid system **60a** of indeno[1,2-*b*]fluorene (**10a**) and heptazethrene (**61a**) was prepared by Wu et al. (Fig. 26) as the respective model compounds of tetraradicaloids with π-conjugation size nearly twice as large (Sect. 5.5) [140]. In **60a** one of the indenyl part in **10a** is replaced by a phenalenyl unit. Because of the pronounced radical-stabilizing ability of phenalenyl as described in Sect. 4, within the series of the three compounds, open-shell character will increase with increasing number of the phenalenyl units in a molecule. Therefore, **61a** is expected to have higher open-shell character than **60a,** which in turn would have higher open-shell character than **10a**. Indeed, DFT calculation at the UCAM-B3LYP/6-31G(d,p) level indicate **61a** has a small diradical character ($y_0 = 0.17$), whereas **60a** and **10a** have negligible y_0 (0.05 and 0, respectively). NICS(1)$_{zz}$ at the central six-membered ring of **61a**, **60a**, and **10a** increases in this order (− 10.8, − 5.1, and 11.3, respectively), showing different electronic character in accord with the diradical character. The bonds lengths determined by X-ray crystallographic structural analysis of mesityl derivatives **61b**, **60b**, and **10e** reveal that bond length alternation of the QDM units increases in this order too. The "*exo*-cyclic" bonds, for example, are 1.382, 1.376 (in the six-membered ring) and 1.382 (in the five-membered ring), and 1.380 Å, respectively. Consequently, **61a** is regarded as a pro-aromatic system, in which the driving force for the formation of diradicals is the aromatic stabilization energy gained by breaking double bonds in a non-aromatic system, such as QDMs, that can surpass the energy required to break the double bonds [22]. On the other hand, **10a** is classified as an antiaromatic system and **60a** is located in between these two. In line with this view, in electronic absorption spectra, **61b** shows an intense band at 613 nm assigned to HOMO → LUMO transition, whereas **10e** exhibits a weak absorption extending to 700 nm due to a forbidden transition characteristic to antiaromatic molecules. In contrast, **60b** displays a broad band extending to NIR region which could originate from intramolecular charge transfer on the basis of the molecular orbitals contributing to this transition assigned by TDDFT calculations. These results demonstrate that replacing a six-membered ring in **61b** with a five-membered ring (i.e., **60b**)

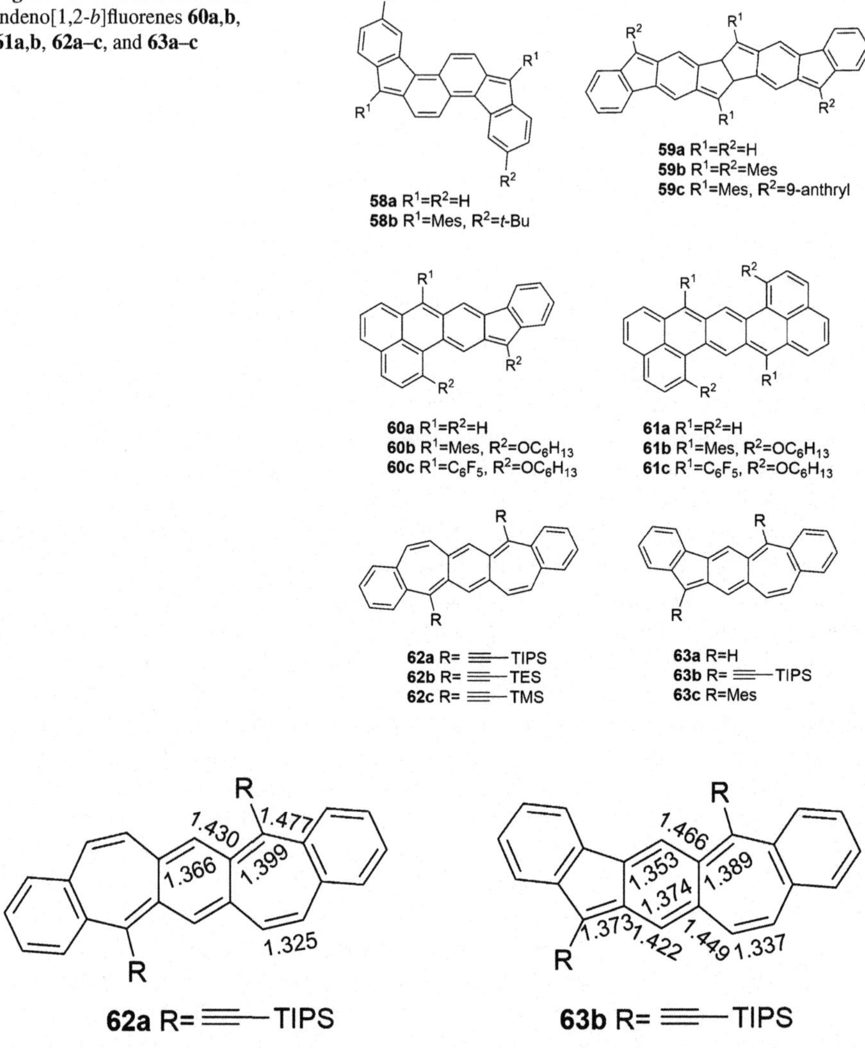

Fig. 26 Structures of modified indeno[1,2-*b*]fluorenes **60a,b**, **61a,b**, **62a–c**, and **63a–c**

58a R^1=R^2=H
58b R^1=Mes, R^2=*t*-Bu

59a R^1=R^2=H
59b R^1=R^2=Mes
59c R^1=Mes, R^2=9-anthryl

60a R^1=R^2=H
60b R^1=Mes, R^2=OC$_6$H$_{13}$
60c R^1=C$_6$F$_5$, R^2=OC$_6$H$_{13}$

61a R^1=R^2=H
61b R^1=Mes, R^2=OC$_6$H$_{13}$
61c R^1=C$_6$F$_5$, R^2=OC$_6$H$_{13}$

62a R= ≡—TIPS
62b R= ≡—TES
62c R= ≡—TMS

63a R=H
63b R= ≡—TIPS
63c R=Mes

62a R= ≡—TIPS

63b R= ≡—TIPS

Fig. 27 Bond lengths (Å) determined by X-ray crystallographic analysis for **62a** (left) and **63b** (right)

gives rise to bond localization and electron-accepting character in the five-membered ring, in addition to changing the total number of electrons involved in conjugation from aromatic to antiaromatic numbers.

Replacing the five-membered ring(s) of the indacene unit with seven-membered ring(s) is expected to modify the electronic properties of the resulting system significantly in view of the potential contribution of aromatic character due to cyclopentadienyl anion and cycloheptatrienyl cation (Fig. 26). Miao et al. synthesized **62a–c** consisting of conjugated 6–7–6–7–6 rings with silylethynyl substituents [141]. Although **62a** and **62b** are stable, TMS derivative **62c** decomposes in solution within a day. As shown in Fig. 27, though the "*exo*-cyclic bond" of the *p*-QDM moiety is relatively elongated (1.399 Å), other bond lengths show considerable bond localization. In particular, C15-C16 bond in the seven-membered ring (1.325 Å) has large double bond character. **62a** shows reversible reduction and oxidation peaks in CV at − 1.66 and 0.12 V, respectively, from which HOMO and LUMO levels are estimated to be − 4.91 and − 3.14 eV, respectively. The HOMO–LUMO gap of 1.77 eV agrees with that estimated from the absorption edge at 649 nm (1.91 eV). Comparison of the HOMO/LUMO levels of **62a** with those of **10j** (− 5.88 and − 4.00 eV, respectively) reveals significantly higher HOMO/LUMO levels of **62a** due to the electron-releasing seven-membered rings. In crystal, **62a** adopts a one-dimensional offset stacking geometry with a interlayer distance of about 3.47 Å, whereas **62b** packs in a herringbone-like geometry with small intermolecular overlap. Thin-film OFET was fabricated by dip-coating or drop-casting method. The film of **62a** deposited on the SiO_2 surface consists of aligned crystalline fibers as revealed by polarized micrograph and X-ray diffraction patterns. As such, the device using **62a** showed the hole mobility of 0.19–0.76 $cm^2V^{-1}s^{-1}$, while those using **62b** and **62c** exhibited low mobilities of 6×10^{-4} and 5×10^{-5} $cm^2V^{-1}s^{-1}$.

Replacing just one five-membered ring of the indacene part with a seven-membered ring results in conjugated 6–5–6–7–6 ring system **63a**, called dibenzoannelated "homoazulene", which is expected to show dipolar character like azulene due to the presence of five- and seven-membered rings (Fig. 26) [142]. The experimental bond lengths of **63c** shown in Fig. 27 indicate quinoid character of this system with marked bond length alternation. Theoretical calculations at the (U)CAM-B3LYP/6-31G* level support the closed-shell electronic configurations ($y_0 = 0$) of the ground states of **63a** and **63c**, whereas C ≡ CTIPS-substituted derivative **63b** is predicted to adopt open-shell singlet ground state with a small y_0 of 0.047. The calculated dipole moments of **63b** and **63c** are 3.18 and 2.65 Debye, respectively, which are larger than that of azulene (1.27 Debye). Experimental HOMO/LUMO levels of **63b** and **63c** are determined by CV to be − 5.36/− 3.80 and − 5.27/− 3.33 eV, respectively. The high HOMO level of **63b** yet lower than that of **63a** and low LUMO level comparable to that of **10j** suggests that oxidation and reduction take place at seven- and five-membered rings of this system, respectively. Similar to azulene, **63b** and **63 cc** exhibit broad electronic absorptions assigned to HOMO → LUMO transitions at 754 and 635 nm, respectively, most likely because of small HOMO/LUMO overlap due to characteristic orbital distribution.

Further extension of the central part of the indenofluorene core may modify the electronic properties, as well as help intermolecular interactions in condensed phase, thereby enhancing charge mobility. In this respect, compounds **64** and its isomer **65** in which a pyrene unit is flanked by two indenyl units are prepared. However, because of their limited stabilities and synthetic accessibility, only fundamental physical properties are reported (Fig. 28) [143]. From the same, but with additional prospects of lowering the HOMO–LUMO gaps, very recently, Feng et al. reported the synthesis, structures, and properties of three antiaromatic hydrocarbons **66a,b**, **67**, and **68**, in which the fluorenyl part of the indacene core of **10a** and its homologue **56a** are fused by ethenylene or triphenylenylene linkage to open additional conjugation paths [144]. The structures revealed by X-ray crystallography and aromaticity assessed by NICS(0) values and current densities calculated at the PBE/

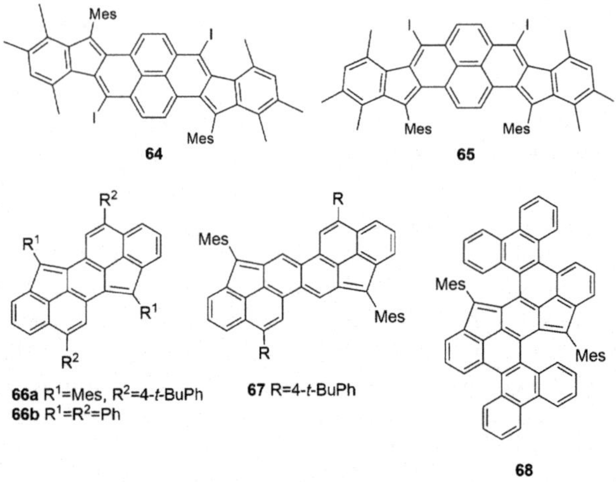

Fig. 28 Structures of compounds **64**, **65**, **66a,b**, **67**, and **68**

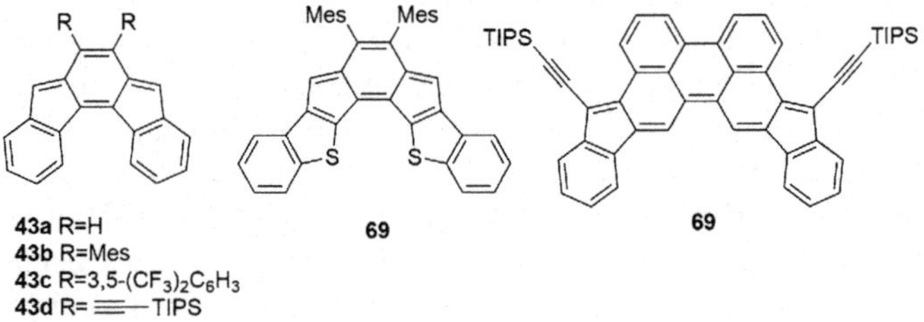

Fig. 29 Structures of indeno[2,1-*c*]fluorenes **43a–d** and its congeners **69** and **70**

TZ2P level of theory indicate that the electronic structures of **66a** and **67** are not much different from the unperturbed systems. In contrast **68** adopts a non-planar curved structure resulting from the steric hindrance at the cove and fjord regions, making the carbon framework chiral. In crystal, the double [4]helicene and double [5]helicene moieties have the same axial chirality. They show excellent redox properties in CV from which HOMO/LUMO levels (and their gaps) are estimated to be − 5.59/− 3.50 (2.01), − 5.25/− 3.70 (1.55), and − 5/43/− 3.55 (1.88) eV, for **66a**, **67**, and **68**, respectively. Thus lateral extension of the π system turned out to be most effective to reduce the HOMO–LUMO gap. FET device using single crystal **66b** was fabricated to exhibit hole mobilities of $(5.0 \pm 1.0) \times 10^{-3}$ cm^{2}V^{-1}s^{-1}.

5.2 Indeno[2,1-*c*]fluorene

Another class of indenofluorene isomer containing a *p*-QDM subunit is indeno[2,1-*c*]fluorene (**43a**), which was much less documented compared with the [1,2-*b*] isomer **10a**. Three stable derivatives **43b–d** were prepared by Haley et al. (Fig. 29)

[145]. X-ray crystallography of **43b** shows slightly twisted carbon frameworks and apparent quinoidal structure of the *as*-indacene portion of the molecules, indicating closed-shell electronic configuration with negligible diradical character. The lengths "*exo*-cyclic bond" of the *p*-QDM unit are 1370–1.371 Å, which are even shorter than those found in the derivatives of [1,2-*b*] isomer. On the basis of reduction/oxidation potentials of **43b** and **43c** in CV, which are − 0.66/1.07 and − 1.05/1.02 V, respectively, the HOMO/LUMO levels are estimated to be − 5.75/− 4.02 and − 5.70/− 3.63 eV, respectively. While these LUMO levels are similar to those of the corresponding [1,2-*b*] isomer, HOMO levels are significantly higher by about 0.15 eV, thus leading to significantly reduced gaps.

Benzothiophene-fused analogue of **69** was reported by Zhu et al. [129]. NICS values at the central six-membered ring and five-membered ring are calculated to be − 9.60 and 22.1 (Though in the paper it is reported by error to be − 22.1.) at the B3LYP/TPZ level, which conflict to each other with respect to the open-shell (favoring diatropic ring current in the six-membered ring) and closed-shell (favoring paratropic ring current in the five-membered ring) character. In CV **69** exhibits reduction and oxidation peaks at − 1.31 and 0.30 V, from which HOMO/LUMO levels are calculated to be − 5.30/− 3.88 eV, indicating significant reduction of HOMO–LUMO gap.

Similar to the design concept for extended π systems such as **64** and **68**, C ≡ CTIPS derivative **70** of diindeno[1,2-*b*:2′,1′-*n*]perylene, in which the six-membered ring of indeno[2,1-*c*]fluorene is replaced by a perylene unit, was prepared [146]. X-ray crystallography of **70** shows a typical closed-shell quinoid structure; the bond lengths of bonds *a*, *c*, and *e* are 1.406, 1.352, and 1.386 Å, respectively, which are typical for double bonds of *p*-QDMs. On the other hand, those of *b*, *d*, *f* −*i* are 1.435, 1.450, and 1.451–1.467 Å, respectively, which are typical for $C(sp^2)$–$C(sp^2)$ single bond. From the first reduction (− 1.1 V) and oxidation (0.5 V) potentials in CV, the HOMO/LUMO levels are estimated to be − 5.15 and − 3.80 eV, respectively, indicating further reduction of gap compared to the parent **10j** due mainly to an increase of the HOMO level. However, an FET device fabricated using **70** exhibits relatively small hole mobility of 1.7×10^{-3} $cm^2V^{-1}s^{-1}$.

In summary, indeno[2,1-*c*]fluorene (**43a**) is characterized by the closed-shell quinoidal configuration and relatively narrow HOMO–LUMO gap. The difficulty in FET application may arise from its U-shaped molecular structure, which is not favorable to construct π-π stacked geometry in the condensed phase.

5.3 Indeno[2,1-*a*]fluorene

Indeno[2,1-*a*]fluorene (**44a**) is regarded as a dibenzo-fused homologue of *o*-QDM, which is under-represented compared to *p*-QDM derivatives because of high reactivity partly due to the *s*-cis 1,3-butadiene moiety different from the *s*-trans butadiene in *p*-QDMs (Fig. 30). In a series of papers during the 1950s, the group of Le Berre reported the synthesis and reactivities of blue-colored 11,12-diphenyl derivative **44b**, which was described to exhibit an absorption maximum at 556 nm, but highly sensitive to air [147–150]. It was too early to determine and discuss molecular

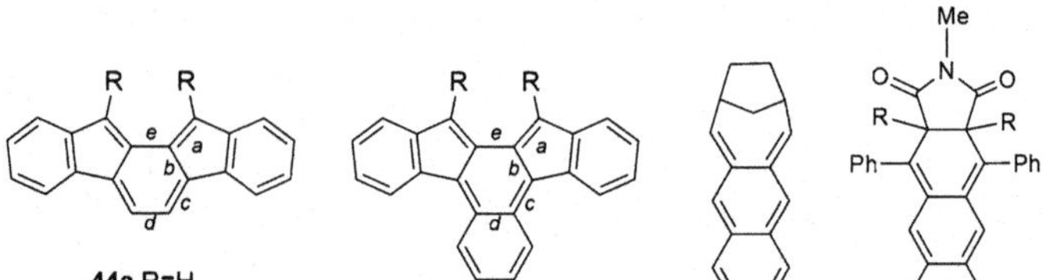

44a R=H
44b R=Ph
44c R= Mes
71a R=H
71b R= Mes
72a
72b R = $(CH_2)_4$

Fig. 30 Structures of indeno[2,1-*a*]fluorenes **44a–c**, their homologues **71a**,**b**, and naphthoquinodimethanes **72a**,**b**

Table 8 Selected bond lengths (Å) etermined by X-Ray crystallographic analysis of **44c** and **71b**

Bond	44c [152]	71b [158]
a	1.391	1.403
b	1.480	1.468
c	1.359	1.398
d	1.431	1.465
e	1.454	1.453
f		1.391
g		1.480
h		1.359

For bond labels see Fig. 30

structure and electronic configuration in detail at the time, and there had been merely one report on the theoretical work on this system [151] before Shimizu and Tobe synthesised air-stable mesityl derivative **44c** more than half a century later in 2011 [152]. The mean bond lengths of the indenofluorene core determined by X-ray crystallographic analysis of **44c** are listed in Table 8 together with those of the higher homologue **71b** discussed in the next section. The structure of **44c** shows significant bond length alternation in the *o*–QDM unit; the lengths of bonds *a* and *c* (1.391 and 1.359 Å, respectively) are short, whereas those of bonds *b*, *d*, and *e* (1.480, 1.431, and 1.454 Å, respectively) are long. The bond lengths of the terminal rings (1.391–1.410 Å) are normal for benzene ring. However, the "*exo*-cyclic bond" of the *o*-QDM unit *a* in is significantly longer than normal $C(sp^2) - C(sp^2)$ double bond (1.349 Å). It is slightly longer than that of the mesityl-substituted [1,2-*b*] isomers (1.38 Å) and the previously isolated *o*-QDM derivative **6** (1.346 Å) [20], indicating higher diradical character. Diradical character (y_0) of **44a** calculated by the broken-symmetry UHF/6-31G(d) method using the Yamaguchi scheme [Eq. (1)] [30] for the geometry optimized by RB3LYP/6-31G(d) level of theory is 0.33, consistent with a closed-shell structure, though this method tends to estimate the diradical factor larger than other methods. For example, calculations at the UB3LYP/6-311 + G**// U(R)B3LYP/6-311 + G** level of theory gave y_0 value of 0.079 [153]. The fact that **44c** exhibits sharp signals in ^{1}H NMR spectrum and did not show line broadening when heated up to 75 °C also supports the closed-shell electronic configuration. The singlet–triplet energy gap ΔE_{ST} of **44a** is estimated theoretically to be − 13.6 kcal/mol. Though experimental chemical shifts do not give useful information about antiaromaticity of **44c** because of the absence of hydrogen on the *as*-indacene core, NICS(1) [NICS(1)$_{zz}$] values of **44a** at the central six-membered ring and five membered ring are calculated to be 4.28 [13.81] and 2.12 [22.02], respectively, indicating significant paratropicity at the *as*-indacene core. On the other hand, the external benzene rings exhibit diatropicity with NICS(1) [NICS(1)$_{zz}$] values of − 6.17 [− 12.62]. In CV **44c** shows reduction and oxidation waves at − 1.51 and 0.59 V, respectively, from which HOMO–LUMO gap is estimated to be 2.10 eV which agrees the theoretical value of 2.25 eV.

Extension of π-conjugation in the vertical direction of **44a** leads to benz[*c*]indeno[2,1-*a*]fluorene (**71a**), which is expected to exhibit larger open-shell character and smaller HOMO–LUMO gap than those of **44a** (Fig. 30). This compound possesses a 2,3-naphthoquinodimethane (2,3-NQDM) substructure, which is known to be more reactive than *o*-QDM and has never been isolated previously despite many attempts. The parent 2,3-NQDM and its bridged derivatives **72a** and pleiadene homologue **26** (Fig. 4) [41] were only spectroscopically characterized in glass matrices at low temperatures [154–156]. Sterically protected NQDM derivatives such as **72b** were reported to be persistent at room temperature for UV–vis measurement and were trapped by a dienophile [157]. However, they had never been isolated due to pronounced air-sensitivity. Tobe et al. [158] synthesized and isolated **71b**, of which half-life is only 77 min under air at room temperature in the dark. Theoretical calculations at the UB3LYP/6-31G(d) level of theory for **71a** indicate that it has larger singlet biradical character ($y_0 = 0.63$) and smaller HOMO–LUMO energy gap (1.66 eV) and single-triplet energy gap (− 7.46 kcal/mol) compared to **44a**. X-ray crystallographic analysis of **71b** showed that the bond lengths of the benzindeno-fluorene core has less quinoidal character than that of **44c** (Table 8). Namely, the bond distance of the "*exo*-cyclic" bond a (1.403Å) of **71b** is longer than that of **44c** (1.391 Å) as is for bond c (1.398 vs. 1.359 Å). On the other hand, bond b is shorter than that of **44c** (1.468 vs. 1.480 Å). Another notable feature is the bond-length alternation in the outer naphthalene ring of **71b**; bonds f (1.441 Å) and h (1.419 Å) are relatively long, whereas bond g is short (1.363 Å). In accord with the theoretical calculations, **71b** exhibited longer wavelength absorption compared to **43c** with a maximum at 697 nm which extends to NIR region (ca. 1050 nm). Similarly, from CV in which the first reduction and oxidation waves are observed at −1.22 and 0.38 V, respectively, and the E^{redox} of 1.60 V which is considerably smaller than that of **44c** (2.10 V). Moreover, in the ^{1}H NMR spectra, **71b** exhibited reversible line broadening at high temperature (100 °C) due to occupation of a thermally excited triplet state, consistent with the calculated small ΔE_{ST} value. All these features indicate that **71a** is regarded as a substantially open-shell singlet species.

Inspection of the spin density distributions of **44a** and **71a** (Fig. 31) indicates the major location of densities at the apical position of the five-membered ring with the second largest densities in the inner naphthalene ring in **71a**. Distribution of HOMO shares the same characteristic. Because of this spin and HOMO density distributions, **71b** exhibits unique chemical behaviour. For example, the product of air

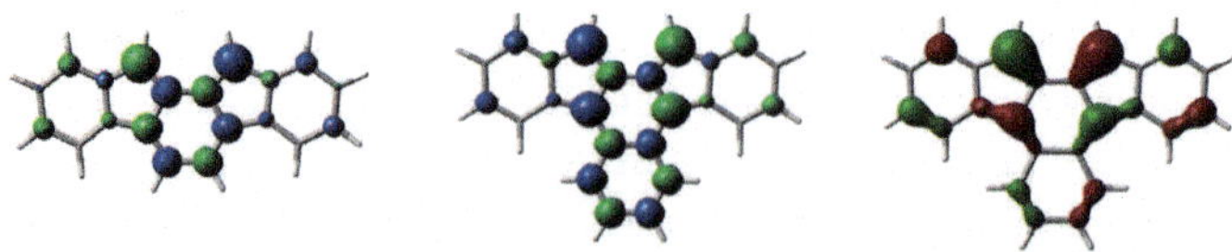

Fig. 31 Spin density distributions in **44a** (left) and **71a** (middle) and HOMO distribution of **71a** (right)

Fig. 32 Structures of adducts **73** and **74** and dimers **75** formed from **71b**

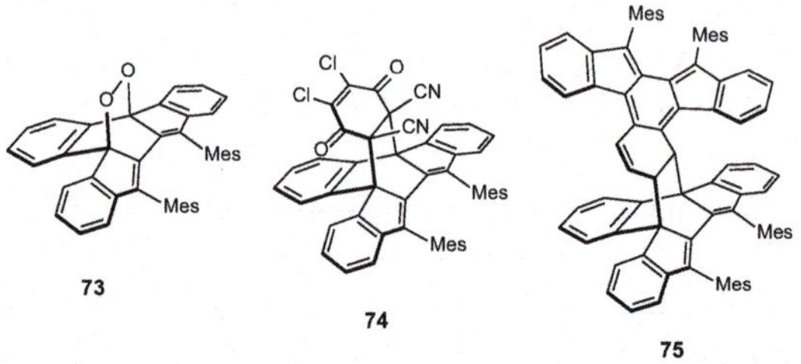

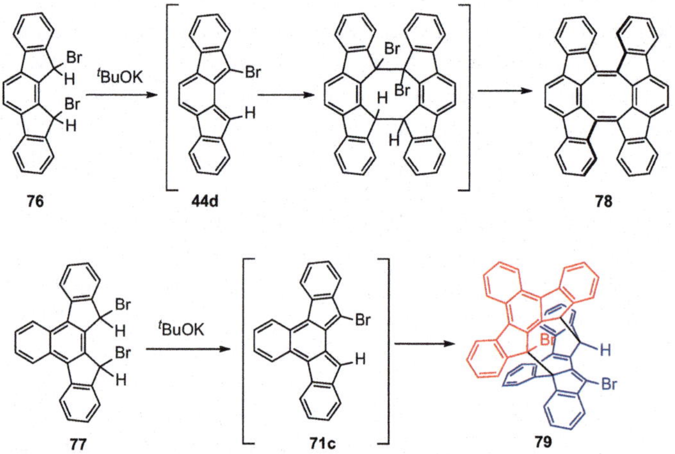

Scheme 4 Dimerization of **44d** and **71c** generated from **76** and **77**, respectively

oxidation was endoperoxide **73**, in which the oxygen bridge attaches at the naphthalene core. Apparently, the most reactive apical positions are protected by steric hindrance due to bulky mesityl groups. Similarly, [4 + 2] cycloaddition of 2,3-dichloro-5,6-dicyano-*p*-benzoquinone (DDQ) took place at the same positions to give adduct **74**. Moreover, heating **71b** at 80 °C resulted in the formation of [4 + 2] dimer **75** in quantitative yield (Fig. 32). Note that in **75** the new bonds were formed at the inner naphthalene (a 4 π-electron component) and the outer naphthalene carbons (a 2 π-electron component), as a consequence of the strong bond length alternation in the outer naphthalene ring of **71b**.

Another interesting behaviour is the dimerization of less protected bromides **44d** and **71c**, which are generated in situ from the corresponding dihydrobromides **76**

Fig. 33 Structures of indeno[2,1-*a*]fluorenes **11a,b** and its homologues **80a,b**

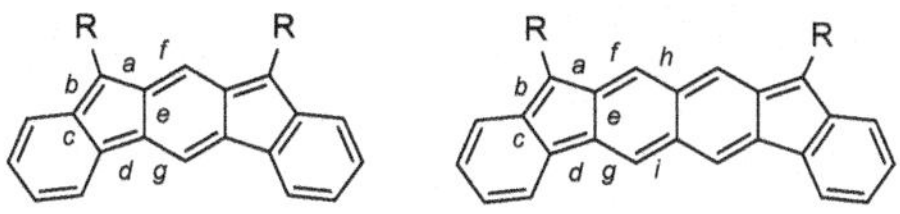

11a R=H
11b R=Mes

80a R=H
80b R= Mes

Table 9 Selected bond lengths (Å) determined by X-ray crystallographic analysis of **11b** and **80b**

Bond	11b [160]	80b [161]
a	1.437	1.437
b	1.420	1.424
c	1.438	1.430
d	1.442	1.445
e	1.438	1.448
f	1.394	1.386
g	1.396	1.370
h		1.417
i		1.412

For bond labels see Fig. 33

and **77** (Scheme 4) [159]. Thus, treatment of **76** with excess potassium *t*-butoxide gave dimer **78**, which is a product of formal [4 + 4] addition of monobromoindenofluorene **44d** followed by elimination of 2HBr. On the other hand, under the same reaction conditions, **77** gave a dimer **79** of complex structure via intermediate **71c** in which one of the new bonds are formed at the inner naphthalene position.

5.4 Indeno[2,1-*b*]fluorene

Indeno[2,1-*b*]fluorene (**11a**) is expected to have a larger singlet diradical character than its constitutional isomers because of the presence of the *m*-QDM unit (Fig. 33). As shown in Fig. 19, the numbers of aromatic sextet structures in the open-shell diradical forms are larger than those closed-shell Kekulé canonical forms by one in [1,2-*b*], [2,1-*c*] and [2,1-*a*] isomers, whereas in [2,1-*b*] isomer **11a** the number of aromatic sextet in the diradical form is larger by two than in the Kekulé form. In fact, the diradical character of **11a** ($y_0 = 0.645$) estimated by the LC-UB3LYP/6-311 + G**//U(R)B3LYP/6-311 + G** level of theory is larger than that of **10a** (0.079) derived at the same method [150]. Air-stable derivative **11b** was prepared by Tobe et al. [157]. As shown in Table 9, X-ray crystal structural analysis showed that the lengths for bonds *a* (1.437 Å) and *b* (1.420 Å) lie between the lengths of the $C(sp^2) - C(sp^2)$ bond in benzene (1.39 Å) and the $C(sp^2) - C(sp^3)$ bond in fluorene (1.468 Å). This observation indicates that **11b** has a substantial contribution of the singlet diradical canonical structure to the ground-state electronic configuration. Theoretical calculations also indicate that **11a** has a narrow HOMO–LUMO energy gap (1.23 eV) and a small single-triplet energy gap (− 2.46 kcal/mol). Experimentally, **11b** showed in CV the first reduction and oxidation peaks at –1.13 and 0.13 V, respectively. The small E^{redox} of 1.26 V is smaller than those of [2,1-*a*] isomer **44c** (2.10 V) and its homologue **71b** (1.60 V) described in the previous section. Moreover, in the ^{1}H NMR spectrum, **11b** did not show signal at 30 °C because of a thermally excited triplet state. By cooling down to –93 °C, broad signals due to the aromatic protons appearing at 5.6–6.9 ppm. Moreover, a triplet species for **11b** was observed by ESR spectroscopy, and the intensity of the ESR signal decreased with decreasing the temperature, indicating a singlet ground state in **11b**. The singlet–triplet energy gap was experimentally estimated to be − 4.21 kcal/mol using variable-temperature SQUID measurements, in agreement with the theoretical energy gap. In contrasted to *m*-QDM whose triplet ground state lies well below the singlet, the singlet state of **11a** is significantly stabilized relative to the triplet by embedding into the indenofluorene framework, by which spin polarization is significantly perturbed.

As in the case of *s*-indacene and indeno[1,2-*b*]fluorene congeners, extension of π-conjugation in the indacene core would increase open-shell character of the [2,1-*a*] isomer too. A derivative **80b** of laterally extended homologue fluoreno[2,3-*b*]fluorene (**80a**) was prepared by the group of Tobe [161]. Although it decomposed with a half-life of 30 h at ambient conditions, it was successfully characterized by spectroscopically and by X-ray crystallography. The diradical character of **80a** ($y_0 = 0.772$) estimated at the same level of theory is even larger than that of **11a** (0.645) [150]. As shown in Table 9, X-ray crystal structural analysis showed that lengths of the critical bonds *a* (1.437 Å) and *b* (1.424 Å) in the five-memberedring of **80b** are virtually the same as those of **11b** despite the larger diradical character, suggesting that the lengths may have reached to the limit. The bond lengths of the central part of the molecule show characteristics of naphthalene bonds. Theoretical calculations also indicate that **80a** has a narrow HOMO–LUMO energy gap (1.10 eV) and a small single-triplet energy gap (− 2.11 kcal/mol). Compound **80b** showed in CV the first reduction and oxidation peaks at − 1.16 and 0.00 V, respectively. The small E^{redox} (1.16 V), which is even smaller than that of **11b** (1.26 V), agrees with the theoretical value estimated by the B3LYP/6-311 + G** level of theory and results mainly from lowered oxidation potential. Moreover, in the ^{1}H NMR spectrum, **80b** did not show signal even at − 60 °C because of a thermally excited triplet state. At –80 °C, only very broad signals centered at ca. 2 and 7 ppm appear. A featureless ESR signal at $g = 2.004$, of which intensity decreases with lowering temperature was observed. By variable-temperature SQUID measurements, $\Delta E_{\text{S-T}}$ value is estimated to be − 2.30 kcal/mol in agreement with the theoretical estimation. These results clearly indicate significant open-shell character of **80a**. This results is also explained in terms of aromaticity of the six-membered rings flanked by the five-membered rings. Figure 34 shows − σ_{yy} maps in the singlet states of **11a** and **80a**, in which the magnetic shielding tensor component (− σ_{yy}) 1 Å above the center of the ring planes are displayed. The negative and positive − σ_{yy} values indicate local aromatic and antiaromatic characters of each ring, respectively. The maps show more aromatic feature in the six-membered rings and less antiaromatic character in the five-membered rings in the singlet states of **80a** compared to **11a** [150].

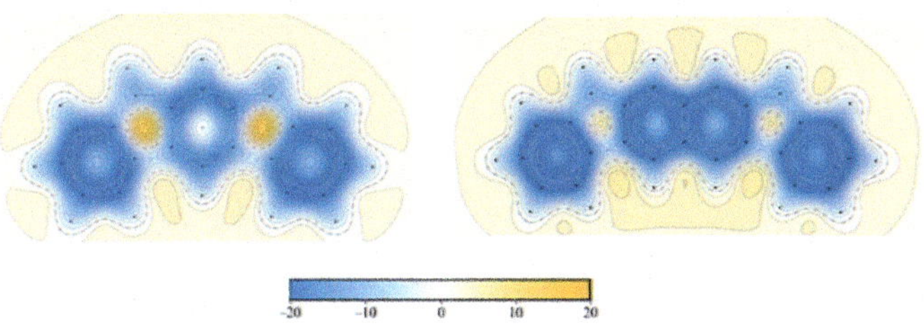

Fig. 34 $-\sigma_{yy}$ Maps in the singlet states of **11a** and **80a**. Reproduced with permission from [153]. Copyright 2015 The American Chemical Society

Fig. 35 Structures of indeno[1,2-*a*]fluorenes **45a,b** (left) and NICS(1) values (right) for **45b**

45a R=H
45b R=Mes

Additionally, an unexpectedly low-energy light absorption of **11b** despite its small conjugation space was recorded. The absorption spectrum for **11b** in CH_2Cl_2 showed a very weak absorption band ascribed to $S_0 \rightarrow S_1$ transition at 1700 nm, which is at longer wavelength than that of the lowest-energy absorption band of its larger homologue **80b** observed at 1293 nm. Theoretical investigations revealed that moderate amplitude of diradical character of **11a**, in addition to relatively small conjugation size (20 π-electrons) is critical for the extraordinary low-energy light absorption [161].

5.5 Indeno[1,2-*a*]fluorene

The last isomer of indenofluorenes **45a** is theoretically predicted to have the largest diradical character ($y_0 = 0.80$) and triplet ground state electronic configuration unlike the other isomers (Fig. 35). Recently, experimental endeavors devoted by Haley et al. [162] revealed the existence of mesityl derivative **45b**, though it was not fully characterized by NMR or ESR owing to its high sensitivity toward oxygen. However, the characteristic absorption bands in UV–vis–NIR spectrum and redox potentials determined by CV are consistent with those expected for **45b**. Moreover, a sample of **45b** was reduced with cesium to give its dianion, which was isolated and fully characterized by spectroscopy and X-ray crystallography, giving further support for the existence of **45b**. It is interesting to note that NICS(1) values shown in Fig. 35 suggest both the strongly diatropic outer benzene ring and also weaker diatropic character of the inner *as*-indacene part of the molecule. ACID calculations also indicate diatropic ring current. These results suggest that **45b** is regarded as a Baird aromatic system [163] due to its 20 π electron conjugation with predicted triplet ground state.

Springer

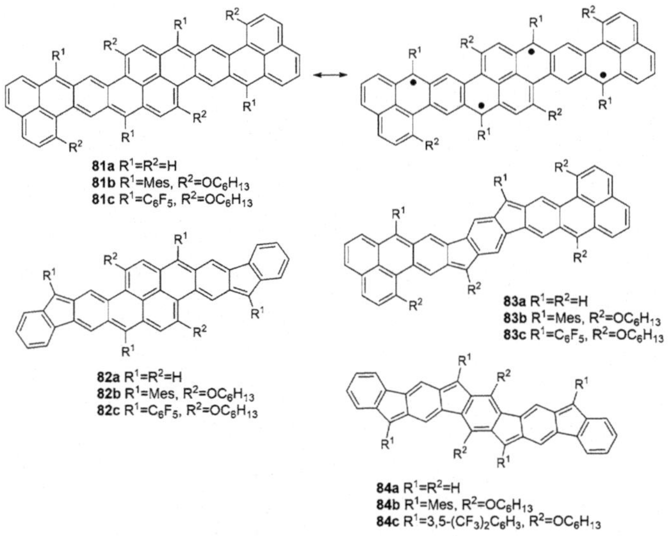

Fig. 36 Structures of compounds **81a–c–84a–c** with resonance structures

5.6 Multiradicaloids

It is appropriate to discuss multiple diradical states here in view of the structural units related to indenofluorenes employed for designing molecules of such extraordinary configurations. Although singlet hydrocarbons with multiradicaloid character are the next targets after diradicaloids in view of scientific challenges in theory, design, synthesis, and characterization of electronic properties, as well as prospective applications in optoelectronic and maybe spintronic devices, only a few studies have been done to date. The design principle is rather simple: either by conjoining diradicaloid units such as linear-shaped indeno[1,2-*b*]fluorene or U-shaped indeno[1,2-*c*]fluorene in a lateral or cyclic fashion, or cyclically conjoining fluorene and carbazole units. From the drawings of resonance structures, they can be regarded as mutiradicaloids such as tetraradicaloids and hexaradicaloids. However, on the basis of their electronic configurations, they may be better called as multiple diradicaloids, such as double or triple diradicaloids, because they have certain occupation numbers in their unoccupied natural orbitals, i.e., y_1, $y_2 > 0$.

To construct linearly extended tetraradicaloids, Wu et al. studied the synthesis of a series of derivatives of bis-heptazethrene **81a** and bis-indeno[1,2-*b*]fluorene **84a** and their hybrids **82a** and **83a** (Fig. 36) [140]. They are related to the modified indenofluorene congeners described in Sect. 5.1.2. As illustrated for **81a–c**, the closed-sell and fully open-shell tetraradical structures can be drawn in addition to their intermediate forms (diradicals). Theoretically estimated diradical factors (y_0 and y_1) of **81a–84a** are listed in Table 10. As can be seen from Table 10, while y_0 of **81a** is large (0.713), its y_1 is small (0.003). In contrast, while the y_0 value of **83a** is not large (0.351), its y_1 is the largest of the four (0.031). This indicates nonlinear

Table 10 Calculated diradical characters (y_0 and y_1) and singlet–triplet energy gaps ($\Delta E_{S\text{-}T}$) of **81a–84a** at the UCAM-B3LYP/6-31G(d,p) level [140]

Compd	y_0	y_1	$\Delta E_{S\text{-}T}$ (kcal/mol)
81a	0.713	0.009	− 3.0
82a	0.635	0.003	− 2.3
83a	0.351	0.031	− 4.2
84a	0.038	0.005	− 8.4

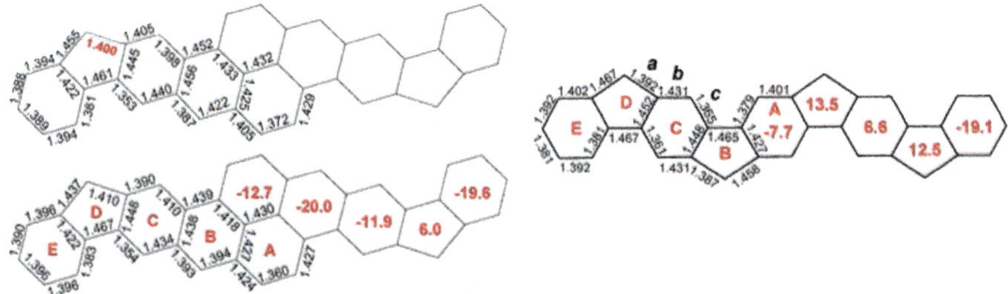

Fig. 37 Bond lengths (Å) and NICS(1)$_{zz}$ values calculated for **82a** (left bottom) and **84a** (right) by the UCAM-B3LYP method and X-ray bond lengths **84b** (left top). Reproduced with permission from [140]. Copyright 2016 The American Chemical Society

relation between the y_0 and y_1 values, thus making the design of multiple diradicaloids complicated. Although compounds **81b,c** and **83b,c** were not isolated and the corresponding diketones derived by air oxidation were obtained, compounds **82b,c** and **84b,c** were isolated and characterized by spectroscopic and crystallographic methods. The above stability trend does not seem to be linearly related to the y_0 values. Rather it seems that the terminal phenalenyl units are responsible for the facile air oxidation. As expected from the large y_0 value (0.635) of **82a**, **82b** shows broad ^{1}H NMR signals even at − 100 °C and broad ESR signal at g = 2.0027, of which intensity decreases by lowering the temperature, indicating a singlet ground state with thermally excited triplet state. The experimental $\Delta E_{S\text{-}T}$ determined by SQUID measurement is − 1.69 kcal/mol in accordance with the theoretical value. The HOMO–LUMO gap of **82b** was estimated to be as small as 0.94 eV, on the basis of the first reduction and oxidation peaks (− 1.45 and − 0.26 V, respectively) in CV. NICS(1)$_{zz}$ of **82a** (Fig. 37) indicate fully aromatic character in all six-membered rings A–C and E, whereas weak antiaromatic character in the five-membered ring D. By X-ray crystallography, it appears that the "*exo*-cyclic" bond lengths in the five-membered ring are significantly elongated (1.410 Å). All these theoretical and experimental results support the open-shell character of this system. In contrast, the corresponding bond lengths in **84b** (1.392 and 1.387 Å) are indicative of quinoidal character, in accord with the NICS(1)$_{zz}$ distribution shown in Fig. 37. The different open-shell character between **82a** and **84a** are interpreted in terms of pro-aromatic (**82a**) and antiaromatic (**84a**) character. In the latter system, charge transfer character may also contribute to the ground state electronic configuration.

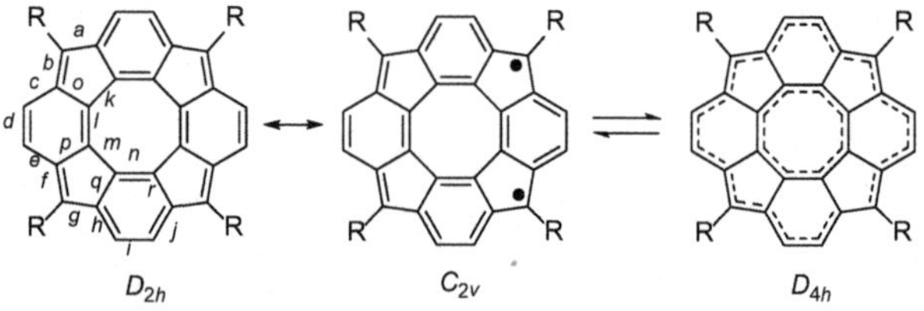

Fig. 38 Structures of **12a–c** with resonance/equilibrium structures

Arrangement of two indeno[2,1-*c*]fluorene units in a cyclic form as in **12a**, tetracyclopenta[*def,jkl,pqr,vwx*]tetraphenylene, may also generate double diradical (tetraradical) character (Fig. 38). Although this hydrocarbon was proposed in 1970 by Hellwinkel as a potential hydrocarbon consisting of doubly antiaromatic circuits, it eluded synthesis for a long time [164]. Synthesis and characterization of its derivative **85b** substittuted with mesityl groups was reported by the group of Tobe [165]. Like many cycloarenes [166], although **85a** can adopt either an annulenoid structure, with internal 8 π-electron and external 20 π-electron circuits, or a benzenoid structure, the latter must be favored because of the aromatic sextets. However, in the benzenoid form, closed-shell D_{2h}, open-shell diradical C_{2v}, and open-shell tetraradical D_{4h} structures exist as shown in Fig. 38. Theoretical calculations based on the spin-flip time-dependent density functional theory [SF-TD-BHHLYP/6-311G(d,p)] indicate that the y_0 and y_1 values for the D_{2h} structure are 0.095 and 0.032, being consistent with a closed-shell nature. In contrast, y_0 and y_1 of the D_{4h} structure are 1.000 and 0.166, indicating that it adopts pure diradical (i.e., occupation number of LUNO = 1) and small tetraradical character (i.e., occupation number of LUNO + 1 = 0.17). However, the latter structure has a higher energy than the former by 8.3 kcal/mol. Experimentally, the D_{2h} structure was confirmed by X-ray crystallography of **12b** as shown in Table 11 which also includes the theoretical bond lengths for the D_{2h} and D_{4h} structures of **12a**. The theoretical bond lengths for the D_{2h} structure agrees well with those of the experimental structure. In ^{1}H NMR spectrum the core hydrogen signal appears as a single peak at 2.85 ppm in toluene-d_8 indicating significant paratropicity. Indeed, NICS(1) values for the D_{2h} structure show nonaromatic to weak antiaromatic characters in the six-membered rings (− 1.45 and 5.51) and antiaromatic character in the five- and eight-membered rings (8.51 and 10.46, respectively). It was not possible to determine whether **12b** adopts rapidly equilibrating D_{2h} structures or a D_{4h} structure in solution, because by decreasing the temperature down to − 80 °C the signal splitting was not observed. At high temperature (50 °C) the signals start to broaden reversibly, indicating the population of thermally excited triplet state, though the theoretical $\Delta E_{\text{S-T}}$ values for the D_{2h} and D_{4h} structures are significantly large (19 and 16.5 kcal/mol). Recently, C ≡ CTIPS-substituted derivative **12c** was prepared and was shown to adopt a

Table 11 Bond lengths (Å) determined by X-ray crystallographic analysis of **12b** and theoretical bond lengths (Å) for D_{2h} and D_{4h} structures of **12a** [165]

Bond	**12b**[a]	D_{2h}-**12a**[b]	D_{4h}-**12a**[b]
a	1.460	1.447	1.397
b	1.371	1.350	1.397
c	1.433	1.437	1.403
d	1.349	1.339	1.366
e	1.440	1.437	1.403
f	1.373	1.350	1.397
g	1.452	1.447	1.397
h	1.391	1.378	1.403
i	1.390	1.390	1.366
j	1.380	1.378	1.403
k	1.479	1.476	1.477
l	1.359	1.344	1.359
m	1.487	1.476	1.477
n	1.379	1.373	1.359
o	1.481	1.469	1.442
p	1.476	1.469	1.442
q	1.433	1.420	1.442
r	1.435	1.420	1.442

For bond labels see Fig. 38

[a]X-ray analysis

[b]SF-TD-BHHLYP/6-311G(d,p) level

D_{4h}-like structure in crystals, thus awaiting further study to demonstrate the tetra-radical character.

Aihara analyzed the stability of **12a** by topological resonance energy (TRE) [167]. TRE ($- 0.1354$ |β|) as well as %TRE ($- 0.342$) indicate that macrocyclic antiaromaticity in **12a** dominates the global anti-aromaticity of the molecule. Dewar resonance energy calculated for the DFT-optimized D_{2h} geometry of **12a**, however, is positive in sign and as large as 37.3 kcal/mol. This is attributed to the presence of tow Clar's sextets in the canonical structure of the D_{2h} geometry.

Although neutral **12a** is theoretically calculated to favor a low-symmetry structure, the corresponding dication **12a**$^{2+}$ and dianion **12a**$^{2-}$ formed by two-electron oxidation and reduction of **12a**, respectively, are expected to adopt high-symmetry D_{4h} geometries [166, 168]. All bond lengths are elongated in both **12a**$^{2+}$ and **12a**$^{2-}$ as a result of repulsion between positive and negative charges, respectively. While the magnitudes of the bond length alternations are similar to that of neutral **12a**, in dication **12a**$^{2+}$ it becomes slightly less significant. Interestingly, NICS(1) of the five-, six- and eight-membered rings of dianion **12a**$^{2-}$ are $- 26.4$, $- 12.2$ and $- 2.7$ ppm, respectively, indicating that peripheral rings are strongly aromatic whereas the central eight-membered ring is almost non-aromatic, thereby forming an "annulene-within-an-annulene" (AWA) system [169] consisting of opposite tro-picity. On the other hand, NICS(1) values of dication **12a**$^{2+}$ are $- 21.6$, $- 15.6$ and $- 8.4$, respectively, indicating that all rings are strongly aromatic. The presence of

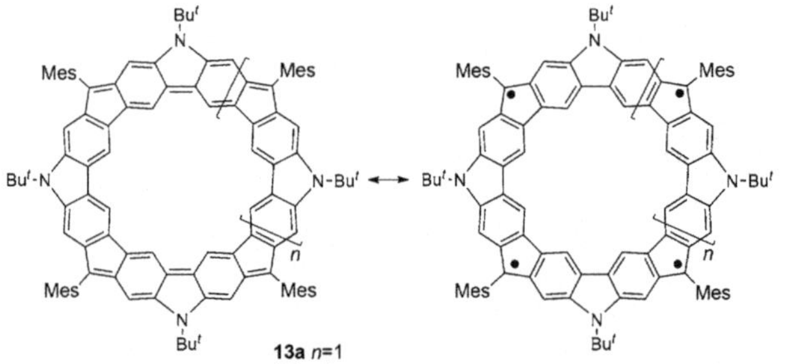

Fig. 39 Structures of macrocycles **13b** and **85** with resonance structures

Table 12 Theoretical multiradical characters and singlet–triplet energy gaps of **13** and **85** [171]

Compd	Point group	Rel energy (kcal/mol)	y_0	y_1	y_2	$\Delta E_{\text{S-T}}$ (kcal/mol)
13b	C_{2v}	0	0.27	0.19		-5.5
	C_{4v}	6.3	1.00	0.27		-2.1
85	C_1	0	0.31	0.31	0.23	-4.0
	C_{2v}	0/.9	0.53	0.52	0.27	-1.8

RAS-SF/6-31G* for conformers optimized by UB3LYP/6-31G*

significant diatropicity at both the outer and inner rings suggests an AWA picture, in which both the outer and inner macrocyclic rings are diamagnetic. Hence, the charged species of derivatives of **12a** would provide real challenges in π-electron conjugation. These predictions together with the generation and characterization of tetra-anion of derivatives of **12a** have been verified experimentally [170].

As the third approach, Wu et al. synthesized large macrocycles **13b** and **85**, which can be viewed as alternately fused fluorene and carbazole units [171]. As shown in Fig. 39, in order to increase the open-shell character, the molecules are designed based on the pro-aromatic concept to stabilize the multi radicaloid canonical forms; for example, in the closed-shell from of **13b** there are four benzenoid sextets, whereas in the fully open-shell tetraradical form there are eight sextets. With the larger increase of the number of sextets, the open-shell from will enjoy more aromatic stabilization, thereby increasing its contribution. In the case of **85** the number of sextets changes from eight to 16. Because of the bond angle constraint, the most stable conformations of **13b** and **85** are not planar. Table 12 shows theoretical multiradical characters (y_0, y_1, and y_2) and singlet–triplet energy gaps ($\Delta E_{\text{S-T}}$) based on restricted active space spin flip method (RAS-SF/6-31G*) for two most stable conformers optimized using the UB3LYP/6-31G* level of theory. Though **13b** and **85** were indeed synthesized, their characterization turned out not easy due to air-sensitivity, occupation of thermally excited triplet, and potential aggregation in solution, all which contribute to make ^{1}H NMR spectra broadened and complicated. By SQUID measurement, $\Delta E_{\text{S-T}}$ of **13b** and **85** are estimated to be as small as -0.30 and -0.25 kcal/mol, which are significantly smaller than the theoretical values. The

HOMO/LUMO energy levels (and HOMO–LUMO gaps) of them are determined by CV to be − 4.01/− 3.51 (0.50) and − 4.14/− 3.51 (0.63) eV, respectively. In accord with the extremely small HOMO–LUMO gap both exhibit electronic absorption longer than 2000 nm. NICS(0) and ACID calculations indicate that the macrocyclic paratropicity in 36 π-electron system **13b** and diatropicity in 54 π-system **85**. Unfortunately no crystal structure analysis for these extraordinary π-systems has been made.

This design principle was nicely advanced to construct molecules by assembling fluorenyl radical units only, either linearly or cyclically, to furnish acyclic radicals **86a–e**, purely fluorenyl macrocyclic radicals **87a–c**, and ethynylene-linked macrocycles **88a–c** (Fig. 40) [172]. Note here that both macrocycles with not only even but also odd numbers of radical centers, **87b** and **88a,c** were prepared and isolated thanks to the stabilizing effect of the anthryl groups which were shown kinetically to block the 9-positions of the fluorenyl radicals in the linear radicals **86a–e** [173].

Theoretical studies combined with X-ray structural analyses for **86a** and **87a** reveal that (i) the linear oligomers **86c–e** adopt *s*-trans conformation, though the crystal structure of dimer **86a** is *s*-cis due to packing force, with substantial twist (~ 32°) between the fluorenyl units, (ii) in the series of macrocycles **87a–c**, tetramer **87a** has most planar geometry, even though the fluorenyl–fluorenyl torsion angles are large (26.5° and 28.9° by X-ray) yet smaller than those of **87b** and **87c** (34°–56°), (iii) ethynylene-bridged macrocycles **88a** and **88b** have nearly planar π-conjugated backbone because of the absence of steric hindrance between the fluorenyl units, and (iv) the relatively long bond lengths between the fluorenyl units in **87a** (1.468–1.463 Å by X-ray) and fluorenyl-ethynyl lengths in **88b** (1.410 Å by theory) indicate that the bonding interaction between the four radical centers are substantially weak. Large positive NICS(0) values at the center of macrocycle of **87a** (+ 11.7 ppm) and **88b** (+ 14.26 ppm) are indicative of moderate anti-aromaticity in their 36 and 44 π-electron circuits, respectively, as is the paratropic ring currents in ACID calculations.

Theoretically estimated open-shell characters by the spin-flip RAS-SF/6-31G* method are summarized in Table 13. As designed, cyclic tetramers **87a** and **88b** are shown to have significant tetraradical character as indicated by large y_0 and y_1 values, as well as the small singlet–triplet energy gap, as is cyclic hexamer **87c** for its hexaradical character. In general, macrocyclic oligomers tend to exhibit narrower HOMO–LUMO energy gaps and smaller excitation energy from the low-spin (LS) ground state to the first high-spin (HS) excited state compared to those of the corresponding linear oligomers. However, there seems to be no obvious relationship between these energy gaps and the number of the fluorenyl units because of the different degree of planarity of the π-conjugated backbone which affects the spin–spin interaction.

6 Dibenzopentalene-Related System

As a last class of QDMs incorporated into conjugated polycyclic frameworks, molecules whose structures are related to yet unknown hydrocarbon, dibenzo[*cd,gh*]

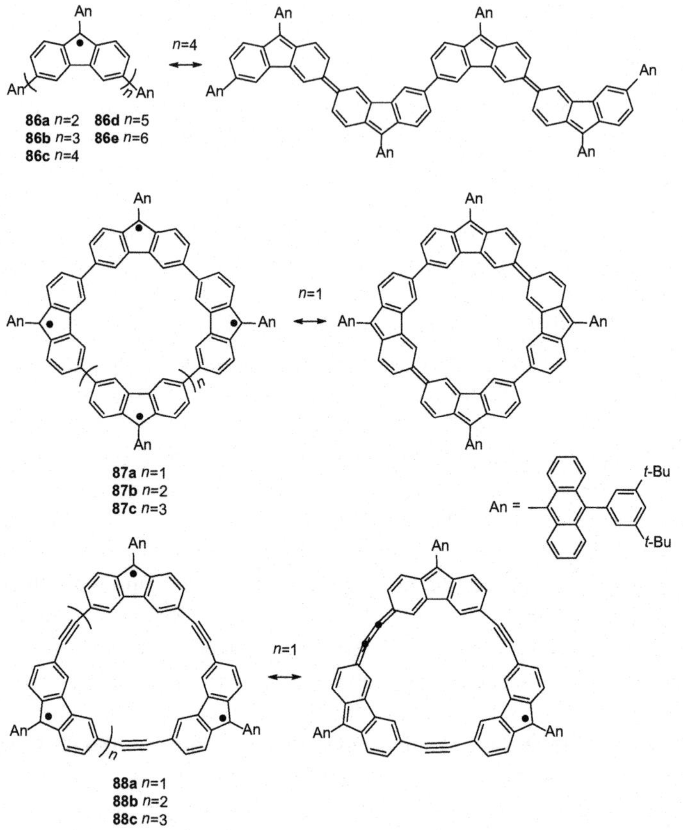

Fig. 40 Structures of fluorenyl radicals **86a–e**, **87a–c**, and **88a–c** with the closed-shell resonance structures for **86c** ($n = 4$) and **87a** ($n = 1$), and a monoradical structure for **88a** ($n = 1$)

pentalene (**89**), are described. These include unknown bis-periazulene (**90**), dibenzo[*a,f*]pentalene (**91a**), dicyclopenta[*ghi,pqr*]perylene (**92**), and cyclopenta[4,5]phenanthro[1,10,9,8-*opqra*]cyclopenta[*ghi*]perylene (**14a**). Synthesis of kinetically stabilized derivatives of **91a** and **14a** were achieved very recently (Fig. 41).

Hydrocarbon **89** is regarded as a perturbed [12]annulene, similar to the known pyracylene (**93**) [174]. PPP-SCF calculations indicate the lowest triplet state is only 0.32 eV above the signet ground state. Despite attempts via several different approaches, synthesis of **89** has not been achieved. However, detailed inspection of the properties of dianion generated by deprotonation of the dihydro derivative and semiquinone radical anion electrochemically generated from the quinone derivative support the peripheral delocalization of electrons.

Replacement of one of the five-membered rings in **89** with a seven-membered ring gives 14 π-electron system **90** called bis-periazulene, which has been attracting continuing interest for a long time, although none of derivatives have been

Table 13 Theoretical multiradical characters, HOMO, LUMO energy levels and their gaps and low-spin ground state to the first high-spin excited state energy gaps of **86b–e**, **87a–c**, and **88a–c** [172]

Compd	number of fluorenyl units	y_0	y_1	y_2	E_{HOMO}^a (eV)	E_{LUMO}^a (eV)	$E_g^{EC\,b}$ (eV)	$\Delta E_{LS\text{-}HS}^c$ (kcal/mol)
86b	3	0.18	0	0	− 4.77	− 3.87	0.90	− 2.19 (D-Q)
88a	3	0.12	0	0	− 4.66	− 4.11	0.55	− 2.10 (D-Q)
86c	4	0.75	0.59	0	− 4.83	− 3.82	1.01	− 2.94 (S-T)
87a	4	0.99	0.51	0	− 4.72	− 3.85	0.87	− 2.80 (S-T)
88b	4	1.00	0.71	0	− 4.65	− 4.02	0.63	− 2.86 (S-T)
86d	5	0.28	0.24	0	− 4.65	− 3.88	0.77	− 1.76 (D-Q)
87b	5	0.33	0.18	0	− 4.70	− 3.95	0.75	− 1.66 (D-Q)
88c	5	0.53	0.14	0	− 4.65	− 3.99	0.66	− 2.02 (D-Q)
86e	6	0.80	0.68	0.59	− 4.74	− 3.82	0.92	− 2.65 (S-T)
87c	6	0.78	0.74	0.56	− 4.77	− 3.82	0.95	− 1.65 (S-T)

[a]Theoretical HOMO and LUMO energy levels

[b]HOMO-LUMO gap determined by electrochemical method

[c]Excitation energy from the low-spin (LS) ground state to the first high-spin (HS) excited state determined by SQUID. S-T and D-Q stand for transition from singlet to triplet and doublet to quartet, respectively

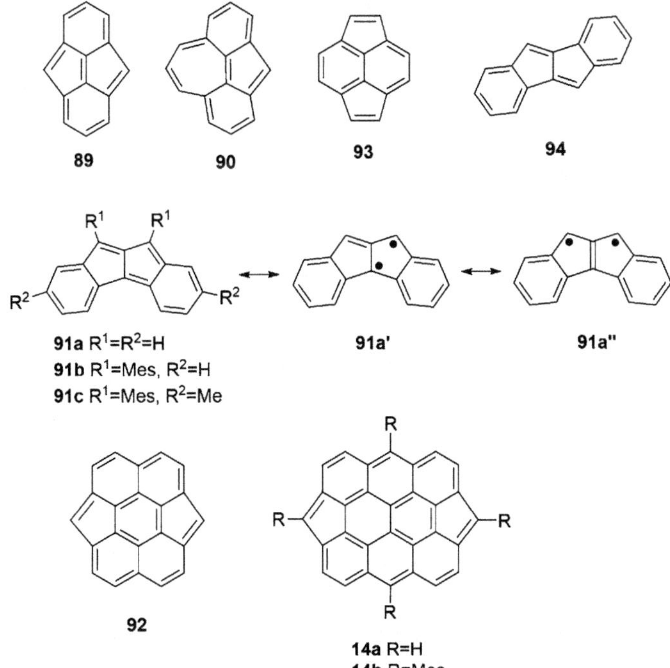

Fig. 41 Structures of compounds **89–90**, **91a–c** with resonance structures of **91a**, **92–94**, and **14a,b**

synthesized yet. An early attempt to synthesize **90** by dehydrogenation failed [175], as theoretically supported for its delocalization energy and stability index based on HMO calculations [176–178]. Earlier it was predicted by Heibronner on the basis of PPP calculations that **90** would have a triplet ground state [179]. It was confirmed by B3LYP/6-31G* calculations that a singlet state of **90** is above the triplet state by 1.7 or 2.3 kcal/mol depending on whether zero-point energy correction is included or not [77, 180]. NICS values for the five-, six- and seven-membered rings of **90** in the singlet state is − 17.29, − 24.97, and 3.69, respectively. Over all, the π system of **90** can be regarded as a biphenyl linked by methyl and allyl radical units. Neither of later attempts via oxidation of the anion of tri-*tert*-butyl derivative nor electrophilic cyclization are successful [181, 182], thus awaiting materialization of this reactive hydrocarbon.

Dibenzo[*a,f*]pentalene (**91a**) and dibenzo[*a,e*]pentalene (**94**) are isomers of **89**. Since the first synthesis of the [*a,e*] isomer in as early as 1912 [183], many derivatives and homologues have been prepared initially in connection with anti-aromaticity [52, 184], then because of their potential for optoelectronic applications supported by their kinetic stability under ambient conditions in recent years [185, 186]. Because of closed-shell character of **94** and in view of the review articles already published recently, special focus is put herein on derivatives of [*a, f*] isomer **91b** and **91c** which were synthesized for the first time recently by the group of Konishi and Yasuda [187].

As illustrated in Fig. 41, there are two open-shell canonical structures, *ortho* diradical **91a′** and trimethylenemethane (TMM)-like diradical **91a″**. Since TMM is a non-Kekulé molecule with a triplet ground state [28], it is of special interest to see how this form would contribute to the structural and physical properties of this system. Observation of temperature-dependent line broadening in the ^{1}H NMR spectra as well as intensity variation of signals due to a triplet state in the ESR spectra of **91c** indicate singlet ground state with a small singlet–triplet energy gaps ($\Delta E_{\text{S-T}}$), which was determined by SQUID to be 4.31 kcal/mol. Two crystallographically independent molecules of C_s- and C_{2v}-symmetry are found in the X-ray crystal structure analysis of **91c**, though quantum chemical calculations (UB3LYP/6-31G**) for **91a** favor the former geometry for singlet and the latter for triplet state. The averaged bond lengths in the five-membered rings of **31c** vary within a range of 1.409–1.458 Å with relatively small bond length alternation. The bond length of the central bond shared by two five-membered rings is 1.458 Å, implying larger contribution of canonical structure **91a′** rather than TMM-like **91a″**. NICS(1) values at the center of the five- and six-membered rings of **91a** are + 25.5 and + 7.48 ppm, indicating paratropicity in not only the 8 π-electron pentalene core but also the entire 16 π-electron system. Consistently, the ^{1}H NMR signals of the anti-aromatic core of **91c** appears at relatively high field (4.49, 5.20 and 5.33 ppm).

Lateral extension of π conjugation of **89** by replacing the benzene rings with naphthalene leads **92**. Although its electronic properties including small HOMO–LUMO gap (1.70 eV) and intermediate diradical character ($y_0 = 0.51$), as well as spin density distribution are theoretically predicted, it has not synthesized yet [27]. However, for further extended 30 π-electron system **14a** with anthracene units on top and bottom of the carbon skeleton, which can be regarded as cyclopenta-fused bisanthene, its derivative **14b** was synthesized by Chi et al. recently [188]. X-ray structure of **14b** shows significant bond length alternation at the outer periphery including the five-membered rings, indicating antiaromatic character of the 20 π-electron circuit. Two bonds (1.502–1.513 Å) around the apical carbons are significantly elongated compared to typical $C(sp^2)$–$C(sp^2)$ bond length (1.47 Å), due to steric constraining. Although the length alternation of the inner 10 π-electron periphery is smaller, because the radial bonds connecting the outer and inner circuits are relatively short (1.409–1.426 Å), they are not decoupled as in an AWA system [169]. In ^{1}H NMR spectrum, the core protons resonate at 4.5 and 4.1 ppm. $NICS(1)_{zz}$ values and ACID calculations indicate significant paratropicity both at local rings and around the outer periphery. The NMR signal exhibits line broadening at elevated temperature, suggesting open-shell character, which is consistent with the moderate y_0 value of 0.22 calculated at the UCAM B3LYP/6-31G(d,p) level of theory and $\Delta E_{S\text{-}T}$ of − 3.17 kcal/mol determined by SQUID measurements. In contrast to neutral **14b**, its dianion **14b^{2-}** generated by reduction with potassium metal and dication **14b^{2+}** generated by oxidation with $NO \cdot SbF_6$ are shown to be diatropic on the basis of ^{1}H NMR spectra, and NICS and ACID calculations. ACID calculations for dication **14b^{2+}** suggest that it adopts an AWA structure with diatropic and paratropic currents at the inner and outer peripheries, respectively.

7 Conclusion and Outlook

Chemistry of QDMs incorporated in non-benzenoid aromatic and antiaromatic frameworks has advanced remarkably during the last decade. Biphenalenyls and indenofluorenes play a central role in the progress of the field due to their kinetic stability enough for isolation, which allows experimental verification, coupled with advances in quantum mechanical investigations, of their structures and electronic configurations relevant to open-shell, diradical character. Material applications such as organic field effect transistors and two-photon absorptions certainly accelerate the advances. It has been shown that incorporation of QDMs into non-benzenoid aromatic or antiaromatic frameworks planarizes the whole π system, thereby facilitating conjugation within the whole π system, and modifies electronic properties inherent to QDMs significantly depending on how the QDM unit is fixed to interact with the rest of the π system. Several guiding principles which govern the open-shell character of a given system have been elucidated including pro-aromatic concept and the balance between antiaromaticity of a whole π system versus aromatic character gained within a portion of a π system. This article also includes achievements during the last century, plaiadene and *s*-indacene, to demonstrate their key contributions of the chemistry in today's still-glowing prosperity.

Careful reading of the literature of the last century would lead us to recognize that there remains many issues that are unsolved or unexplored at this time, which must be answered or settled based on the knowledge accumulated during recent decades and the current state-of-the-art technology in chemical science. Certainly, there remain many possibilities for future directions such as design for high spin-state control and multiradicaloid molecules, which are challenging in communities in synthesis and in theory and materials as well. Supramolecular chemistry of open-shell singlet molecules would be a possible approach to this subject, as already demonstrated for biphenalenyls, which is waiting challenges.

Acknowledgements The author is grateful to for continued support by Grants-in-Aid from the Ministry of Education, Culture, Sports, Science and Technology.

References

1. Platz M (1982) In: Borden WT (ed) Diradicals. Wiley, New York, p 195
2. Flynn CR, Michl J (1974) J Am Chem Soc 96:3280
3. Karafiloglou P (1989) J Chem Educ 66:816
4. Segura JL, Martín N (1999) Chem Rev 99:3199
5. Thiele J, Balhorn H (1904) Ber Dtsch Chem Ges 37:1463
6. Tschitschibabin AE (1907) Ber Dtsch Chem Ges 40:1810
7. Waring RK Jr, Sloan GJ (1964) J Chem Phys 40:772
8. Montgomery LK, Huffman JC, Jurczak EA, Grendze MP (1986) J Am Chem Soc 108:6004
9. Abe M (2013) Chem Rev 113:7011
10. Schmidt R, Brauer H-D (1971) Angew Chem Int Ed 10:506
11. Müller E, Pfanz H (1941) Ber Dtsch Chem Ges 74:1051
12. Schlenk W, Brauns M (1915) Ber Dtsch Chem Ges 48:716
13. Dowd P (1972) Acc Chem Res 5:242
14. Berson J (1982) In: Borden WT (ed) Diradicals. Wiley, New York, p 151
15. Iwamura H, Koga N (1993) Acc Chem Res 26:346
16. Rajca A (1994) Chem Rev 94:871
17. Rajca A (2005) Adv Phys Org Chem 40:153
18. Quinkert G, Wiersdorff WW, Finke M, Opitz K (1966) Tetrahedron Lett 7:2193
19. Quinkert G, Wiersdorf WW, Finke M, Opitz K, von der Haar F-G (1968) Chem Ber 101:2302
20. Iwashita S, Ohta E, Higuchi H, Kawai H, Fujiwara K, Ono K, Takenaka M, Suzuki T (2004) Chem Commun 10:2076
21. Suzuki T, Sakano Y, Iwai T, Iwashita S, Miura Y, Katoono R, Kawai H, Fujiwara K, Tsuji Y, Fukushima T (2013) Chem Eur J 19:117
22. Zeng Z, Shi X, Chi C, López Navarrete JT, Casado J, Wu J (2015) Chem Soc Rev 44:6578
23. Sun Z, Ye Q, Chi C, Wu J (2012) Chem Soc Rev 41:7857
24. Sun Z, Zeng Z, Wu J (2013) Chem Asian J 8:2894
25. Sun Z, Wu J (2014) In: Siegel JS, Wu Y-T (eds) Top Curr Chem 349, Polyarenes I. Springer, Heidelberg, p 197
26. Kubo T (2015) Chem Lett 44:111
27. Shimizu A, Nobusue S, Miyoshi H, Tobe Y (2014) Pure Appl Chem 86:517
28. Borden WT (1982) In: Borden WT (ed) Diradicals. Wiley, New York, p 1
29. Lineberger WC, Borden WT (2011) Phys Chem Chem Phys 13:11792
30. Yamaguchi K (1975) Chem Phys Lett 33:330
31. Jung Y, Head-Gordon M (2003) ChemPhysChem 4:522
32. Snyder JP (ed) (1969) Nonbenzenoid aromatics I. Academic Press, New York
33. Snyder JP (ed) (1971) Nonbenzenoid aromatics II. Academic Press, New York
34. Fieser LF, Fieser M (1933) J Am Chem Soc 55:3010
35. Binsch G, Tamir I (1969) J Am Chem Soc 91:2450
36. Boekelheide V, Vick GK (1956) J Am Chem Soc 78:653
37. Boekelheide V, Langeland WE, Liu C-T (1951) J Am Chem Soc 73:2432
38. Cava MP, Schlessinger RH (1965) Tetrahedron 21:3051
39. Cava MP, Schlessinger RH (1965) Tetrahedron 21:3065
40. Cava MP, Schlessinger RH (1965) Tetrahedron 21:3073
41. Kolc J, Michl J (1973) J Am Chem Soc 95:7391
42. Kawai H, Takeda T, Fujiwara K, Wakeshima M, Hinatsu Y, Suzuki T (2008) Chem Eur J 14:5780

43. Kolc J, Michl J (1970) J Am Chem Soc 92:4147
44. Michl J, Kolc J (1970) J Am Chem Soc 92:4148
45. Downing J, Vdořák V, Kolc J, Manzara A, Michl J (1972) Chem Phys Lett 17:70
46. Kolc J, Downing JW, Manzara AP, Michl J (1976) J Am Chem Soc 98:930
47. Steiner RP, Michl J (1978) J Am Chem Soc 100:6413
48. Castellan A, Kolc J, Michl J (1978) J Am Chem Soc 100:6687
49. Kertesz M, Choi CH, Yang S (2005) Chem Rev 105:3448
50. Gellini C, Salvi PR (2010) Symmetry 2:1846
51. Hafner K, Häfner KH, König C, Kreuder M, Ploss G, Schluz G, Strum E, Vöpel KH (1963) Angew Chem Int Ed 2:123
52. Hafner K (1964) Angew Chem Int Ed 52:3165
53. Hafner K (1982) Pure Appl Chem 54:939
54. Hafner K, Krimmer H-P (1980) Angew Chem Int Ed 19:199
55. Hafner K, Stowasser B, Krimmer H-P, Fischer S, Böhm MC, Lindner HJ (1986) Angew Chem Int Ed 25:630
56. Dunitz JD, Krüger C, Irngartinger H, Maverick EF, Wang Y, Nixdorf M (1988) Angew Chem Int Ed 27:387
57. Wang C-C, Tang T-H, Wu L-C, Wang Y (2004) Acta Cryst A60:488
58. Klann R, Bäuerle RJ, Laermer F, Elsaesser T, Niemeyer M, Lüttke W (1990) Chem Phys Lett 169:172
59. Gellini C, Salvi PR, Hafner K (1993) J Phys Chem 97:8152
60. Gellini C, Angeloni L, Salvi PR, Marconi G (1995) J Phys Chem 99:85
61. Beer M, Longuet-Higgins HC (1955) J Chem Phys 23:1390
62. Rentzepis PM (1969) Chem Phys Lett 3:717
63. Murata S, Iwanaga C, Toda T, Kokubun H (1972) Chem Phys Lett 15:152
64. Bearpark MJ, Celani P, Jolibois F, Olivucci M, Robb MA, Bernardi F (1999) Mol Phys 96:645
65. Arnold BR, Michl J (1990) In: Platz MS (ed) Kinetics and spectroscopy of carbenes and biradicals. Plenum Press, New York, p 1
66. Whitman DW, Carpenter BK (1980) J Am Chem Soc 102:4272
67. Whitman DW, Carpenter BK (1982) J Am Chem Soc 104:6473
68. Carpenter BK (1983) J Am Chem Soc 105:1700
69. Kozuch S (2014) RSC Adv 4:21650
70. Nakajima T, Saijo T, Yamaguchi H (1964) Tetrahedron 20:2119
71. Hafner K, Stowasser B, Krimmer H-P, Fischer S, Böhm MC, Lindner HJ (1963) Angew Chem Int Ed 2.123
72. Gellini C, Cardini G, Salvi PR, Marconi G, Hafner K (1993) J Phys Chem 97:1286
73. Heilbronner E, Yang Z-Z (1987) Angew Chem Int Ed 26:360
74. Hertwig RH, Holthausen MC, Koch W, Maksić ZB (1994) Angew Chem Int Ed 33:1192
75. Hertwig RH, Holthausen MC, Koch W, Maksić ZB (1995) Int J Quantum Chem 54:147
76. Nendel M, Goldfuss B, Houk KN, Hafner K (1999) J Mol Struct 461–462:23
77. Nendel M, Goldfuss B, Beno B, Houk KN, Hafner K, Lindner HJ (1999) Pure Appl Chem 71:221
78. Moroni L, Gellini C, Salvi PR (2004) J Mol Struct 667:1
79. Makino M, Aihara J (2008) Phys Chem Chem Phys 10:591
80. Julg A, François P (1967) Theo Chim Acta 8:249
81. Jartín RS, Ligabue A, Soncini A, Lazzeretti P (2002) J Phys Chem A 106:11806
82. Havenith RWA, Engelberts JJ, Fowler PW, Steiner E, van Lenthe JH, Lazzeretti P (2004) Phys Chem Phys Chem 6:289
83. Motomura S, Nakano M, Fukui H, Yoneda K, Kubo T, Carion R, Champagne B (2011) Phys Chem Chem Phys 13:20575
84. Prinzbach H, Knothe L (1986) Pure Appl Chem 58:25
85. Wiersum UE, Jenneskens LW (1993) Tetrahedron Lett 34:6615
86. Nakajima T (1971) Pure Appl Chem 28:219
87. Hess BA Jr, Shaad LJ (1971) J Org Chem 36:3418
88. Toyota A (1975) Bull Chem Soc Jpn 48:1152
89. Böhm MC, Schütt J (1992) J Phys Chem 96:3674
90. Randić M (2003) Chem Rev 103:3449
91. Tobe Y (2015) Chem Rec 15:86
92. Prinzbach H, Babsch H, Hunkler D (1978) Tetrahedron Lett 19:649

165

93. Prinzbach H, Bingmann H, Beck A, Hunkler D, Sauter H, Hädicke E (1981) Chem Ber 114:1697
94. Prinzbach H, Schneider H-W (1973) Angew Chem Int Ed 12:1007
95. Prinzbach H, Knothe L, Schneider H-W (1973) Angew Chem Int Ed 12:1009
96. Kuroda S, Asao T, Funamizu M, Kurihara H, Kitahara Y (1976) Tetrahedron Lett 17:251
97. Beck A, Bingmann H, Kagabu S, Knothe L, Hädicke E, Prinzbach H (1983) Chem Ber 116:1963
98. Kubo T (2015) Chem Rec 15:218
99. Motita Y, Nishida S (2010) In: Hicks R (ed) Stable radicals. Wiley, Chichester, p 81
100. Goto K, Kubo T, Yamamoto K, Nakasuji K, Sato K, Shiomi D, Takui T, Kubota M, Kobayashi T, Yakusi K, Ouyang J (1999) J Am Chem Soc 121:1619
101. Kubo T, Shimizu A, Sakamoto M, Uruichi M, Yakushi K, Nakano M, Shiomi D, Sato K, Takui T, Morita Y, Nakasuji K (2005) Angew Chem Int Ed 44:6564
102. Kubo T, Shimizu A, Uruichi M, Yakushi K, Nakano M, Shiomi D, Sato K, Takui T, Morita Y, Nakasuji K (2007) Org Lett 9:81
103. Shimizu A, Hirao Y, Matsumoto K, Kurata H, Kubo T, Uruichi M, Yakushi K (2012) Chem Commun 48:5629
104. Shimizu A, Uruichi M, Yakushi K, Matsuzaki H, Okamoto H, Nakano M, Hirao Y, Matsumoto K, Kurata H, Kubo T (2009) Angew Chem Int Ed 48:5482
105. Shimizu A, Kubo T, Uruichi M, Yakushi K, Nakano M, Shiomi D, Sato K, Takui T, Hirao Y, Matsumoto K, Kurata H, Morita Y, Nakasuji K (2010) J Am Chem Soc 132:14421
106. Chikamatsu M, Mikami T, Chisaka J, Yoshida Y, Azumi R, Yase K, Shimizu A, Kubo T, Morita Y, Nakasuji K (2007) App Phys Lett 91:043506
107. Mou Z, Uchida K, Kubo T, Kertesz M (2014) J Am Chem Soc 136:18009
108. Uchida K, Mou Z, Kertesz M, Kubo T (2016) J Am Chem Soc 138:4665
109. Nakano M, Kishi R, Nitta T, Kubo T, Nakasuji K, Kamada K, Ohta K, Champagne B, Botek E, Yamaguchi K (2005) J Phys Chem A 109:885
110. Ohta S, Nakano M, Kubo T, Kamada K, Ohta K, Kishi R, Nakagawa N, Champagne B, Botek E, Umezaki S-Y, Takebe A, Takahashi H, Furukawa S-I, Morita Y, Nakasuji K, Yamaguchi K (2006) Chem Phys Lett 420:432
111. Nakano M, Kubo T, Kamada K, Ohta K, Kishi R, Ohta S, Nakagawa N, Takahashi H, Furukawa S-I, Morita Y, Nakasuji K, Yamaguchi K (2006) Chem Phys Lett 418:142
112. Kamada K, Ohta K, Kubo T, Shimizu A, Morita Y, Nakasuji K, Kishi R, Ohta S, Furukawa S-I, Takahashi H, Nakano M (2007) Angew Chem Int Ed 46:3544
113. Kamada K, Ohta K, Shimizu A, Kubo T, Kishi R, Takahashi H, Botek E, Champagne B, Nakano M (2010) J Phys Chem Lett 1:937
114. Frederickson CK, Rose BD, Haley MM (2017) Acc Chem Res 50:977
115. Fix AG, Chase DT, Haley MM (2014) In: Siegel JS, Wu Y-T (eds) Polyarenes I, Top Curr Chem 349:159
116. Haley MM (2015) Chem Rec 15:1140
117. Zhou Q, Carroll PJ, Swager TM (1994) J Org Chem 59:1294
118. Scherf U (1993) Macromol Rapid Commun 14:575
119. Chase DT, Rose BD, McClintock SP, Zakharov LN, Haley MM (2011) Angew Chem Int Ed 50:1127
120. Chase DT, Fix AG, Kang SJ, Rose BD, Weber CD, Zhong Y, Zakharov LN, Lonergan MC, Nuckolls C, Haley MM (2012) J Am Chem Soc 134:10349
121. Rose BD, Zakharov LN, Haley MM (2013) Acta Cryst E69:o890
122. Chase DT, Fix AG, Rose BD, Weber CD, Nobusue S, Stockwell CE, Zakharov LN, Lonergan MC, Haley MM (2011) Angew Chem Int Ed 50:11103
123. Fu X, Zhao D (2015) Org Lett 17:5694
124. Nishida J-I, Tsukaguchi S, Yamashita Y (2012) Chem Eur J 18:8964
125. Frederickson CK, Zakharov LN, Haley MM (2016) J Am Chem Soc 138:16827
126. Young BS, Chase DT, Marshall JL, Vonnegut CL, Zakharov LN, Haley MM (2014) Chem Sci 5:1008
127. Marshall JL, Uchida K, Frederickson CK, Schütt C, Zeidell AM, Goetz KP, Finn TW, Jarolimek K, Zakharov LN, Risko C, Heges R, Jurchescu OD, Haley MM (2016) Chem Sci 7:5547
128. Marshall JL, O'Neal NJ, Zakharov LN, Haley MM (2016) J Org Chem 81:3674
129. Ren L, Liu C, Wang Z, Zhu X (2016) J Mater Chem C 4:5202
130. Gershoni-Poranne R, Stanger A (2014) Chem Eur J 20:5673
131. Anthony JE, Brooks JS, Eaton DL, Parkin SR (2001) J Am Chem Soc 123:9482

132. Ryno SM, Risko C, Bredas J-L (2014) J Am Chem Soc 136:6421
133. Lim Y-F, Shu Y, Parkin SR, Anthony JE, Malliars GG (2009) J Mater Chem 19:3049
134. Rose BD, Summer NJ, Filatov AS, Peters SJ, Zakharov LN, Petrukhina MA, Haley MM (2014) J Am Chem Soc 136:9181
135. Barker JE, Frederickson CK, Jones MH, Zakharov LN, Haley MM (2017) Org Lett 19:5312
136. Rudebush GE, Zafra JL, Jorner K, Fukuda K, Marshall JL, Arrechea-Marcos I, Espejo GL, Ortiz RP, Gómes-García CJ, Zakharov LN, Nakano M, Ottosson H, Casado J, Haley MM (2016) Nat Chem 8:753
137. Rose BD, Vonnegut CL, Zakharov LN, Haley MM (2012) Org Lett 14:2426
138. Rudebush GE, Espejo GL, Zafra JL, Peña-Alvarez M, Spisak SN, Fukuda K, Wei Z, Nakano M, Petrukhina M, Casado J, Haley MM (2016) J Am Chem Soc 138:12648
139. Maekawa T, Ueno H, Segawa Y, Haley MM, Itami K (2016) Chem Sci 7:650
140. Hu P, Lee S, Herng TS, Aratani N, Gonçalves TP, Qi Q, Shi X, Yamada H, Huang K-W, Ding J, Kim D, Wu J (2016) J Am Chem Soc 138:1065
141. Yang X, Liu D, Miao Q (2014) Angew Chem Int Ed 53:6786
142. Yang X, Shi X, Aratani N, Gonçalves TP, Huang K-W, Yamada H, Chi C, Miao Q (2016) Chem Sci 7:6176
143. Hibi D, Kitabayashi K, Fujita K, Takeda T, Tobe Y (2016) J Org Chem 81:3735
144. Liu J, Ma J, Zhang K, Ravat P, Machata P, Avdoshenko S, Hennersdorf F, Komber H, Pisula W, Weigand JJ, Popov AA, Berger R, Müllen K, Feng X (2017) J Am Chem Soc 139:7513
145. Fix AG, Deal PE, Vonnegut CL, Rose BD, Zakharov LN, Haley MM (2013) Org Lett 15:1362
146. Sbargoud K, Mamada M, Marrot J, Tokito S, Yassar A, Frigoli M (2015) Chem Sci 6:3402
147. Étienne A, Le Berre A (1956) C R Hebd Seances Acad Sci 242:1493
148. Étienne A, Le Berre A (1956) C R Hebd Seances Acad Sci 242:1899
149. Le Berre A (1956) C R Hebd Seances Acad Sci 242:2365
150. Le Berre A (1957) Ann Chim 13:371
151. Chalvet O, Peltier J (1956) Bull Soc Chim Fr 1667
152. Shimizu A, Tobe Y (2011) Angew Chem Int Ed 50:6906
153. Fukuda K, Nagami T, Fujiyoshi JY, Nakano M (2015) J Phys Chem A 119:10620
154. Gisin M, Wirz J (1976) Helv Chim Acta 59:2273
155. Steiner RP, Miller RD, Dewey HL, Michl J (1979) J Am Chem Soc 101:1820
156. Cofino WP, Engelsma M, Kamminga DA, Hoornweg PG, Gooijer C, MacLean C, Velthost NH (1984) Spectrochim Acta 40A:269
157. Jones WD, Pomfret A, Eife RL (1983) J Chem Soc Perkin Trans 1459
158. Miyoshi H, Nobusue S, Shimizu A, Hisaki I, Miyata M, Tobe T (2014) Chem Sci 5.163
159. Nobusue S, Tobe Y (2016) Synlett 21:2140
160. Shimizu A, Kishi R, Nakano M, Shiomi D, Sato K, Takui T, Hisaki I, Miyata M, Tobe Y (2013) Angew Chem Int Ed 52:6076
161. Miyoshi H, Miki M, Hirano S, Shimizu A, Kishi R, Fukuda K, Shiomi D, Sato K, Takui T, Hisaki I, Nakano M, Tobe Y (2017) J Org Chem 82:1380
162. Dreeler JJ, Zhou Z, Marshall JL, Kishi R, Takamuku S, Wei Z, Spisak SN, Nakano M, Petrukhina MA, Haley MM (2017) Angew Chem Int Ed 56:15363
163. Baird NC (1972) J Am Chem Soc 94:4941
164. Hellwinkel D, Reiff G (1970) Angew Chem Int Ed 82:516
165. Nobusue S, Miyoshi H, Shimizu A, Hisaki I, Fukuda K, Nakano M, Tobe Y (2015) Angew Chem Int Ed 54:2090
166. Miyoshi H, Nobusue S, Shimizu A, Tobe Y (2015) Chem Soc Rev 44:6560
167. Aihara J (2016) Bull Chem Soc Jpn 89:1425
168. Baryshnikov GV, Valiev RR, Karaush NN, Sundholm D, Minaev BF (2016) Phys Chem Chem Phys 18:8980
169. Dickens TK, Mallion RB (2013) Phys Chem Chem Phys 15:8245
170. Miyoshi H, Nishinaga T, Hisaki I, Nakano M, Petrukhina M, Tobe Y unpublished results
171. Das S, Herng TS, Zafra JL, Burrezo PM, Kitano M, Ishida M, Gopalakrishna TY, Hu P, Osaka A, Casado J, Ding J, Casanova D, Wu J (2016) J Am Chem Soc 138:7782
172. Lu X, Lee S, Hong Y, Phan H, Gopalakrishna TY, Herng TS, Tanaka T, Sandoval-Salinas ME, Zeng W, Ding J, Casanova D, Osuka A, Kim D, Wu J (2017) J Am Chem Soc 139:13173
173. Lu X, Lee S, Kim JO, Gopalakrishna TY, Phan H, Herng TS, Lim Z, Zeng Z, Ding J, Kim D, Wu J (2016) J Am Chem Soc 138:13048

174. Trost BM, Kinson PL (1975) J Am Chem Soc 97:2438
175. Reid DH, Stafford WH, Ward JP (1955) J Chem Soc 1193
176. Zahradník R, Michl J, Pancíř J (1966) Tetrahedron 22:1355
177. Hess BA Jr, Schaad LJ (1971) J Org Chem 36:3418
178. DasGupta A, DasGupta NK (1974) Theoret Chim Acta 33:177
179. Baumgartner P, Weltin E, Wagnière G, Heilbronner E (1965) Helv Chim Acta 48:751
180. Nendel M, Goldfuss B, Houk KN, Hafner K, Grieser U (1999) Theor Chem Acta 102:397
181. Grieser U, Hafner K (1994) Chem Ber 127:2307
182. Das S, Wu J (2015) Org Lett 17:5854
183. Brand K (1912) Ber Dtsch Chem Ges 45:3071
184. Bergmann ED (1959) In: Ginsburg D (ed) Non-Benzenoid Aromatic Compounds. Interscience, New York, p 141
185. Parkhurst RR, Swager TM (2014) In: Siegel JS, Wu YT (eds) Polyarenes II. Springer, Heidelberg, p 141
186. Rudebusch GE, Haley MM (2016) In: Miao Q (ed) Polycyclic Arenes and Heteroarenes. Wiley, Weinheim, p 37
187. Konishi A, Okada Y, Nakano M, Sugisaki K, Sato K, Kakui T, Yasuda M (2017) J Am Chem Soc 139:15284
188. Wang Q, Gopalakrishna TY, Phan H, Herng TS, Dong S, Ding J, Chi C (2017) Angew Chem Int Ed 56:11415

Top Curr Chem (Z) (2017) 375:68
DOI 10.1007/s41061-017-0154-3

REVIEW

Heterocyclic Quinodimethanes

Xueliang Shi[1] · **Chunyan Chi**[1]

Received: 29 January 2017 / Accepted: 30 May 2017 / Published online: 20 June 2017
© Springer International Publishing Switzerland 2017

Abstract This chapter describes a series of heterocyclic quinodimethanes mainly containing O, N, S and Si atoms in their frameworks with either closed- or open-shell electronic structures in the ground state. Their syntheses, structural characterizations and chemical and physical properties are comprehensively reviewed. Some of them are used as materials for dyes/pigments and semiconductors for organic electronics. Some of them show an open-shell singlet diradical character with unusual properties. The purpose of this chapter is to provide fundamental understanding on the structure-property relationships of quinoidal π-conjugated compounds and give some insight into the rational design of such molecules with desirable properties.

Keywords Heterocyclic · Quinodimethane · Diradicaloid · Aromaticity · Anti-aromaticity · Pro-aromaticity

Abbreviations

ACID	Anisotropy of the induced current density
2,6-AQDM	2,6-Anthraquinodimethane
DCM	Dichloromethane
DDQ	2,3-Dichloro-5,6-dicyano-1,4-benzoquinone
DFT	Density functional theory
DHTAP	Dihydro-5,7,12,14-tetraazapentacene

Chapter 4 was originally published as Shi, X. & Chi, C. Top Curr Chem (Z) (2017) 375: 68. DOI 10.1007/s41061-017-0154-3.

✉ Chunyan Chi
chmcc@nus.edu.sg

[1] Department of Chemistry, National University of Singapore, 3 Science Drive 3, Singapore 117543, Singapore

 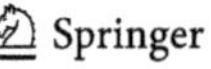

ESR   Electron spin resonance
FET   Field-effect transistor
HOMO   Highest occupied molecular orbital
LUMO   Lowest unoccupied molecular orbital
NBS   *N*-Bromosuccinimide
NICS   Nucleus-independent chemical shift
NIS   *N*-Iodosuccinimide
NMR   Nuclear magnetic resonance
2,6-NQDM   2,6-Naphthoquinodimethane
p-QDM   *para*-Quinodimethane
SOMO   Singly occupied molecular orbital
SQUID   Superconducting quantum interference device
TCNQ   Tetracyanoquinodimethane
TIPSE   Triisopropylsilylethynyl
TMEDA   Tetramethylethylene diamine
TPA   Two-photon absorption

1 Introduction

Quinodimethane (QDM) has two exocyclic double bonds resembling the corresponding quinones in structure (Fig. 1a). Compared with conventional aromatic molecules, quinoidal compounds containing QDM units have attracted considerable

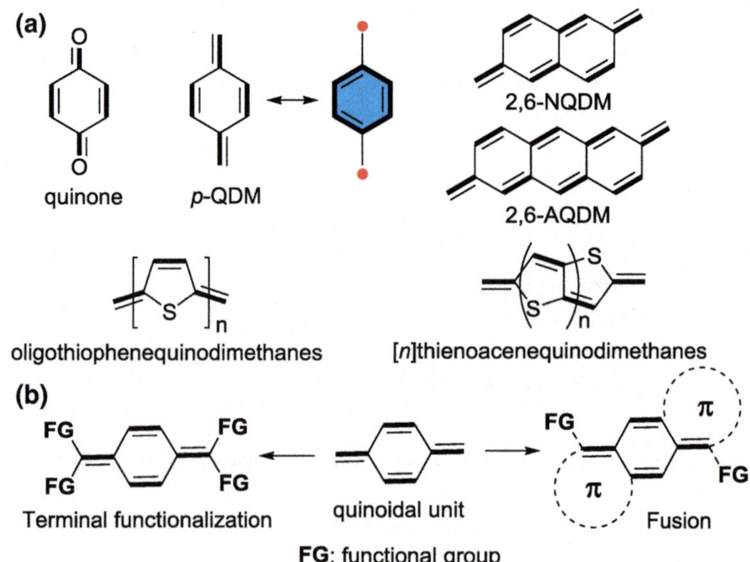

Fig. 1 **a** Structures of quinone and some typical quinoidal structures. **b** Two stabilizing strategies toward stable quinoidal compounds

interests because of their intrinsic open-shell diradical character. The parent *para*-quinodimethane (*p*-QDM) is extremely unstable and has an irresistible wish to be diradical [1, 2]. The main reason is that the quinoidal structure could recover the aromaticity of the benzenoid ring in the diradical form (Fig. 1a, the aromatic sextet ring is shaded in blue color). Therefore, parent quinoidal compounds readily undergo dimerization, oxidation, polymerization and decomposition, which have hindered their investigations and practical applications. Hence, different strategies have been developed by chemists to stabilize such meta-stable quinoidal compounds during the last decades. Generally, there are two general stabilizing strategies, leading to two types of quinoidal compounds (Fig. 1b). The first method, named terminal functionalization, is to end cap the terminal methylene sites by some functional groups. The functional groups, which are usually electron-withdrawing cyano or ester groups or aryl groups, can effectively block the reactive sites and delocalize the spin, hence, stabilizing the quinoidal compounds. The second approach, developed recently, is to embed the quinoidal unit into an aromatic π-conjugated framework (i.e., by fusion). This can enhance the structural rigidity and at the same time lead to more efficient spin delocalization. In addition, kinetic blocking of the most reactive sites with bulky groups can further improve the stability and solubility. Moreover, the diversity of quinoidal structures [e.g., *p*-QDM, 2,6-naphthoquinodimethane (2,6-NQDM), 2,6-anthraquinodimethane (2,6-AQDM), oligothiophenequinodimethanes and [*n*]thienoacenequinodimethanes, etc., in Fig. 1], gives chemists plenty of space to design various quinoidal compounds. For example, terminal functionalized quinoidal oligothiophenes [3, 4], Thiele's hydrocarbon, Tschitschibabin's hydrocarbon [5–8] and aromatic ring-fused bisphenalenyls [9, 10], zethrenes [11, 12] and indenofluorenes [13–16] have been intensively studied. The structural diversity of these quinoidal compounds allows us to investigate their fundamental structure-property relationships. Generally speaking, most of these quinoidal compounds display an open-shell singlet diradical character to a certain extent. The preliminary studies also demonstrate that different factors including the length of the quinoidal units (aromatic stabilization energy), fusion motifs as well as the planarity of molecules (steric strain) could affect the ground-state electronic structures of these quinoidal compounds and consequently their physical properties. Due to the intriguing diradical character, they display unique optical, electronic and magnetic properties and thus have potential applications for organic electronics, nonlinear optics, organic spintronics and energy storage [1–21].

The diverse quinoidal subunits together with the different fusion motifs give rise to various quinoidal compounds. Among them, quinoidal π-conjugated polycyclic hydrocarbons (benzenoid quinodimethanes and non-benzenoid quinodimethanes) are thoroughly summarized and highlighted in other chapters; thus, in this chapter we will particularly focus on the heterocyclic quinodimethanes that contain heteroatoms such as O, N, S and Si in (or linked to) the quinoidal units. The existence of heteroatoms usually stabilizes the quinoidal structures and also allows us to approach the stable charged species. The syntheses, structural characterizations and chemical and physical properties of such heterocyclic quinoidal compounds will be comprehensively reviewed. We intend to give some insight into the fundamental structure-property relationships, which are of importance for rational design of stable heterocyclic quinoidal compounds with desirable properties.

2 *O*-Heterocyclic Quinodimethanes

Benzodifuranone dyes, the representative *O*-heterocyclic quinodimethanes, were developed about 40 years ago (Fig. 2) [22]. Because of their bright red color and excellent wash and sublimation fastness properties, they are one of the most successful classes of commercial disperse dyes [23]. The deep red color was believed to originate from the quinoidal structure of the central core unit.

Symmetrical benzodifuranones **1a** were prepared upon condensation of 1,4-dihydroxybenzene **2** with mandelic acid or its derivatives **3** followed by oxidation (Scheme 1a) [24, 25]. However, it was found that symmetrical benzodifuranones had much poorer build up properties on polyester fibers than asymmetrical benzodifuranones, and the poor substantivity was ascribed to the high degree of symmetry and good crystallinity of the molecules. Thus, the method to synthesize asymmetrical benzodifuranones was developed (Scheme 1b). It involves a half-condensed intermediate **5**, which undergoes a condensation reaction with **6** to form **7**, and **7** is then oxidized by using chloranil or nitrobenzene to give **1b** [26]. Very recently, chemists have renewed their interest in benzodifuranone dyes and demonstrated that both *p*-benzodifuranone (**1a**, R = Br) and *o*-benzodifuranone **10** (Scheme 1c) could serve as good building blocks (acceptor) for novel π-conjugated polymers suitable for electronic applications [25, 27].

Benzodifuranone exhibits extremely good stability though one quinodimethane unit is incorporated in its structure. The good stability could be ascribed to the stabilizing effect of the electron-deficient lactone moieties as the benzofuranyl radical itself has been demonstrated to be an air-persistent monoradical [28, 29]. The singlet diradical resonance form is expected to contribute to the ground state structure of benzodifuranone dyes, but experimentally they behave like closed-shell compounds (Fig. 3). The extended naphthadifuranone **11** and anthradifuranone **12** are supposed to exhibit increased diradical characters and the related work is in progress in our laboratory.

Two other types of *O*-heterocyclic quinodimethanes **13** [30, 31] and **14** [32] fused by six-membered rings were developed recently (Fig. 4), and the synthetic route is outlined in Scheme 2. The synthesis of **13** commenced with the dilithiation

Fig. 2 Structures of benzodifuranone dyes

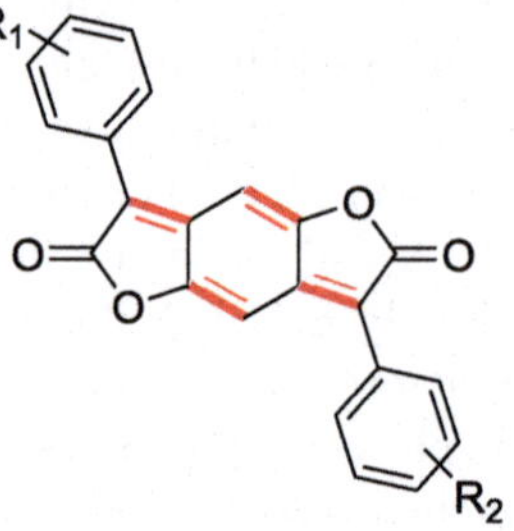

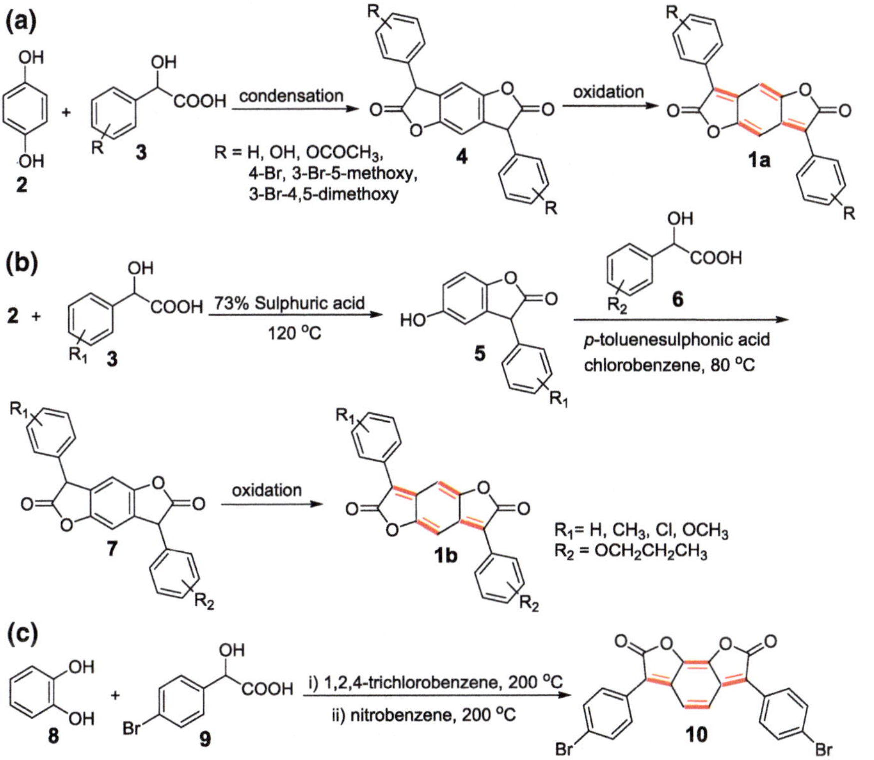

(a)

Scheme 1 Synthetic routes to symmetrical (**a**), asymmetrical (**b**) benzodifuranone dyes and *o*-benzodifuranone (**c**)

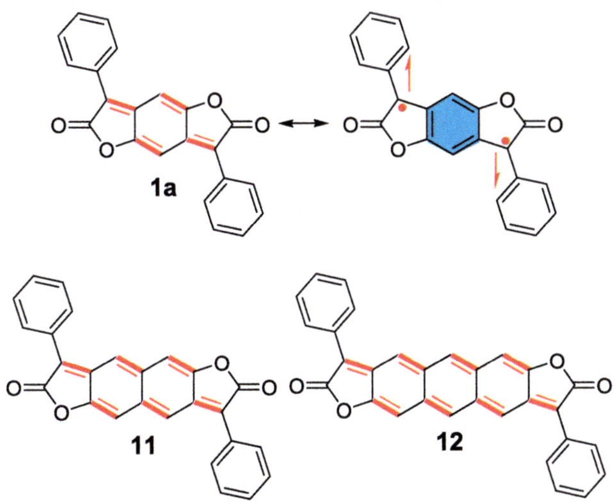

Fig. 3 Resonance forms of benzodifuranone and structures of extended naphthadifuranone and anthradifuranone

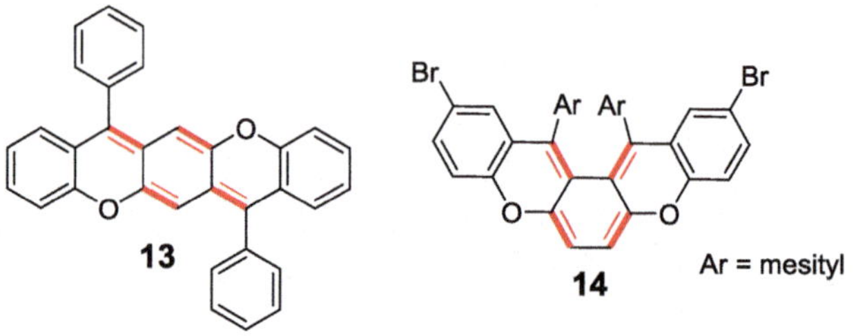

Fig. 4 Structures of *O*-heterocyclic quinodimethanes **13** and **14**

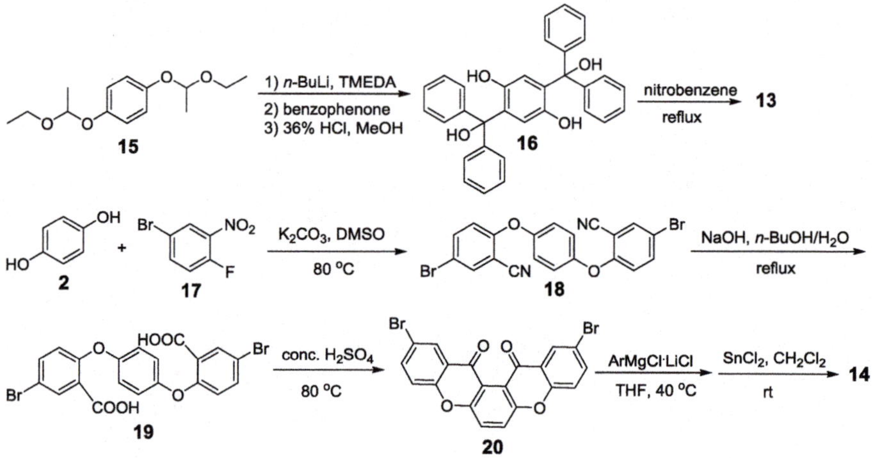

Scheme 2 Synthetic routes to compounds **13** and **14**

of **15** with *n*-butyllithium (*n*-BuLi) in the presence of tetramethylethylene diamine (TMEDA) followed by treatment with benzophenone and subsequent 36% HCl/MeOH-mediated deprotection to afford the key intermediate **16**. The final ring closure and aromatization reaction in refluxing nitrobenzene gave greenish red crystals of **13** [30]. Compound **14** was synthesized in five steps from 1,4-hydroquinone **2** [32]. First, the nucleophilic substitution of 5-bromo-2-fluorobenzonitrile with 1,4-hydroquinone in the presence of K_2CO_3 gave dinitrile **18**, which underwent hydrolysis under basic conditions to form the corresponding dicarboxylic acid **19**. Interestingly, the acid-catalyzed intramolecular electrophilic cyclization of **19** in hot concentrated H_2SO_4 regio-selectively afforded the *ortho*-diketone **20** as the only product in good yield. The origin of this regio-selectivity can be explained by the shape of the HOMO in the (9-xanthone)oxidanium cation intermediate, which has a large coefficient at the 8-position (*ortho*-carbon with the respect to the carbonyl group) rather than the 6-position (*para*-carbon); thus, subsequent cyclization in the cation intermediate prefers to attack *ortho*-carbon. Finally, compound **14** was obtained as a dark bluish-purple solid by addition of mesityl Grignard reagent to the diketone **20** followed by reductive dehydroxylation with $SnCl_2$.

Figure 5 shows the X-ray crystallographic structures of **13** and **14**. The backbone of **13** in the crystal is essentially flat, and the dihedral angle between the phenyl ring and the backbone is 70.22°. In contrast to acenes such as pentacene, which is known to crystallize in a herringbone structure, the molecules of **13** form slipped stacks. It showed a p-type field effect transistor (FET) behavior with hole mobility around 0.16 cm^2 V^{-1} s^{-1} in single-crystal devices [31]. Compared to **13**, molecule **14** has a significantly twisted π-framework due to steric repulsion between the two mesityl groups. The large bond length alternation of the central quinoidal unit in both of the two molecules (p-QDM in **13**, o-QDM in **14**) unambiguously revealed their quinoidal structure. The quinoidal conjugation in **14** is consistent with its negligible diradical character ($y_0 = 9\%$ calculated for **14** by the CASSCF(2,2)/6-31G//RHF/6-31G** method) [32].

3 *N*-Heterocyclic Quinodimethanes

Tetrabenzodiazadiketoperylene **21** and its derivatives, containing a *p*-QDM in their framework, are widely used pigments and dyes for polymeric substrates (Fig. 6) [33]. Compound **21** was normally prepared by reacting aminobenzophenone **26** with a dimethyldisuccinylsuccinate **27** to form an intermediate **28**, which was then heated in refluxing Therminol VP-1 (a high-boiling aromatic ether-based solvent) (Scheme 3) [33]. "Pigment Violet 23" (P.V. 23) **22** could also be regarded as a heterocyclic quinoidal compound containing both O and N atoms (Fig. 6) [34]. It exhibits a unique blue-shade violet color with exceptional intensity. In contrast to **21** and other heterocyclic quinodimethanes, methylene moieties of the quinoidal unit in

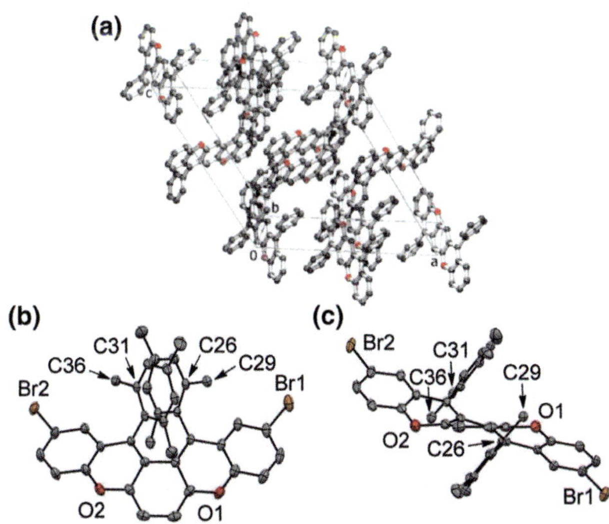

Fig. 5 Crystal structure of **13** and **14**: **a** 3D packing structure of **13**[a], **b** top and **c** side views of **14**[b]. [a]Adapted with permission from Ref. [31]. Copyright 2009 American Institute of Physics. [b]Adapted with permission from Ref. [32]. Copyright 2016 American Chemical Society

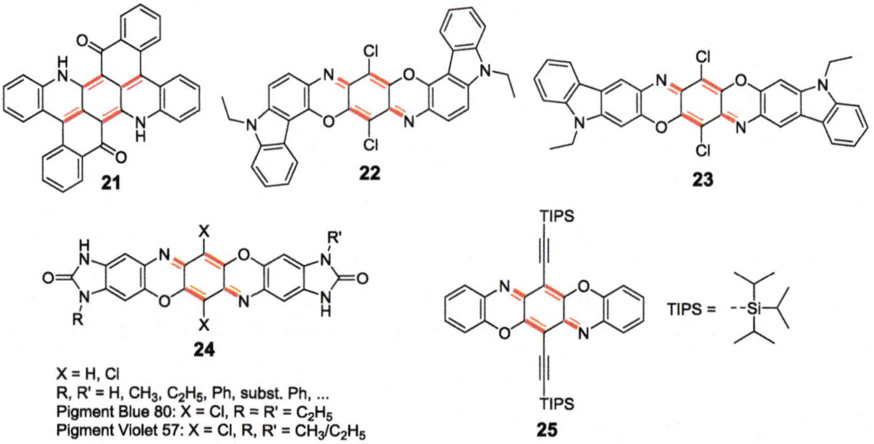

Fig. 6 Structures of *N*-heterocyclic quinodimethanes **21-25**

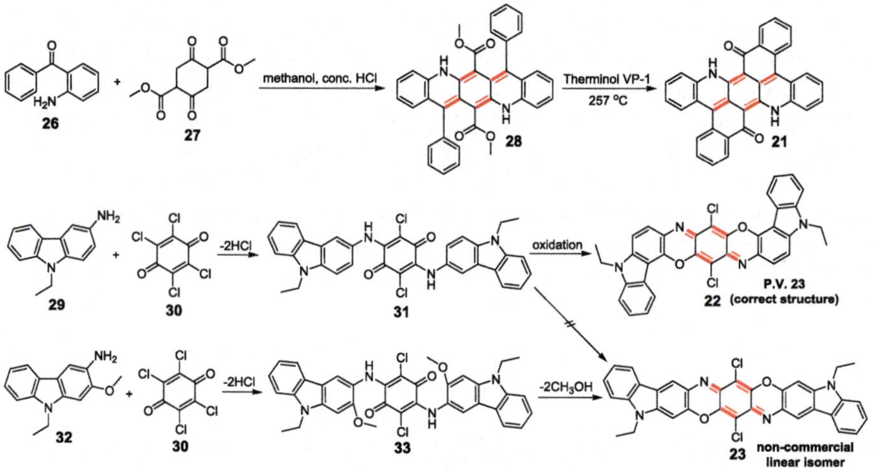

Scheme 3 Synthetic routes to compounds **21**, **22** and **23**

22 and **23** are replaced by imine-type nitrogen atoms. Pigment Violet 23 was assumed to be molecule **23** with a linear structure until in 1987 a single-crystal X-ray analysis proved the actual S-shape of molecule **22** (Fig. 6) [35]. The synthesis of both compounds is outlined in Scheme 3, and the linear molecular structure of **23** was confirmed by X-ray crystallographic analyses [35]. Compound **23** has a blue shade and is not produced industrially. Some other dyes such as imidazolone-annellated triphenedioxazine pigments **24** [34, 36, 37] and **25** [38] also contain the same quinoidal unit as in Pigment Violet 23. Compounds **24** are a class of violet pigments with very good fastness properties and extremely high color strength. Compound **25** exhibits red color and is also a n-type organic semiconductor [38]. All these dyes display good stability and deep color. These quinoidal compounds are mainly used as pigments and dyes, and very few other studies or applications were reported.

Another type of *N*-heterocyclic quinodimethanes is based on an old and long-debated molecule, dihydro-5,7,12,14-tetraazapentacene (DHTAP) **34**, which was not only used as a small-molecule model for ladder polymers [39–42], but also developed as an organic semiconductor in FETs [43]. The dispute about this molecule exists in its two possible structures, benzenoid one and quinoidal one (Fig. 7). Though later the correct structure of DHTAP was determined to be benzenoid by ^{1}H NMR spectroscopy [44–46], the quinoidal structure was still misused in the literature.

To further address the debate, Miao's group reported that the methylation of DHTAP allowed not only *N*-alkylation but also *C*-alkylation and afforded both benzenoid and quinoidal molecules, as shown in Scheme 4 [47]. The formation of different alkylated products **37a–c** was ascribed to the multiple resonance forms of the dianion of DHTAP, where the negative charges can delocalize not only on nitrogen atoms but also on the neighboring carbon atoms. X-ray crystallography analysis revealed the structures of these three molecules (Fig. 8). The backbones of **37a** and **37c** bend at the methylated nitrogen atoms, and yet **37b** is essentially planar. The large bond length alternation found in the central quinoidal unit in **37b** revealed its quinoidal form. In addition, the bond length analysis of **37a–c** indicated minor contributions from potential zwitterionic resonance forms. Single-crystal FET devices were fabricated based on both benzenoid and quinonoid molecules **37a–b**, and they showed p-type FET behavior with mobility of 1–$5 \times 10^{-4} \ cm^2 \ V^{-1} \ s^{-1}$.

The dihydroindolocarbazoles refer to a family of heterocyclic compounds containing an indole unit fused to one of the benzenoid rings of a carbazole group [48]. Five isomers exist in the indolocarbazole family based on the position and orientation of the indole-carbazole ring fusion (Fig. 9), which is very similar to the indenofluorene family [13–16]. Though dihydroindolocarbazoles and *N*-alkylation indolocarbazoles are well known because of their potential biological activity as anti-cancer drugs [49, 50] and active materials for organic electronics [51–53], indolocarbazole derivatives with imine nitrogens are rarely studied and reported because of their intrinsic instability. Apparently, the bare indolocarbazoles are extremely unstable because of the contribution of the diradical structure (Fig. 9). Unlike substituted indenofluorene derivatives recently developed by Tobe's and Haley's groups [13–16], conventional protecting substituents such as triisopropy-lsilylethynyl (TIPSE) or aryl groups cannot directly install to the most reactive site of indolocarbazoles because of the trivalent character of N atom. In addition, the

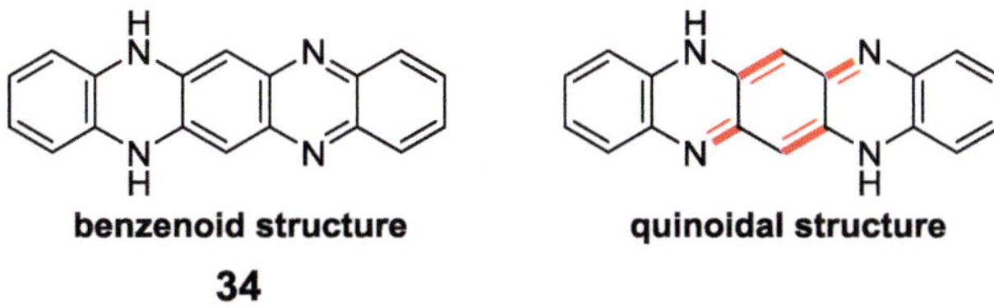

Fig. 7 Two possible structures of dihydro-5,7,12,14-tetraazapentacene **34**

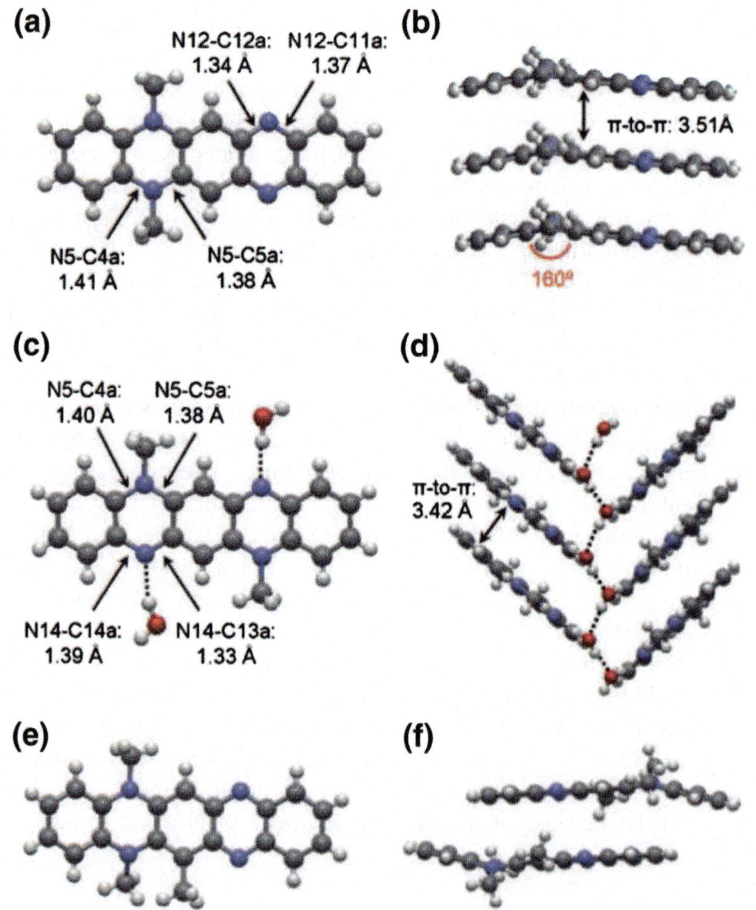

Scheme 4 Synthetic routes to compounds **37a–c**

Fig. 8 X-ray crystallographic structures and packing structures of **a–b 37a**, **c-d 37b** and **e–f 37c**. Adapted with permission from Ref. [47]. Copyright 2009 Wiley-VCH Verlag GmbH & Co.

poor solubility of bare indolocarbazoles is another issue that hinders the synthesis of such meta-stable compounds. All these issues impose restrictions on the synthesis, characterization and functionalization of indolocarbazole-based molecules and

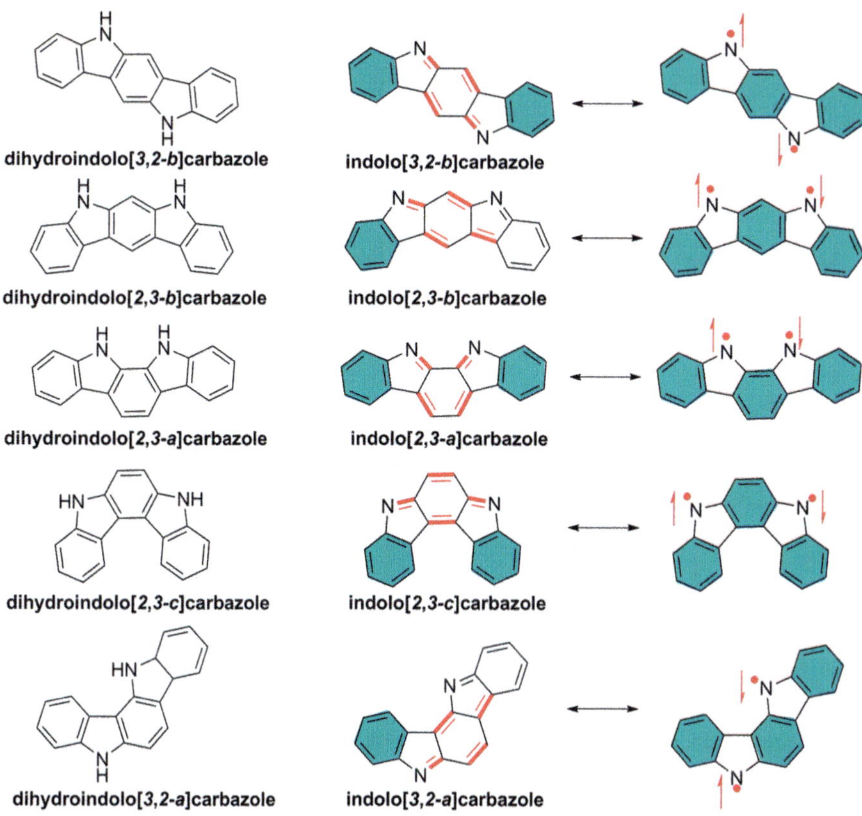

Fig. 9 Chemical structures of five isomers of dihydroindolocarbazoles and indolocarbazoles

materials. Nevertheless, some isomers in the indolocarbazole family have been successfully synthesized and their intriguing properties have been disclosed.

Indolo[*3,2-b*]carbazole was reported to be generated through dehygrogenation of dihydroindolo[*3,2-b*]carbazole in the presence of 2,3-dichloro-5,6-dicyano-1,4-benzoquinone (DDQ) in 2000 by Yudina et al. [54]. However, Dehaen and co-workers examined this reaction under the same condition, and they found that the formation of corresponding indolo[*3,2-b*]carbazole was not successful. In 2008, Dehaen and co-workers first prepared the fully characterized indolo[*3,2-b*]carbazole derivatives through a deprotonation approach (Scheme 5) [55]. Interestingly, they found that the formation of indolo[*3,2-b*]carbazole derivatives **39** could be directly achieved with some strong bases such as *t*-BuOK and *t*-BuLi in the absence of oxidant DDQ. No stable **39a** was formed under this condition, presumably because **39a** was insufficiently protected. Thus, the result demonstrates that space shielding of the imine sites by bulky groups is crucial to obtain stable compounds. The authors also found that Michael addition of *n*-BuLi to the formed indolo[*3,2-b*]carbazole could happen and generated **40** in good yield (Scheme 5) [55]. Compound **40** could act as a selective chemosensor in aprotic solvents, for either F^-

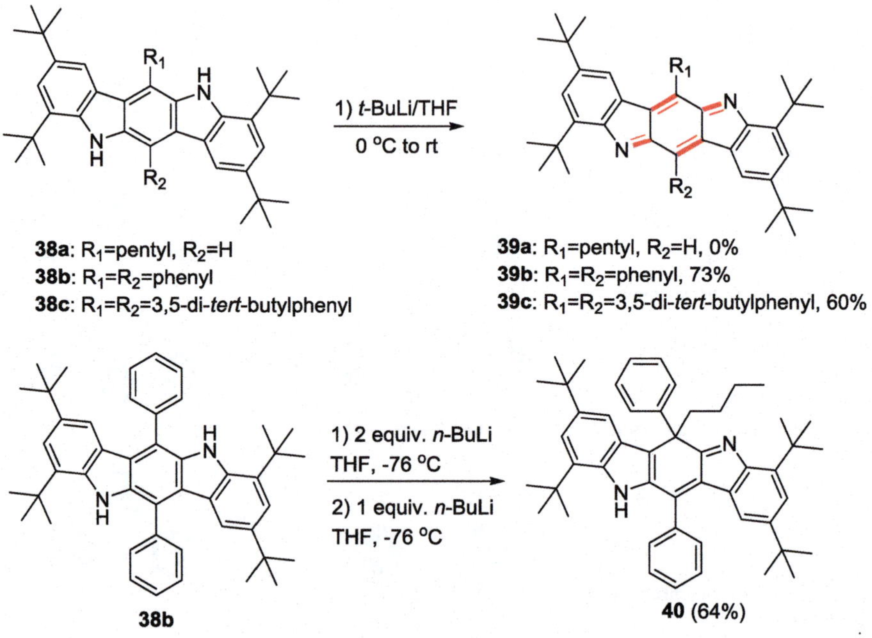

Scheme 5 Synthetic routes to compounds **39a–c** and **40**

or Brønsted acids. However, except for the X-ray crystallographic structure of **39b**, the basic information such as stability, electrochemical and optical properties of indolo[*3,2-b*]carbazole derivatives **39** were not mentioned, and their potential diradical character was not studied.

Indolo[*2,3-b*]carbazole is supposed to be more reactive and has larger singlet diradical character than indolo[*3,2-b*]carbazole based on recent study of indenofluorene family (Fig. 9) [13–16]. Thus, the synthesis of indolo[*2,3-b*]carbazole derivatives is even more challenging. In 2014, Wu and co-workers reported the first example of quinoidal indolo[*2,3-b*]carbazole derivatives. They synthesized stable/persistent indolo[*2,3-b*]carbazole derivatives **41** and **42** with tunable ground states (Fig. 10) [56]. The precursors **43** and **44** were obtained by multi-step synthesis and subsequently converted to corresponding **41** and **42** through deprotonation followed by oxidation or by direct oxidative dehydrogenation, respectively (Scheme 6).

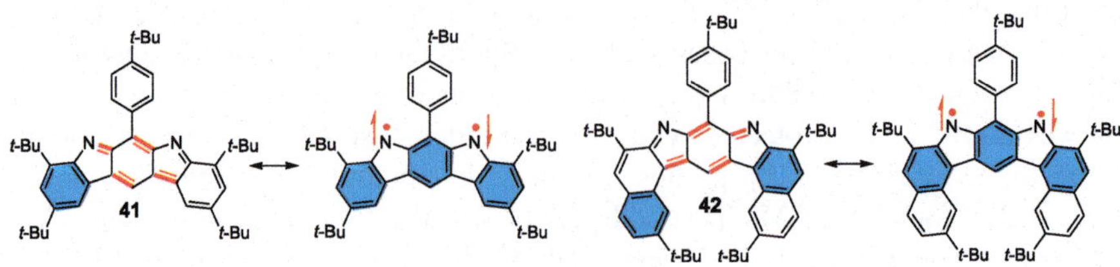

Fig. 10 Chemical structures of indolo[*2,3-b*]carbazole derivatives **41** and **42** and their diradical resonance forms

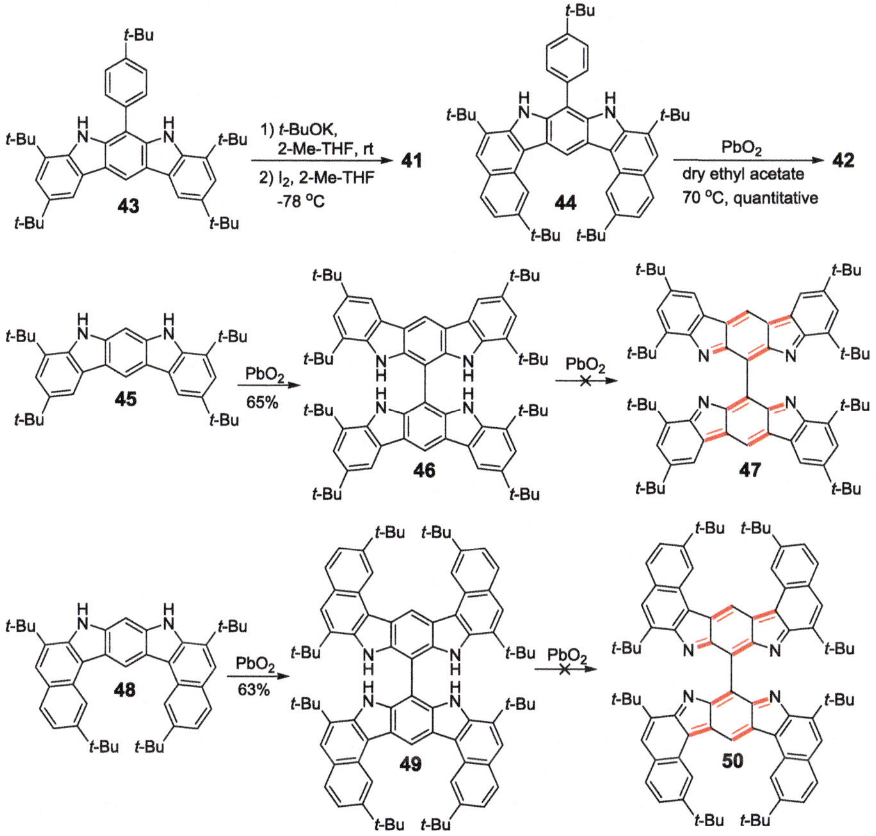

Scheme 6 Synthetic routes to compounds **41**, **42**, **46** and **49**

Similar to the design and synthesis of indolo[*3,2-b*]carbazole derivatives **39**, the bulky *tert*-butyl and 4-*tert*-butylphenyl groups are installed to protect and stabilize such reactive molecules. Consequently, the obtained compounds **41** and **42** were reasonably stable under an inert atmosphere. In contrast, the oxidative dehydrogenation of the unblocked precursors **45** and **48** gave the homo-coupled dimers **46** and **49** in good yields. No predicted quinoidal products **47** and **50** or dihydrazine products (via intramolecular N–N bond formation) were formed with increased reaction time.

The ground states of **41** and **42** were studied by variable-temperature NMR/ESR measurements and density functional theory (DFT) calculations. It was found that **41** is a persistent singlet diradicaloid in the ground state with a moderate diradical character ($y_0 = 0.269$) (UCAM-B3LYP/6-31G*), while **42** exhibits a quinoidal closed-shell ground state [56]. This difference was explained by Clar's aromatic sextet rule [57] for singlet diradicaloids, i.e., the more aromatic sextet rings can be gained, the larger the character expected, which has been demonstrated to be a very useful qualitative rule to interpret the reactivity/stability of diradicaloids. Another notable property of **41** and **42** is their very low-lying LUMO energy levels that makes these indolo[*3,2-b*]carbazole derivatives subject to attack by active protons from either protic solvents or silica gel. Thus, the high reactivity of indolocarbazoles is better ascribed to their intrinsic singlet diradical character as well as their inherently low-lying LUMO energy level. The design and synthesis of three other isomers of the indolocarbazole family are interesting but also very challenging.

4 *S*-Heterocyclic Quinodimethanes

4.1 Terminal Functionalized *S*-Heterocyclic Quinodimethanes

Tetracyano-substituted α-oligothiophenequinodimethanes and thienoacenequinodimethanes, which are the most widely studied *S*-heterocyclic quinodimethanes, have been demonstrated to be very interesting molecules and materials. In this part, we will provide a summary of recent representative studies on the chemistry and applications of these types of quinoidal oligothiophenes and thienoacenes. For more details on this topic, interested readers are directed to some excellent review articles [3, 4].

The shortest quinoidal thiophene **51a** (Fig. 11) was synthesized and developed as the alternative to the tetracyanoquinodimethane (TCNQ) in 1974 by Gronowitz and Uppström, but failed to be the electron accepter as the role of TCNQ [58]. Later, Ogura and co-workers developed the extended bi- and ter-quinoidal thiophenes **51b** and **51c** [59]. Both **51b** and **51c** could serve as electron accepter for organic metals and were considered good TCNQ alternatives because of the better electrical conductivities. Compounds **51a–c** were prepared from 2,5-dibromo-oligothiophenes with tetracyanoethylene oxide in dry 1,2-dibromoethane under reflux (Scheme 7). However, the development of even longer dibromo-oligothiophenes encountered a poor solubility problem. This issue again hampered the synthesis of the longer quinoidal thiophenes. To overcome this issue, Otsubo and co-workers developed a series of oligothiophenes with β-positions fused with a bis(butoxymethyl)cyclopentane ring, which significantly improved the solubility [60]. Based on this work, in 2005 the same group reported a series of dicyanomethylene-capped quinoidal oligothiophenes **52a–52f** (Fig. 11). Compounds **52b–f** were synthesized by using Pd(0)-promoted Takahashi coupling with sodium dicyanomethanide in refluxing tetrahydrofuran (Scheme 7) followed by air oxidation to give the quinoidal oligothiophenes [61]. However, this approach could not be applied to the synthesis of **52a**, because the thiophene monomer was less reactive to *N*-iodosuccinimide (NIS), and the dibromothiophene instead obtained with *N*-bromosuccinimide (NBS) did not undergo the Takahashi reaction. Alternatively, **52a** was prepared by a similar procedure to **51a** from the corresponding dibromide treated with tetracyanoethylene oxide.

This family of quinoidal oligothiophenes **52a–f** showed highly amphoteric redox behavior. They displayed strong electronic absorptions in the visible to near infrared region and their energy gap decreased with the extension of chain length. Interestingly, the ^{13}C NMR spectra of **52e** and **52f** showed no signals of the tetracyanothienoquinodimethane and cyclopentane carbons, though definite signals assignable to the pendant butoxy carbons were observed. This indicated the

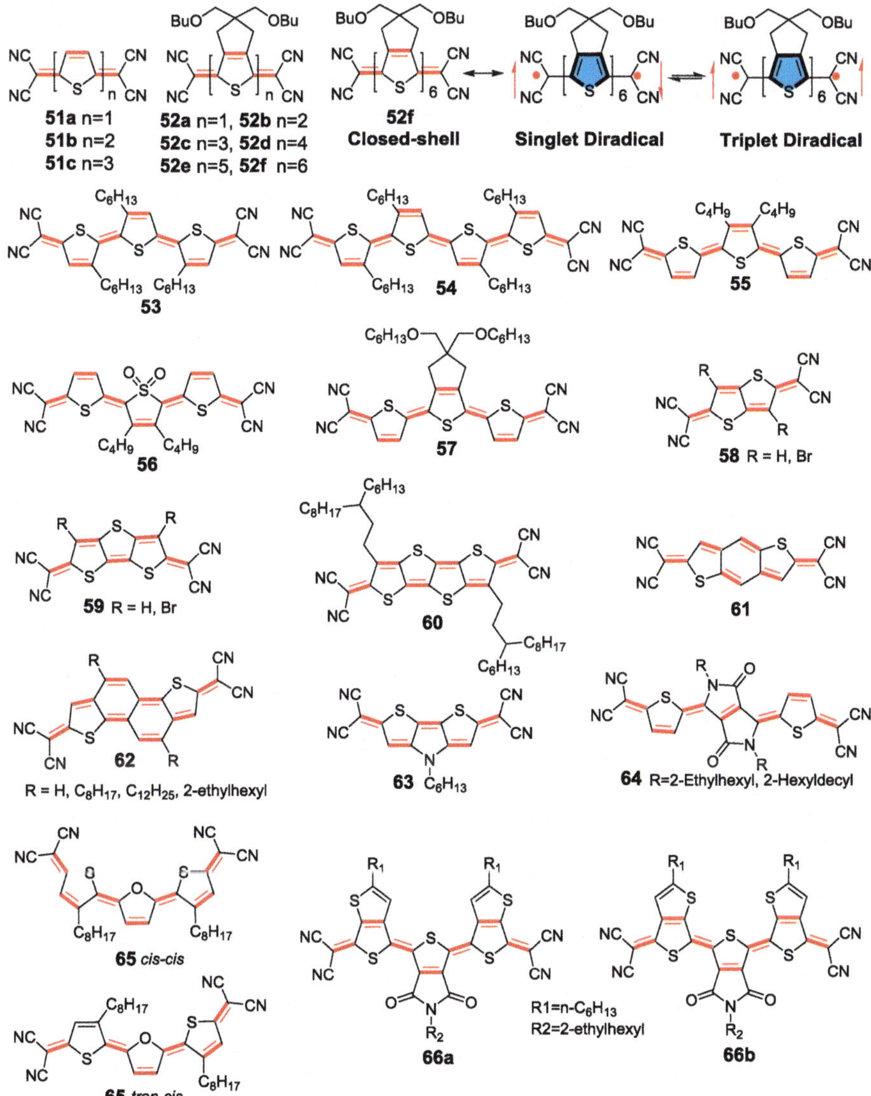

Fig. 11 Representative examples of quinoidal oligothiophenes with two terminal dicyanomethylene units

contribution of paramagnetic triplet biradical species in **52e** and **52f**, which was further confirmed by ESR measurement. This system nicely demonstrates the chain length-dependent properties and the increased diradical characters of quinoidal compounds with the extension of the central quinoidal units. Ortiz et al. in 2007 established the open-shell singlet nature of these quinoidal oligomers on the basis of the Raman spectra as well as the presence of a low-lying triplet excited state on 1–4 kcal mol^{-1} for the longer oligomers responsible for the ESR activity [62].

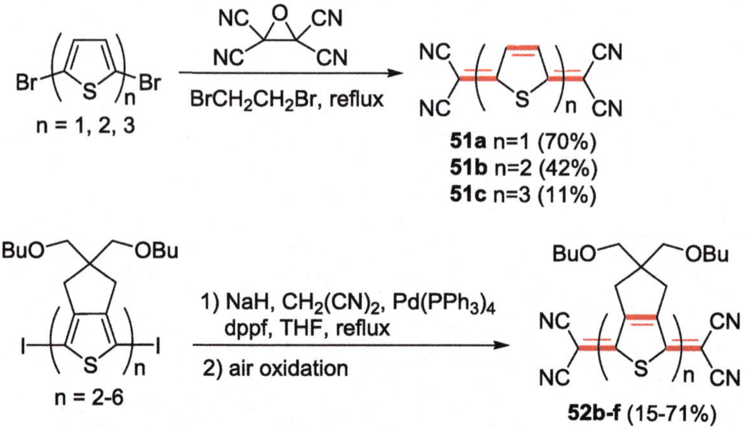

Scheme 7 Synthetic routes of **51** and **52** from oligothiophene dihalides

In addition, a series of quinoidal oligothiophenes with linear alkyl chains at the β-position of thiophene rings were also synthesized, and their physical properties were studied [63–65]; some examples (**53–55**) are listed in Fig. 11. The alkylated terthiophene and quaterthiophene analogs of TCNQ showed amphoteric redox behavior. The quaterthiophene derivative forms four stable redox species, the dianion, neutral, cation radical and dication. This indicated that they could act as both electron acceptors and electron donors [63, 64]. This behavior makes these molecules a promising class of active materials for organic electronics, especially for complementary circuits and n-type/ambipolar FETs. For example, the complementary logic circuits and gate-modulated lateral p–n diodes based on **54** were made by the solution-processing technique by Ribierre et al. [66]. The FET device based on **55** was fabricated and displayed n-type behavior with electron mobilities as high as 0.005 and 0.002 $cm^2\ V^{-1}\ s^{-1}$ for vapor and solution-deposited films, respectively [67]. The device could be further optimized and displayed a saturation electron mobility of 0.2 $cm^2\ V^{-1}\ s^{-1}$ at room temperature [68]. The devices based on **55** could be ambipolar when the films of **55** were grown at higher temperature. Later, Casado et al. converted **55** to an *S,S*-dioxide derivative **56** (Fig. 11) [69]. Sulfonation of terthiophene de-aromatized the central thiophene, which became an electron acceptor and gave a competing intramolecular charge transfer (i.e., from the central **55** backbone toward the outermost dicyanomethylene groups and from the outer thiophene rings toward the central one in **56**). Electron affinity was improved in **56** with respect to **55**. Takimiya and co-workers further optimized the structure of quinoidal terthiophenes by changing the butyl groups in **55** with a bis(butoxymethyl)cyclopentane ring in **57**. The solution-processed FETs based on **57** showed electron mobility as high as 0.16 $cm^2\ V^{-1}\ s^{-1}$ [70]. Qunoidal thienoacenes such as **58–60** were also prepared [71, 72]. Compounds **58** and **59** have been found to be superior electron acceptors with very small on-site Coulomb repulsion. They formed a variety of molecular complexes with π-electron donors, most of which exhibited very high electrical conductivities up to the metallic region

[71]. Compound **60** displayed n-type FET behavior with electron mobility up to 0.9 cm^2 V^{-1} s^{-1} under ambient conditions [72].

Quinoidal mixed or co-oligomer tetracyanoquinodimethanes such as **61** [73, 74], **62** [75] and **63** [76] (Fig. 11) have also been prepared by combining thiophene with benzene, naphthalene and pyrrole units, respectively. They exhibited a closed-shell structure and could also be used as n-type organic semiconductors. Recently, compound **64** [77] was synthesized by Zhu's group. It also showed n-type FET behavior with electron mobility of 0.35 cm^2 V^{-1} s^{-1}, by solution-processing techniques. The good device performance may be attributed to the utilization of the rigid diketopyrrolopyrrole group in **64**. A furan–thiophene-based quinoidal compound **65** was reported by Li and co-workers [78]. Interestingly, **65** existed as a mixture of *cis–cis* and *trans–cis* isomers (in terms of the double bonds between the heterocyclic rings) in solution and as a *cis–cis* structure in the solid state. The thin film of **65** deposited from the nonchlorinated solvent displayed a high electron mobility of 1.11 cm^2 V^{-1} s^{-1}, and the micro-sized ribbon of **65** showed an electron mobility as high as 7.7 cm^2 V^{-1} s^{-1}, which surpassed the corresponding thiophene-based compounds. 2D π-expanded quinoidal terthiophenes **66a–b** with outstanding semiconducting properties were reported by Zhu's and Di's groups [79]. It was found that the thiophene orientation in **66a** and **b** significantly affected their film morphology and as a consequence different OFET performances of **66a** (3.0 cm^2 V^{-1} s^{-1}) and **66b** (0.44 cm^2 V^{-1} s^{-1}) were achieved in ambient conditions by the solution processing technique. The superior electron mobility of **66a** was attributed to its higher crystallinity, longer crystalline coherence lengths and lower orientational disorder compared to **66b**.

In 2010, Takimiya and co-workers synthesized a series of thienoquinodimethanes **67–70** (Fig. 12) with a newly employed terminal group, the ((alkyloxy)carbonyl)-cyanomethylene moiety [80]. This terminal group plays two important roles in these compounds: i.e., as an electron-withdrawing group to keep the LUMO energy level

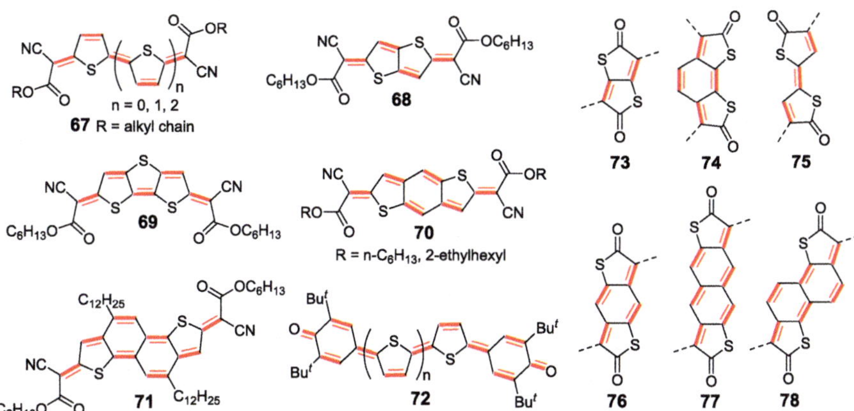

Fig. 12 Representative examples of thienoquinoidal compounds with terminal ((alkyloxy)carbonyl)cyanomethylene moieties (**67–71**) and benzoquinone units (**72**) and a quinoidal thienothiophene-dione (**73–78**) building block

sufficiently low for acting as an n-channel organic semiconductor and as a solubilizing group to facilitate solution processability. As a result, these thieno-quinodimethanes exhibited good solubility in normal organic solvents, high stability in the atmosphere, electron-accepting properties and solution processability. Solution-processed FETs based on the terthienoquinoid derivative **67** ($n = 2$) exhibited good electron mobilities ($\mu_e \sim 0.015$ cm^2 V^{-1} s^{-1}) and $I_{on}/I_{off} \approx 10^5$ under ambient conditions.

Very recently, Takimiya, Casado and co-workers reported the quinoidal naphthodithiophene **71** with cyano-alkyloxycarbonyl-methylene functionalities [81], which could undergo reversible diradical σ-dimerization/σ-polymerization triggered by mild stimuli such as concentration, temperature and pressure (Fig. 13a, c). The reaction reversibility was caused by the quinoidal molecule of **71**, which changes its structure to an aromatic species by forming weak and long intermolecular C–C single bonds (Fig. 13a, c). The reaction was accompanied by a big chromic effect of about 2.5 eV. The two opposite but complementary quinoidal and aromatic tautomers provided the Janus faces of the reactants and products, which produced the observed chromic effect. For comparison, no such changes were detected for the quinoidal naphthodithiophene **62** with dicyanomethylene terminal groups since it can only form π dimers (Fig. 13b).

Aside from the cyano end-caps, quinoidal oligothiophenes with benzoquinone as terminal groups such as **72** (Fig. 12) have also been reported and exhibited a low band gap and amphoteric redox behavior [82–85]. However, the study of their

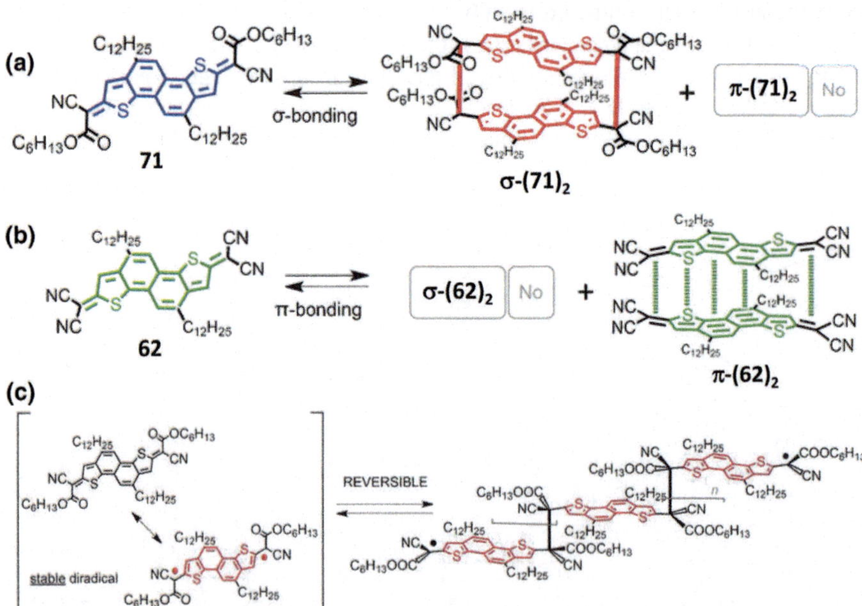

Fig. 13 a Reversible σ-dimerization of **71**; **b** reversible π-dimerization of **62**; **c** reversible σ-polymerization of **71** Adapted with permission from Ref. [81]. Copyright 2016 Wiley-VCH Verlag GmbH & Co.

diradical character is rare and the FET devices based on such structures were even less reported. Very recently, quinoidal thienothiophene-diones such as **73–78** (Fig. 12) have been developed by Takimiya et al. and demonstrated to be versatile building blocks for π-conjugated polymers for FETs applications [86–89].

4.2 Fused *S*-Heterocyclic Quinodimethanes

The above-discussed quinoidal oligothiophenes are typical terminal functionalized quinoidal compounds, and so far there are very few reports on the fused *S*-heterocyclic quinodimethanes. Incorporation of a quinoidal oligothiophene or [*n*]thienoacenequinodimethane unit into a polycyclic hydrocarbon framework would lead to new types of fused quinoidal compounds, which are supposed to exhibit unique physical properties. With this design in mind, our group developed several new types of quinoidal compounds containing S atoms. First, we targeted bisindeno-[*n*]thienoacene, which consists of a quinoidal thienoacene unit annulated with two indene rings. There are different isomers in the bisindeno-[*n*]thienoacenes family. Taking the simplest bisindenothiophene as an example, because of the different position and orientation of the indeno–thiophene ring fused in, three isomers could exist with different numbers of aromatic sextet rings, which is illustrated in Fig. 14. The ground-state geometry and aromaticity of this system are of great interest. Taking molecule **a** as an example, two resonance forms (closed-shell quinoidal one and open-shell diradical one) are supposed to contribute to the ground-state structure. Moreover, if looking into this system from another angle, it can be regarded as dibenzannulated anti-aromatic system containing $4n$ π electrons if two π electrons are counted for each sulfur atom (Fig. 14).

In view of the stability of these three isomers, we can make a simple judgment intuitively according to Clar's sextet rule [57] that isomer **a** should be the most stable one. It was further experimentally validated that only isomer **a** with substituent (**a'** in Scheme 8) could be synthesized, whereas during the attempted synthesis of **b'** and **c'**, the final reduction step gave a complicated mixture because of their high reactivity, and the mixture could not be isolated (Scheme 8). The high reactivity of isomers **b** and **c** is likely due to their inherent singlet diradical character as discussed above. These results indicated that for the system of bisindeno-

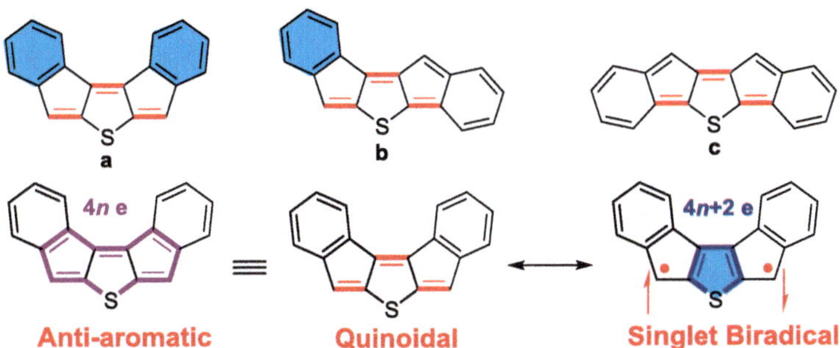

Fig. 14 Chemical structures of three isomers of bisindenothiophene and their diradical resonance forms

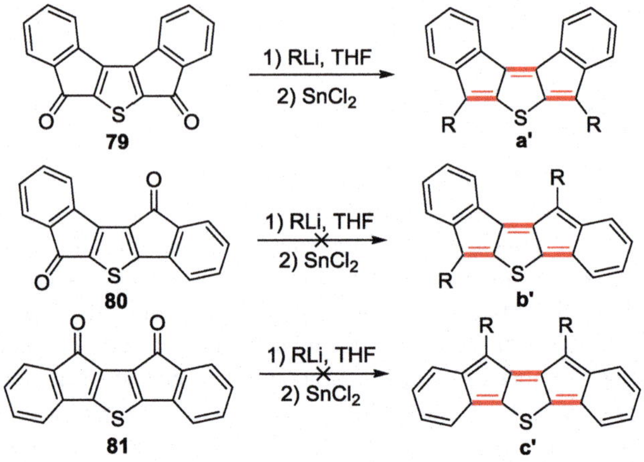

Scheme 8 Synthetic route to compounds **a'**, **b'** and **c'**

[*n*]thienoacenes only molecules with a fusion mode like **a** could be synthesized in a reasonably stable form. With this information in mind, we set about synthesizing the extended analogs of isomer **a**, namely bisindeno-[*n*]thienoacenes **82a–d** (Fig. 15), and in 2014 our group and Haley's group independently reported this system [90, 91].

The synthetic route was quite straightforward (Scheme 9) and mainly involved multi-step synthesis of diacids **83** and **85**, which were then treated with an excess of thionyl chloride to afford dicarbonyl chloride, and subsequent double Friedel–Crafts acylation with $AlCl_3$ gave the diketone precursors **84** and **86**. The target bisindeno-[*n*]thienoacenes **82a–d** were obtained by nucleophilic addition of the diketone with lithiated triisopropylsilylacetylene followed by reduction of the intermediate diols with $SnCl_2$. Compounds **82a**, **b** and **c** were all successfully isolated; however, **82d** could be only prepared in situ in an inert atmosphere because it was extremely unstable. The reduction reaction was followed by UV–Vis-NIR absorption spectroscopic measurements, and the formation of **82d** was accompanied by simultaneous decomposition of the product. Attempted separation of **82d** by column

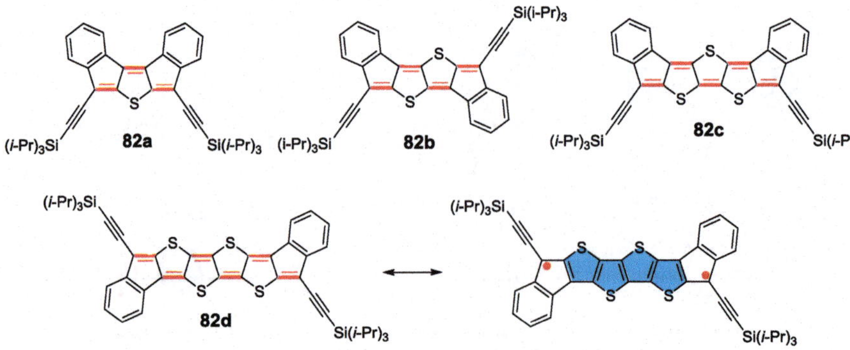

Fig. 15 Chemical structures of bisindeno-[*n*]thienoacenes **82a–d**

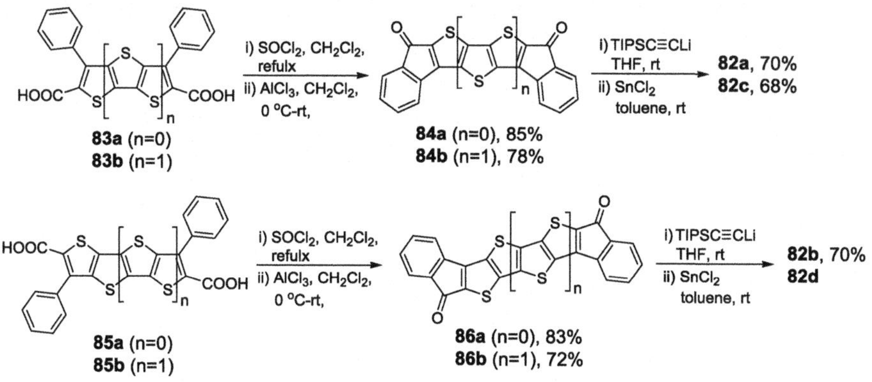

Scheme 9 Synthetic routes to bisindeno-[n]thienoacenes **82a–d**

chromatography or recrystallization all failed because of its high reactivity to oxygen, protonated reagents and silica gel. The high reactivity of **82d** was hypothesized to be associated with its anti-aromatic character and intrinsic diradical character, which was evidenced by variable-temperature ESR measurement and DFT calculations.

The structures of **82a–c** were all unambiguously confirmed by X-ray crystallographic analysis as shown in Fig. 16. The π-frameworks (bisindeno[n]thienoacene) of all molecules are almost planar, and large bond length alternation was found in all molecules, suggesting that the typical quinoidal resonance form dominates in the ground state. However, spin-unrestricted DFT calculations predicted that the molecules showed a gradual increase of the singlet diradical character with an extended quinoidal thienoacene unit, finally leading to a highly reactive bisindeno [4] thienoacene (**82d**) with an open-shell singlet ground state. Meanwhile, the positive NICS(1)zz values for the central quinoidal thienoacene rings (Fig. 16) implied the anti-aromaticity of this system, which is similar to the reported indenofluorenes. However, from **82a** to **82d**, the NICS(1)zz values for the central thiophene ring (ring A1 in Fig. 16) become more negative, indicating an increase of aromaticity, which is in accordance with the increased diradical character. The optical and electronic properties of compounds **82a–d** were also carefully investigated by various experimental techniques including one-photon absorption, two-photon absorption (TPA), femtosecond transient absorption, Raman spectroscopy and cyclic voltammetry measurements and revealed chain length-dependent behavior. Some data are listed in Table 1. They have a short singlet excited state lifetime (7–12 ps), moderate TPA cross section (340–520 GM) and amphoteric redox behavior, which all are related to their anti-aromaticity and singlet diradical character. The transformation from anti-aromatic to pseudo-aromatic/ aromatic systems was conducted using electrochemical oxidation and reduction followed by UV–Vis-NIR spectroscopic measurements. Their radical cations/anions and dianions are achievable, while dications with two anti-aromatic cyclopentadienyl rings are unstable and difficult to attain using electrochemistry. All these results fully demonstrated that these bisindeno[n]thienoacenes could be regarded as

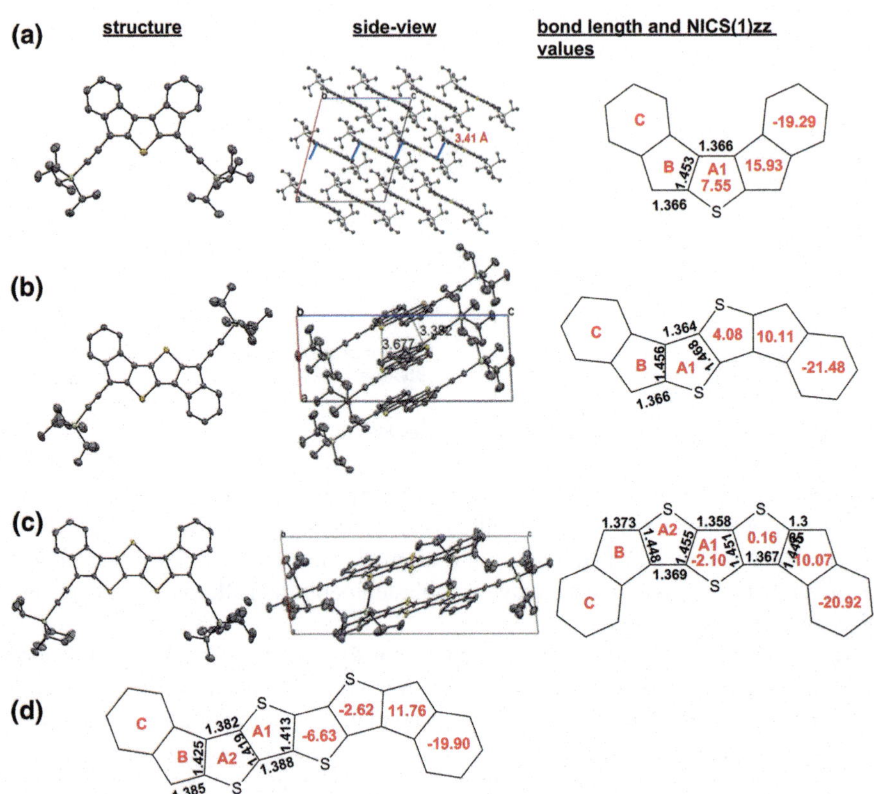

Fig. 16 X-ray crystallographic structures, packing structures and selected bond lengths of **a 82a**, **b 82b**, **c 82c** and **d 82d** (calculated data). The *red numbers* in the rings denote the calculated NICS(1)zz values. Adapted with permission from Ref. [90]. Copyright The Royal Society of Chemistry 2014

an anti-aromatic system with a quinoidal resonance form dominating in the ground state, at the same time, with a small singlet diradical character.

Inspired by the work of bisindeno-[*n*]thienoacenes, another relevant system based on benzo-thia-fused [*n*]thienoacenequinodimethanes was designed and published by our group in 2016 (Fig. 17) [92]. Compared to bisindeno-[*n*]thienoacenes, benzo-thia-fused [*n*]thienoacenequinodimethanes (**87a–c** and **88**) have a very similar quinoidal unit (central quinoidal thienoacene) but different fusion modes (benzo-thia versus indeno). In view of the above-mentioned structural similarities and differences, the ground-state geometry, aromaticity and physical properties of this system are expected to be distinct compared with those of the bisindeno-[*n*]thienoacenes system and thus are of great interest. The synthetic strategy of **87a–c** and **88** is very similar to bisindeno-[*n*]thienoacenes **82a–d**, mainly via nucleophilic addition of the corresponding diketone with lithiated TIPSE followed by reduction of the intermediate diols with $SnCl_2$.

Compounds **87a–c** and **88** were stable in both solution and solid state, and their X-ray crystallographic structures and 3D packing structures are shown in Fig. 18.

Table 1 Summary of the optical and electrochemical data of molecules reported by Chi's group

Compound	λ_{abs} (nm)	ε (10^4 M^{-1}cm^{-1})	HOMO (eV)	LUMO (eV)	E_g^{EC} (eV)	E_g^{opt} (eV)
82a	417	3.61	−5.41	−3.69	1.72	1.51
	444	4.62				
	684	0.76				
82b	450	2.75	−5.35	−3.75	1.60	1.58
	480	2.50				
	606	3.10				
82c	514	3.45	−5.30	−3.68	1.62	1.37
	554	4.65				
	687	2.33				
82d	561	3.36	−	−	−	1.27
	610	4.63				
	729	4.41				
87a	570	7.32	−4.77	−3.06	1.71	2.05
87b	640	7.46	−4.56	−3.17	1.39	1.90
87c	666	5.23	−4.42	−3.35	1.07	1.65
88	702	5.52	−4.39	−3.40	0.99	1.46
89	682	10.8	−4.65	−3.46	1.19	1.65
90a	318	5.90	−4.67	−3.06	1.61	1.88
	625	5.76				
90b	310	8.90	−4.52	−2.43[a]	−	2.09
	558	3.83				
90c	304	6.26	−4.80	−2.88	1.92	2.13
	555	2.71				
91	620	2.20	−4.70	−3.14	1.56	1.86
92	368	2.63	−4.61	−3.24	1.37	1.62
	662	2.42				
101	360	3.58	−4.79	−3.10	1.69	1.34
	650	0.77				
103	413	5.65	−4.72	−3.20	1.52	1.28
	694	3.07				

HOMO and LUMO are estimated from the onset potentials of the first oxidative and reductive waves, respectively

λ_{abs} absorption maxima, ε molar extinction coefficient for the corresponding absorption maximum, E_g^{EC} electrochemical band gap, E_g^{opt} optical band gap estimated from the absorption onset

[a] LUMO of **90b** was estimated according to the equation: LUMO = HOMO + E_g^{opt}

For all compounds, two molecules form an anti-parallel packed dimer via π–π interactions with a distance less than 3.57 Å. In all cases, large bond length alternation was observed for the central thienoacenequinodimethane unit, indicating a dominant quinoidal structure, which is similar to bisindeno-[*n*]thienoacenes **82a–c**. However, the different aromaticity of these two systems was evidenced by their

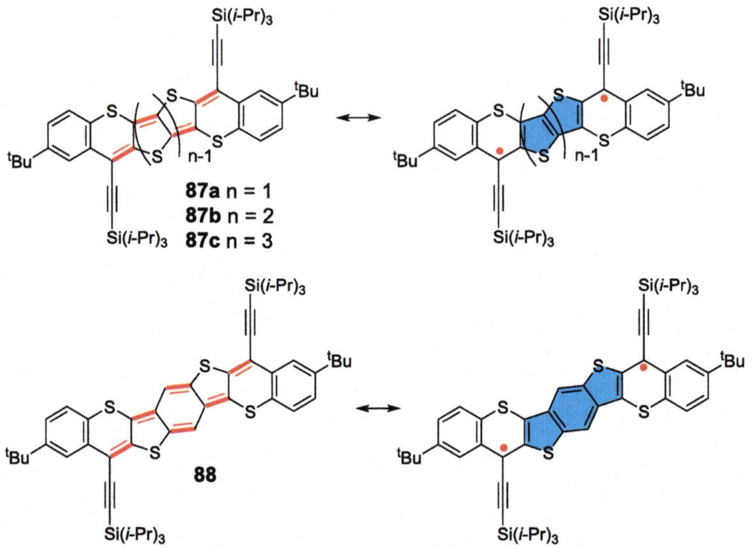

Fig. 17 Resonance structures of benzo-thia-fused [*n*]thienoacenequinodimethanes **87a–c** and **88** and their diradical resonance forms

distinctively different electronic absorption spectra. That is, **87a–c** and **88** have a characteristic p-band, which is very similar to that of many closed-shell polycyclic aromatic hydrocarbons such as acenes [93] and rylenes [94], while **82a–d** have a broad absorption band attributed to their anti-aromatic character.

DFT calculation indeed revealed the distinctive difference between these two closely related systems in the matter of aromaticity and singlet diradical character. Bisindeno-[*n*]thienoacene is a typical anti-aromatic system [NICS(1)zz values of the cyclopenta- and the central thiophene rings are significantly positive as shown in Fig. 16] with a small singlet diradical character, whereas benzo-thia-fused [*n*]thienoacenequinodimethane is a pro-aromatic system [the NICS(1)zz values of the central thiophene/benzene rings are significantly negative, as shown in Fig. 18] with an enhanced singlet diradical character. A "pro-aromatic" compound refers to a nonaromatic molecule containing a quinodimethane-like structure that has a high tendency to become an open-shell diradical and/or zwitterionic form containing one or more aromatic sextets. Consistent with the calculated data, the experimental results also revealed that **87a–c** and **88** have an enhanced singlet diradical character compared to **82a–d**. In particular, **88** displays temperature-dependent magnetic susceptibility and showed a broadened ^{1}H NMR spectrum and ESR signal at room temperature because of the presence of thermally populated triplet species. In addition, the physical properties of these two systems were also intensively studied and systematically compared. It was found that the pro-aromatic compounds (**87a–c** and **88**) show much longer singlet excited state lifetimes, larger TPA cross-section values and better redox amphotericity compared with their anti-aromatic counterparts (**82a–d**). This detailed study provided a comprehensive understanding of how the pro-aromaticity and anti-aromaticity affect the ground-state electronic structure,

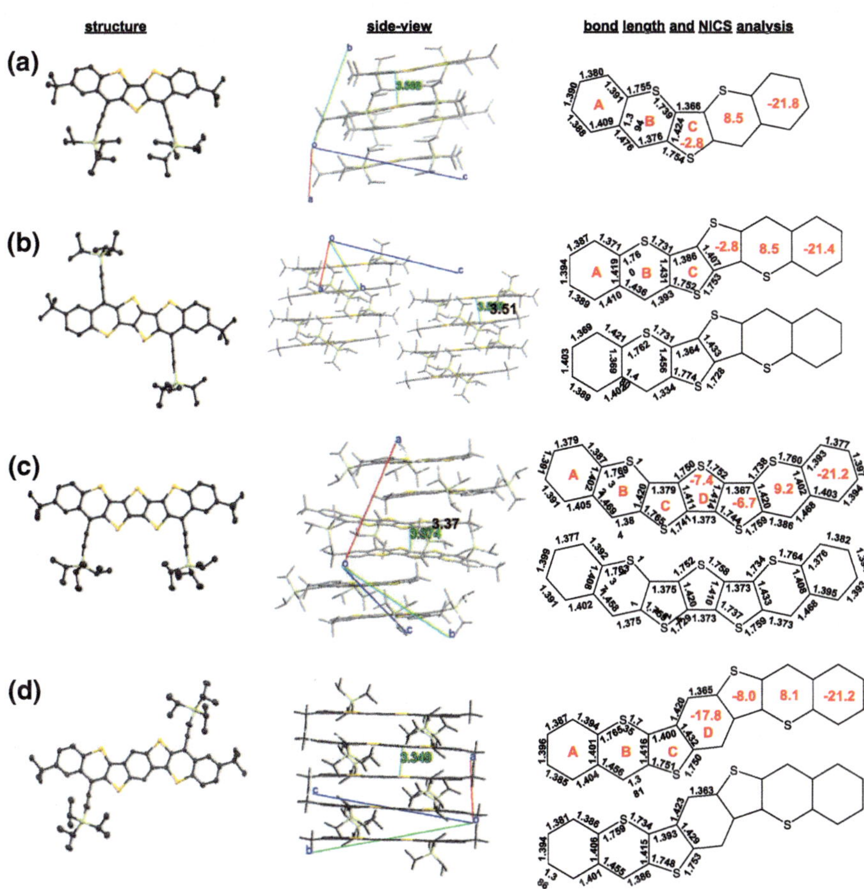

Fig. 18 X-ray crystallographic structures, 3D packing structures (side view) and selected bond lengths of **a 87a**, **b 87b**, **c 87c** and **d 88**. The *red numbers* in the rings denote the calculated NICS(1)zz values. Adapted with permission from Ref. [92]. Copyright The Royal Society of Chemistry 2016

diradical character and consequently the physical properties of quinoidal compounds.

Another pro-aromatic system based on bisphenaleno-thieno[3,2-*b*]thiophene (**89**) was also developed in our group in 2015 [95]. Compound **89** displays good solubility and stability in common organic solvents. The X-ray crystallographic analysis revealed its typical quinoidal resonance form with a surprisingly wave-shaped structure presumably because of the strong intermolecular phenalenyl–phenalenyl interaction in crystals (Fig. 19). Similar to the above-demonstrated results, pro-aromatic **89** also showed a larger diradical character, stronger absorption, longer excited state lifetime and better redox amphotericity than its anti-aromatic analog **82b**. It is notable that molecules **89**, **82b** and **87b** have the same central quinoidal thienoacene unit with similar molecular length, but the diradical character of **89** ($y_0 = 18.6\%$) is larger than both **82b** ($y_0 = 0\%$) and **87b** ($y_0 = 2.5\%$) under the same calculation method. Normally, a larger segregation

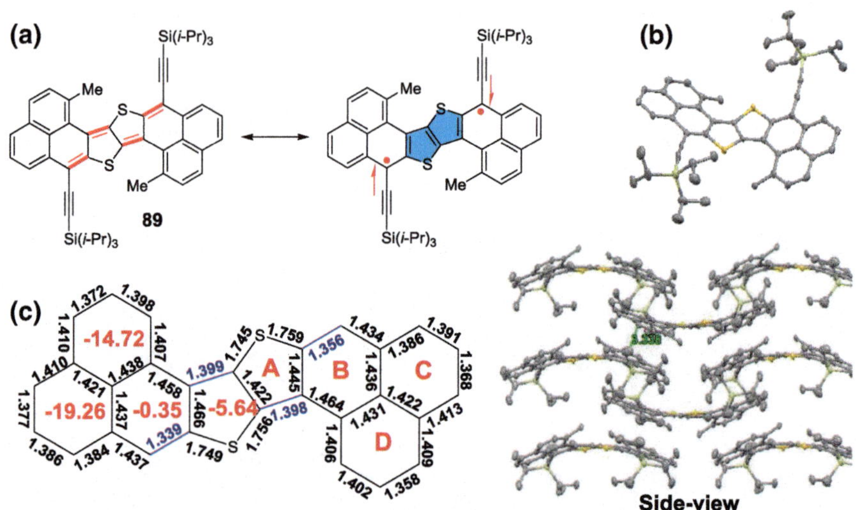

Fig. 19 Chemical resonance forms, X-ray crystallographic structures, 3D packing structures (side view) and bond lengths of bisphenaleno-thieno[3,2-*b*]thiophene **89**. The *red numbers* in the rings denote the calculated NICS(1)zz values. Adapted with permission from Ref. [95]. Copyright The Royal Society of Chemistry 2015

between SOMO-α and SOMO-β indicates a larger diradical character. Considering that quinoidal compounds containing the phenalenyl moiety such as bisphenalenyls [9, 10] and zethrenes [11, 12] usually have a significant diradical character, the larger diradical charater of **89** likely originates from the thermodynamic stabilizing effect of the delocalized phenalenyl radical. In **87b**, SOMO-α and SOMO-β showed less segregation than that of **89**. For **82b**, it is an anti-aromatic compound and showed a closed-shell structure. In addition, FETs based on solution-processed **89** thin films exhibited p-type operation with a high charge carrier mobility of 0.26 cm^2 V^{-1} s^{-1}. The relatively high charge carrier mobility together with the ordered packing structure of this type of diradicaloid molecule would be a potential material for spin-dependent charge transport studies in the future.

In parallel with above anti- and pro-aromatic quinoidal compounds, another research interest of our group focused on the design and synthesis of acene analogs with quinoidal conjugation. Here, the term "quinoidal conjugation" refers the π-conjugation in the *p*-QDM framework. Acenes are the polycyclic aromatic hydrocarbons containing linearly fused benzene rings, which were first introduced by Clar in 1939 (Fig. 20) [96]. To date, almost all the published acenes or acene analogs have a diene conjugation mode (conjugation in the diene form), and the stability of acenes decreases with the increment of molecular length. Longer acenes with a typical diene character are highly reactive and readily undergo oxidation [97, 98] or Diels-Alder reactions [99, 100]. Incorporation of two hetero-atoms (S, O, N–R) into the acene framework at the *para* position of the central benzenoid ring will result in quinoidal acene analogs (Fig. 20). As a consequence, the diene character of the molecule decreases and at the same time the quinoidal character

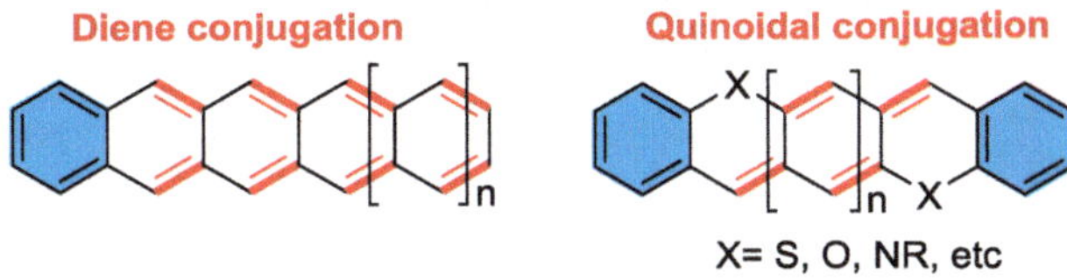

Fig. 20 Chemical structures of normal acenes and quinoidal heteroacenes showing the diene and quinoidal conjugation, respectively

increases. The designed molecules have one more aromatic sextet ring; therefore, they are expected to exhibit improved stability and distinctive properties compared to the normal acenes with diene conjugation (Figs. 20, 21). More interestingly, their dications obtained by oxidation are isoelectronic structures of corresponding acenes (Fig. 21).

With this concept in mind, in 2014 our group reported the synthesis of the stable bis(benzothia)-fused quinodimethane **90a–b** (Fig. 21 and Scheme 10) [101]. They were prepared from the diacid intermediate **93**, which was converted to diacyl chloride in the presence of thionyl chloride, and a subsequent intramolecular Friedel-Craft acylation reaction of diacyl chloride with $TiCl_4$ generated the key intermediate dithioquinacridone **94**. The TIPSE and phenyl substitution groups were introduced to the 7,14-positions by nucleophilic addition of **94** with the corresponding lithium reagents. Subsequent reduction of the generated diols with tin (II) chloride afforded the target compounds **90a** and **b**.

As expected from our designed concept, **90a–b** have superior stability in solution compared with their pentacene derivatives, e.g., the chloroform solution of **90a** was stored at ambient conditions for over 2 months without any indication of decomposition. The substantially increased stability of these quinoidal molecules would most likely originate from the disruption of the normal diene conjugation of acenes and hence interruption of the decomposition pathways. X-ray

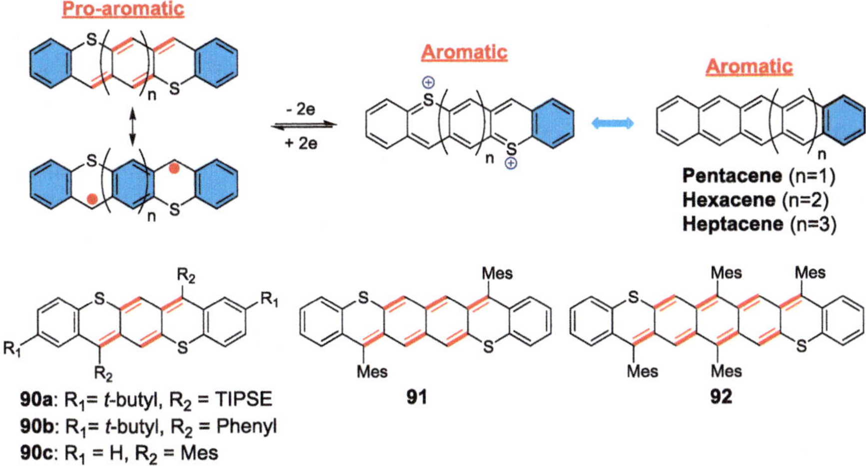

Fig. 21 Structures of the quinoidal bis(benzothia)quinodimethanes, their dications and acene analogs

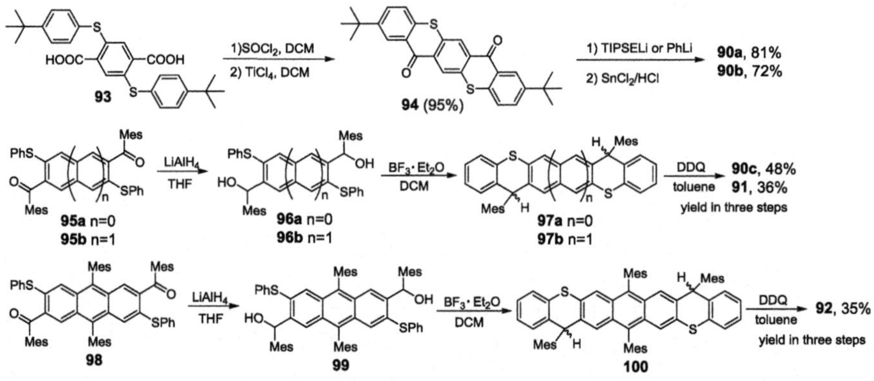

Scheme 10 Synthetic routes to compounds **90–92**

crystallographic analysis of **90a–b** revealed typical quinoidal conjugation with a large bond length alternation. Similar to pentacene, both **90a** and **90b** pack in a herringbone motif. FETs based on solution-processed thin films of **90a** and **90b** exhibited a hole mobility of up to 0.032 cm^{-2} V^{-1} s^{-1}. **90a–b** have a closed-shell ground state with a negligible singlet diradical character.

We hypothesize that the extension of the central *para*-quinodimethane unit would enhance the diradical character. Therefore, a series of extended bis(benzothia)-quinodimethanes **90c**, **91** and **92** were prepared [102]. The key intermediates for the synthesis of extended bis(benzothia)quinodimethanes **90c**, **91** and **92** are the thiophenol- and mesitylketone-substituted compounds **95a–b** and **98** (Scheme 10). They were reduced by $LiAlH_4$ to give diol **96a–b** and **99**, which then underwent $BF_3 \cdot Et_2O$-mediated intramolecular Friedel-Crafts alkylation to generate the dihydro compounds **97a–b** and **100**, and the subsequent oxidative dehydrogenation by DDQ afforded the target compounds **90c**, **91** and **92**. All three compounds exhibited excellent photostability with a half-life time of more than 1 month in dichloromethane under ambient light and air conditions. Like **90a** and **90b**, **90c** has a closed-shell ground state with a very small diradical character [$y_0 = 0.004$, at UCAM-B3LYP/6-31G(d,p) level of theory]. Similar calculations suggest that both **91** and **92** have an open-shell singlet ground state, with a diradical character of 0.195 and 0.384, respectively. NMR spectral broadening was observed for both **91** and **92** at elevated temperatures, which can be ascribed to a thermally populated triplet species. They also showed a broad ESR signal in solution. The bond length alternation in the central quinodimenthane units becomes smaller from **90c** to **92** based on X-ray crystallographic analysis (Fig. 22). Calculations show that NICS(1)zz values of the central quinodimethane units gradually change from +1.65 in **90c** to −9.86 ppm (ring C) in **91** and −10.24 ppm (ring C)/−15.85 ppm (ring D) in **92** (Fig. 22). All of the calculations and experimental results indicated an increase of aromaticity and diradical character with extension of the molecular length. The singlet-to-triplet energy gap (ΔE_{S-T}) of −5.3 kcal/mol was estimated by a superconducting quantum interference device (SQUID) measurement for **92**. The increase of the diradical character with extension of the central quinodimethane unit

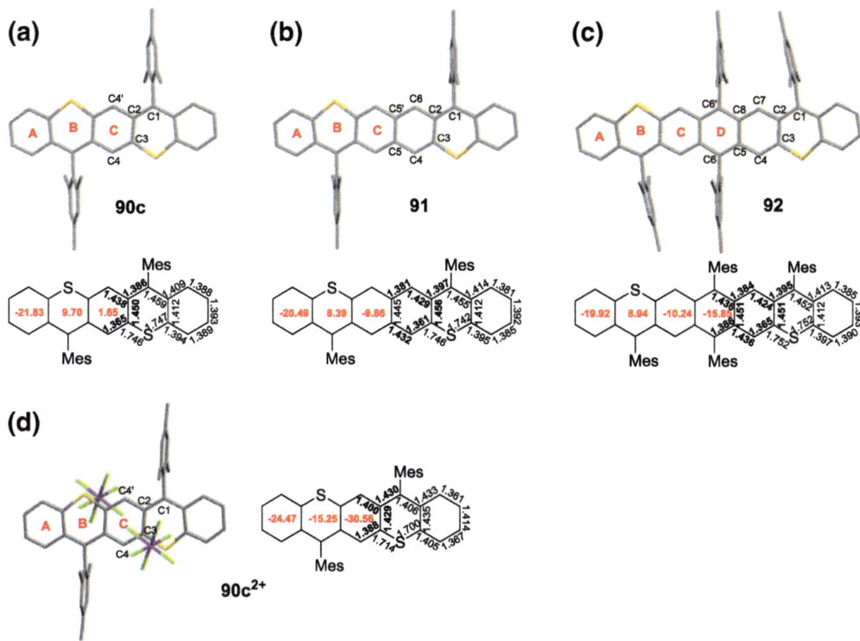

Fig. 22 X-ray crystallographic structures, selected bond lengths from single-crystal structures and calculated NICS(1)zz values of **a 90c**, **b 91**, **c 92** and **d 90c^{2+}**. Hydrogen atoms are omitted for clarity. Adapted with permission from Ref. [102]. Copyright 2016 Wiley-VCH Verlag GmbH & Co.

can be simply explained by the increase of the aromatic resonance energy from benzene to naphthalene and further to anthracene.

Electrochemical measurements revealed that all three compounds **90c**, **91** and **92** exhibited two reversible oxidations. Chemical oxidation by one and two equivalents of NO·SbF$_6$ gave the corresponding radical cations and dications, which can be reversibly reduced back to neutral compounds via addition of triethylamine. ^{1}H NMR spectra of the three dications (**90c^{2+}** as an example shown in Fig. 23) showed largely downfield-shifted resonances compared to their neutral compounds, indicating the formation of aromatic molecules. Consistent with the NMR results, the bond length alternation of the original *p*-QDM unit in dications becomes much smaller. In addition, calculations showed large negative NICS(1)zz values for all the dithiapentacene framework of the dications (Fig. 22d). Anisotropy of the induced current-density (ACID) plots of the dications clearly shows clockwise ring currents delocalized along the periphery (Fig. 23), which is similar to the corresponding acenes. However, for the neutral quinoidal compounds, the clockwise ring currents are mainly localized on the lateral benzene rings. All of experimental results and calculation data indicate that the dications indeed have an acene-like structure, and thus they can be regarded as the genuine isoelectronic structures of acenes. This study provides a new way to design and synthesize stable longer acene analogs with similar electronic structures to the traditional all-carbon acenes.

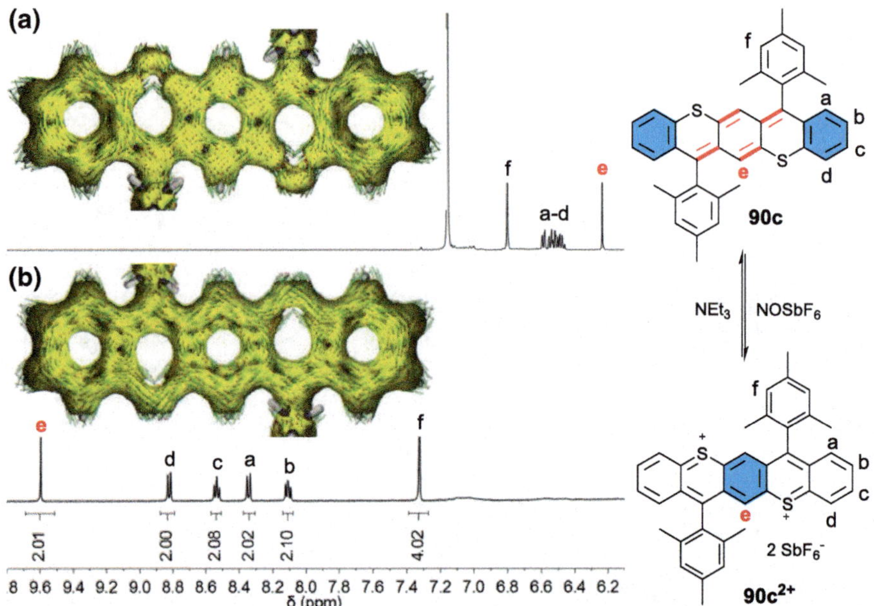

Fig. 23 ^{1}H NMR spectra of **a** neutral **90c** and **b** its dication in the aromatic region. Insert is the corresponding ACID plots, and the external magnetic field is applied orthogonal to the plane. Adapted with permission from Ref. [102]. Copyright 2016 Wiley-VCH Verlag GmbH & Co.

Although bis(benzothia)-fused quinodimethane resembles pentacene in topology, it possesses two more π electrons than pentacene [e.g., 24 πe for bis(benzothia)-fused quinodimethane while 22 πe for pentacene; Fig. 24]. Thus, our other interest is to make isoelectronic structures of acenes with dipolar character, i.e., asymmetrical quinodimethanes with different fusion motifs at two sides. For example, if one sulfur atom in bis(benzothia)-fused quinodimethane is removed, the obtained molecule **A** can be regarded as the isoelectronic structure of pentacene (Fig. 24). Similarly, the molecules **B** and **C** are isoelectronic structures of nonacene. Both five-membered rings (indene) and hetero-atoms (benzothia) are incorporated into the central quinodimethane unit, which leads to a new hybrid system with an asymmetrical quinodimethane framework. So far, there are very limited reports about the synthesis and properties of such asymmetrical quinodimethanes [103, 104]. The ground-state geometric and electronic structures of **A–C** are of particular interest because besides the closed-shell quinoidal structure, one open-shell diradical resonance form and one dipolar zwitterionic resonance form are also supposed to contribute to the ground-state structure because of the recovery of one more aromatic sextet ring in the latter two forms (Fig. 24) [105].

The parent compounds **A–C** are predicted to be unstable and insoluble. Therefore, their derivatives **101–103** in which the most reactive sites are kinetically blocked by the mesityl group (Fig. 24) were designed [105]. Similar to **90c**, an intramolecular Friedel-Crafts alkylation followed by an oxidative dehydrogenation strategy was utilized to construct the target molecules from the key intermediates of

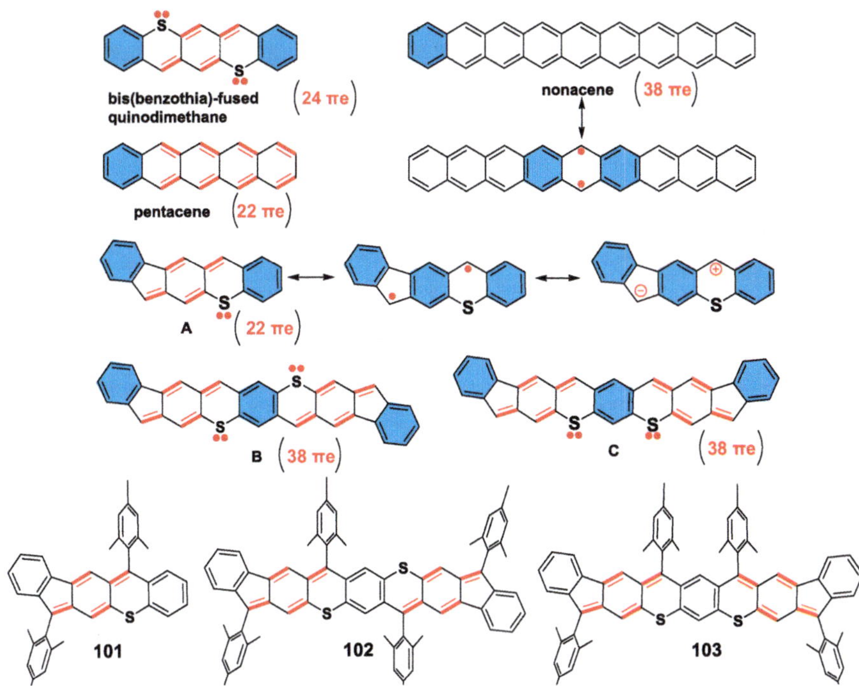

Fig. 24 Structures of bis(benzothia)-fused quinodimethane, pentacene, nonacene and their isoelectronic quinoidal molecules

diol **105** and tetrol **108** (Scheme 11). Compound **101** was successfully obtained by this strategy in a very good yield. Following a similar strategy, the synthesis of the mesityl-substituted **B** derivative was carried out. However, interestingly, instead of the targeted compound **102**, the mesityl substituted **C** derivative **103** was obtained exclusively. The formation of the *meta*-dithia compound was likely due to 1,2-sulfur migration via a spirocyclic cationic intermediate during the ring cyclization reaction in the presence of $BF_3 \cdot Et_2O$ (Scheme 11). Compounds **101** and **103** were extremely stable in air in contrast to their corresponding reactive pentacene and nonacene derivatives.

Both **101** and **103** displayed an intense absorption band at 310–450 nm and a broad band from 520 nm to 900 (for **101**) or 1080 nm (for **103**) (see Table 1), which was distinctly different from that of their corresponding acene derivatives and **90** [105]. The unique absorption bands in **101** and **103** could be ascribed to the weak intramolecular donor-acceptor interactions in a quinoidal skeleton. DFT calculation also predicted that **101** and **103** have a significant dipole moment of 3.14 and 2.87 D, respectively. Their dipole moments point to the different directions, which cause the different molecular packing structure in the solid state. Single crystals of **101** and **103** revealed their typical quinoidal-type conjugation with a large bond length alternation (Fig. 25). Molecules of **101** are packed into a square-like tetrameric structure via intermolecular dipole–dipole interactions and [C-H···π] interactions

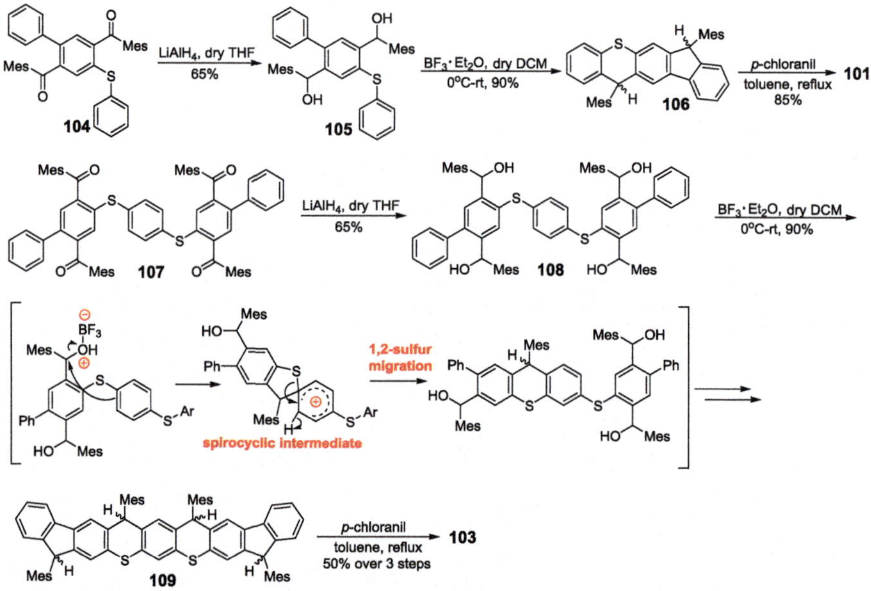

Scheme 11 Synthetic routes to compounds **101** and **103**

between the adjacent molecules. Interestingly, **103** packs in a slipped face-to-face manner mainly via [C-H···S] interactions and [C-H···π] interactions between adjacent molecules. The dipole moments of all **103** molecules in this case point in the same direction. The preliminary theoretical study on molecules **101** and **103** revealed that they have a negligible diradical character. Instead, closed-shell quinoidal conjugation mostly dominates their ground-state structures with a small dipolar/zwitterionic character.

5 Fused Silole Quinodimethanes

Quinoidal compounds containing Si are rarely reported because of the difficulty to construct silicon-bridged skeletons [106–108]. Very recently, Shintani and co-workers reported the synthesis of quinoidal-fused oligosiloles via a concise stitching reaction with rhodium catalysis [109]. They first devised a reaction between two oligo(silylene-ethynylene)s **110** and **111**, one of which (**110**) contains an arylmetal moiety (aryl boronic acid ester in Scheme 12) on one end and a haloarene moiety on the other end, under rhodium catalysis to stitch them together. Up to five siloles **112a–c** have been fused in quinoidal form. The structures of **112a–c** were unambiguously confirmed by X- ray crystallographic analysis (Fig. 26). Generally, all the structures are completely flat, and large bond length alternation was found in

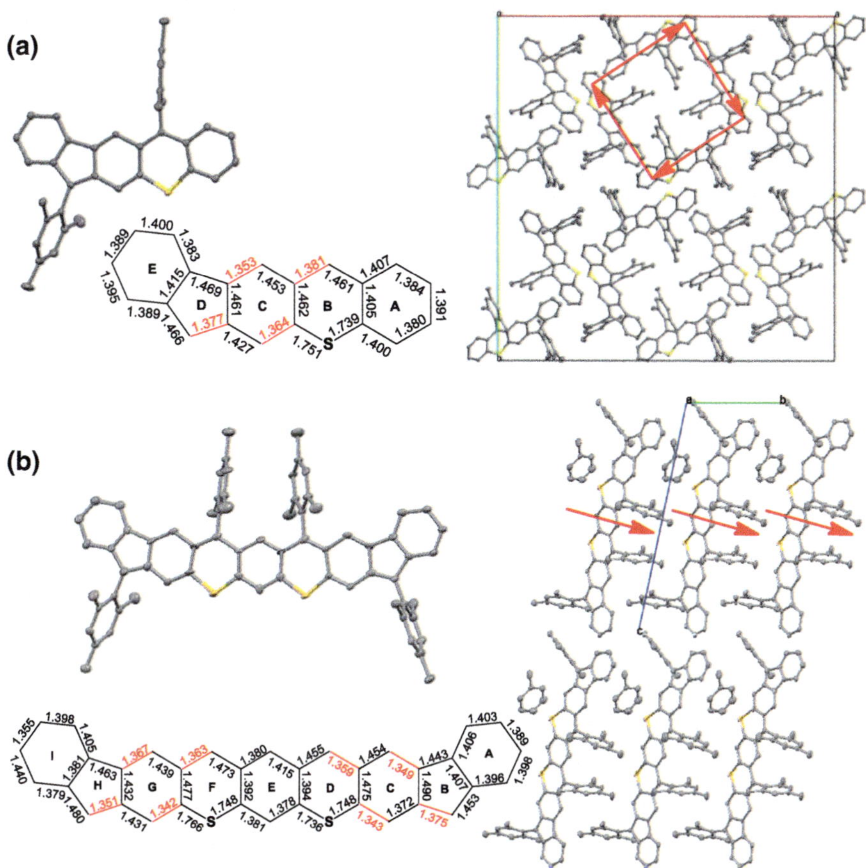

Fig. 25 Single-crystal structures, 3D packing and bond length analysis of **a 101** and **b 103**. The *arrows* roughly denote the dipole moments. Adapted with permission from Ref. [105]. Copyright 2015 Wiley-VCH Verlag GmbH & Co.

the central silicon-bridged five-membered rings, i.e., vertical carbon–carbon bonds (C1–C2, C3–C4, C5–C6, C7–C8) have a typical single-bond character (1.473–1.488 Å), and horizontal carbon–carbon bonds [C2–C3, C4–C5, C6–C7(C6′), C8–C8′] have a typical double-bond character (1.356–1.375 Å). They showed gradually red-shifted absorption from **112a** to **112c**. Both HOMO and LUMO energy levels increase with extension of molecular length, and their band gaps are larger than 2 eV. Compared to our previously reported compounds **82a–d**, **112a–c** are more stable and have larger energy gaps. The difference could be explained by the fact that S atoms in **82** participate in the π-conjugation and lead to an antiaromatic system while these quinoidal-fused oligosiloles are nonaromatic, bridged *cis–transoid* oligoacetylenes.

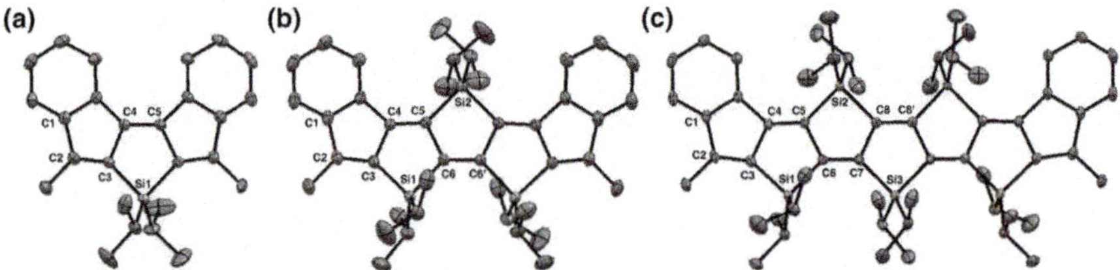

Scheme 12 Synthetic route to bisindeno-[*n*]thienoacenes **112a–c**

Fig. 26 X-ray crystallographic structures of **a 112a**, **b 112b** and **c 112c** Adapted with permission from Ref. [109]. Copyright 2016 American Chemical Society

6 Conclusion

In summary, a large number of heterocyclic quinodimethanes containing O, N, S and Si have been reviewed in this chapter. The early interests in this kind of quinoidal compound were mainly related to their applications as high-performance pigments/dyes and semiconductors for organic electronics. The recently emerging interests in quinoidal compounds have accelerated the development of some novel heterocyclic quinodimethanes with unique properties. Particularly, they have been demonstrated to be a good system to fundamentally correlate their unique properties to their inherent open-shell singlet diradical character and aromaticity/pro-aromaticity/anti-aromaticity. Without doubt, together with other quinoidal poly-cyclic hydrocarbons, heterocyclic quinodimethanes still remain a largely unexplored area of research. We believe that this kind of molecule will eventually become a real and useful material in the near future.

Acknowledgements We acknowledge financial support from the MOE Tier 1 Grant (R-143-000-623-112) and Tier 3 program (MOE2014-T3-1-004).

References

1. Zeng Z, Shi X, Chi C, López Navarrete JT, Casado J, Wu J (2015) Pro-aromatic and anti-aromatic p-conjugated molecules: an irresistible wish to be diradicals. Chem Soc Rev 44:6578–6596
2. Kubo T (2015) Recent progress in quinoidal singlet biradical molecules. Chem Lett 44:111–122
3. Casado J, Ortiz RP, López Navarrete JT (2012) Quinoidal oligothiophenes: new properties behind an unconventional electronic structure. Chem Soc Rev 41:5672–5686
4. Burrezo PM, Zafra JL, López Navarrete JT, Casado J (2017) Quinoidal/Aromatic Transformation in pi-conjugated oligomers: vibrational Raman studies on the limit of rupture of pi-bonds. Angew Chem Int Ed 56:2250–2259
5. Thiele J, Balhorn H (1904) Ueber einen chinoïden kohlenwasserstoff. Chem Ber 37:1463–1470

6. Tschitschibabin AE (1907) Über einige phenylierte derivate des p, p-ditolyls. Chem Ber 40:1810–1819

7. Montgomery LK, Huffman JC, Jurczak EA, Grendze MP (1986) The molecular structures of Thiele's and Chichibabin's hydrocarbons. J Am Chem Soc 108:6004–6011

8. Zeng Z, Sung YM, Bao N, Tan D, Lee R, Zafra JL, Lee BS, Ishida M, Ding J, Lûpez Navarrete JT, Li Y, Zeng W, Kim D, Huang KW, Webster RD, Casado J, Wu J (2012) Stable tetrabenzo-Chichibabin's hydrocarbons: tunable ground state and unusual transition between their closed-shell and open-shell resonance forms. J Am Chem Soc 134:14513–14525

9. Kubo T (2015) Phenalenyl-based open-shell polycyclic aromatic hydrocarbons. Chem Rec 15:218–232

10. Uchida K, Kubo T (2016) Recent advances in the chemistry of phenalenyl. J Synth Org Chem Jpn 74:1069–1077

11. Sun Z, Zeng Z, Wu J (2014) Zethrenes, extended p- quinodimethanes, and periacenes with a singlet biradical ground state. Acc Chem Res 47:2582–2591

12. Sun Z, Wu J (2014) Zethrene and expanded zethrenes with tunable ground states and physical properties. Pure Appl Chem 86:529–537

13. Shimizu A, Nobusue S, Miyoshi H, Tobe Y (2014) Indenofluorene congeners: biradicaloids and beyond. Pure Appl Chem 86:517–528

14. Fix AG, Chase DT, Haley MM (2014) Indenofluorenes and derivatives: syntheses and emerging materials applications. Top Curr Chem 349:159–196

15. Haley MM (2015) Origins of the indenofluorene project: serendipity and other surprises. Chem Rec 15:1140–1143

16. Tobe Y (2015) Non-alternant non-benzenoid aromatic compounds: past, present, and future. Chem Rec 15:86–96

17. Sun Z, Wu J (2012) Open-shell polycyclic aromatic hydrocarbons. J Mater Chem 22:4151–4160

18. Shimizu A, Hirao Y, Kubo T, Nakano M, Botek E, Champagne B (2012) Theoretical consideration of singlet open-shell character of polyperiacenes using Clar's aromatic sextet valence bond model and quantum chemical calculations. AIP Conf Proc 1504:399–405

19. Sun Z, Ye Q, Chi C, Wu J (2012) Low band gap polycyclic hydrocarbons: from closed-shell near infrared dyes and semiconductors to open-shell radicals. Chem Soc Rev 41:7857–7889

20. Sun Z, Zeng Z, Wu J (2013) Benzenoid polycyclic hydrocarbons with an open–shell biradical ground state. Chem Asian J 8:2894–2904

21. Abe M (2013) Diradicals. Chem Rev 113:7011–7088

22. Annen O, Egli R, Henzi B, Jacob H, Matzinger P (1987) Replacement of disperse anthraquinone dyes. Rev Prog Color 17:72–85

23. Greenhalgh CW, Carey JL, Hall N, Newton DF (1994) The benzodifuranone chromogen and its application to disperse dyes. J Soc Dyes Colour 110:178–184

24. Greenhalgh CW, Carey JL, Newton DF (1980) The synthesis of quinodimethanes in the benzodifuranone and benzodipyrrolidone series. Dyes Pigm 1:103–120

25. Zhang K, Tieke B (2011) Low-bandgap benzodifuranone-based polymers. Macromolecules 44:4596–4599

26. Hallas G, Yoon C (2001) The synthesis and properties of red and blue benzodifuranones. Dyes Pigm 48:107–119

27. Zhang K, Tieke B, Forgie JC, Vilela F, Skabara PJ (2012) Donor-acceptor conjugated polymers based on *p*- and *o*-benzodifuranone and thiophene derivatives: electrochemical preparation and optical and electronic properties. Macromolecules 45:743–750

28. Font-Sanchis E, Aliaga C, Cornejo R, Scaiano JC (2003) Reactivity toward oxygen of isobenzofuranyl radicals: effect of nitro group substitution. Org Lett 5:1515–1518

29. Frenette M, Aliaga C, Font-Sanchis E, Scaiano JC (2004) Bond dissociation energies for radical dimers derived from highly stabilized carbon-centered radicals. Org Lett 6:2579–2582

30. Berens U, Stassen A, Schmidhalter B, Kalb W, Bienewald F (2007) Quinoid systems as organic semiconductors. PCT:WO 118799 A1

31. Kalb WL, Stassen AF, Batlogg B, Berens U, Schmidhalter B, Bienewald F, Hafner A, Wagner T (2009) Quinoid heteropentacenes as promising organic semiconductors for field-effect transistor applications. J Appl Phys 105:043705

32. Sato C, Suzuki S, Kozaki M, Okada K (2016) 2,11-Dibromo-13,14-dimesityl-5,8-dioxapentaphene: a stable and twisted polycyclic system containing the *o*-quinodimethane skeleton. Org Lett 18:1052–1055

33. Cole DT, Sarver JE, Campbell CD (2007) Tetrabenzodiazadiketoperylene pigments and dyes. US 0179222 A1
34. Faulkner EB, Schwartz RJ (2009) High performance pigments. Wiley, Weinheim
35. Dietz E, Kroh A, Paulus EF, Prokschy F, Lincke G (1991) Chinacridone und Dioxazine—Neues über bekannte Pigmente 11. Internationales Farbensymposium, Montreux
36. Boeglin P (1998) Synthèse de nouveaux pigments et colorants heterocycliques de structures triphenodioxazine, phtaloperinone et perimidophtalone. Ph.D. Thesis, Université de Haute Alsace
37. Born R (2002) In: Presentation on 9th international conference COLORCHEM Špindlerův Mlýn, Czech Republic
38. Nicolas Y, Castet F, Devynckb M, Tardy P, Hirsch L, Labrugère C, Allouchi H, Toupance T (2012) TIPS-triphenodioxazine versus TIPS-pentacene: enhanced electron mobility for n-type organic field-effect transistors. Org Electron 13:1392–1400
39. Manassen J, Khalif Sh (1966) Organic polymers. Correlation between their structure and catalytic activity in heterogeneous systems. II. Models for dehydrogenation catalysts. J Am Chem Soc 88:1943–1947
40. Stille JK, Mainen EL (1968) Thermally stable ladder polyquinoxalines. Macromolecules 1:36–42
41. Jenekhe SA (1991) Electroactive ladder polyquinoxalines. 1. Properties of the model compound 5,12-dihydro-5,7,12,14-tetraazapentacenaene and its complexes. Macromolecules 24:1–10
42. Goyette MA, Leclerc M (1995) Structure-property relationships in poly(o-phenylenediamine) derivatives. J Electroanal Chem 382:17–23
43. Ma Y, Sun Y, Liu Y, Gao J, Chen S, Sun X, Qiu W, Yu G, Cui G, Hu W, Zhu D (2005) Organic thin film transistors based on stable amorphous ladder tetraazapentacenes semiconductors. J Mater Chem 15:4894–4898
44. Armand J, Boulares L, Bellec C, Pinson J (1987) Chemical and electrochemical reduction of pyrazino[2,3-g]quinoxalines and of their benzo and dibenzo derivatives; the structure of fluorindine and the formation of tetraanion. Can J Chem 65:1619–1623
45. Sawtschenko L, Jobst K, Neudeck A, Dunsch L (1996) Electrochemical and spectroelectrocemical studies of dihydro-tetra-azapentacene as a model of polyazaacene. Electrochim Acta 41:123–131
46. Seillan C, Brisset H, Siri O (2008) Efficient synthesis of substituted dihydrotetraazapentacenes. Org Lett 10:4013–4016
47. Tang Q, Liu J, Chan HS, Miao Q (2009) Benzenoid and quinonoid nitrogen-containing heteropentacenes. Chem Eur J 15:3965–3969
48. Janosik T, Wahlstroem N, Bergman J (2008) Recent progress in the chemistry and applications of indolocarbazoles. Tetrahedron 64:9159–9180
49. Long BH, Rose WC, Vyas DM, Matson JA, Forenza S (2002) Discovery of antitumor indolocarbazoles: rebeccamycin, NSC 655649, and fluoroindolocarbazoles. Curr Med Chem Anti Cancer Agents 2:255–266
50. Sánchez C, Méndez C, Salas J (2006) Indolocarbazole natural products: occurrence, biosynthesis, and biological activity. Nat Prod Rep 23:1007–1045
51. Wakim S, Bouchard J, Simard M, Drolet N, Tao Y, Leclerc M (2004) Organic microelectronics: design, synthesis, and characterization of 6,12-dimethylindolo[3,2-b]carbazoles. Chem Mater 16:4386–4388
52. Wu YL, Li YN, Gardner S, Ong BS (2005) Indolo[3,2-b]carbazole-based thin-film transistors with high mobility and stability. J Am Chem Soc 127:614–618
53. Li YN, Wu YL, Ong BS (2006) Polyindolo[3,2-b]carbazoles: a new class of p-channel semiconductor polymers for organic thin-film transistors. Macromolecules 39:6521–6527
54. Yudina LN, Preobrazhenskaya MN, Korolev AM (2000) Transformation of 5H,11H-indolo[3,2-b]carbazole through 5,11-didehydroindolo[3,2-b]carbazole. Chem Heterocycl Compd 36:1112–1113
55. Gu R, Robeyns K, Meervelt LV, Toppet S, Dehaen W (2008) Facile synthesis of novel indolo[3,2-b]carbazole derivatives and a chromogenic-sensing 5,12-dihydroindolo[3,2-b]carbazole. Org Biomol Chem 6:2484–2487
56. Luo D, Lee S, Zheng B, Sun Z, Zeng W, Huang KW, Furukawa K, Kim D, Webster RD, Wu J (2014) Indolo[2,3-b]carbazoles with tunable ground states: how Clar's aromatic sextet determines the singlet biradical character. Chem Sci 5:4944–4952
57. Clar E (1972) The aromatic sextet. Wiley-VCH, London
58. Gronowitz S, Uppström B (1974) On the reaction of 2,5-dihalothiophenes with tetracyanoethylene oxide. Acta Chem Scand Ser B 28:981–985

59. Yui K, Aso Y, Otsubo T, Ogura F (1987) New electron acceptors for organic metals: extensively conjugated homologues of thiophene-7,7,8,8-tetracyanoquinodimethane (TCNQ). J Chem Soc Chem Commun (24):1816–1817

60. Izumi T, Kobashi S, Takimiya K, Aso Y, Otsubo T (2003) Synthesis and spectroscopic properties of a series of β-blocked long oligothiophenes up to the 96-mer: revaluation of effective conjugation length. J Am Chem Soc 125:5286–5287

61. Takahashi T, Matsuoka K, Takimiya K, Otsubo T, Aso Y (2005) Extensive quinoidal oligothiophenes with dicyanomethylene groups at terminal positions as highly amphoteric redox molecules. J Am Chem Soc 127:8928–8929

62. Ponce Ortiz R, Casado J, Hernández V, López Navarrete JT, Viruela PM, Ortí E, Takimiya K, Otsubo T (2007) On the biradicaloid nature of long quinoidal oligothiophenes: experimental evidence guided by theoretical studies. Angew Chem Int Ed 46:9057–9061

63. Janzen DE, Burand MW, Ewbank PC, Pappenfus TM, Higuchi H, da Silva Filho DA, Young VG, Brédas J-L, Mann KR (2004) Preparation and characterization of & #x03C0;-stacking quinodimethane oligothiophenes. Predicting semiconductor behavior and bandwidths from crystal structures and molecular orbital calculations. J Am Chem Soc 126:15295–15308

64. Casado J, Miller LL, Mann KR, Pappenfus TM, Higuchi H, Ortí E, Milián B, Pou-Amérigo R, Hernández V, López Navarrete JT (2002) Quinonoid oligothiophenes as electron-donor and electron-acceptor materials. a spectroelectrochemical and theoretical study. J Am Chem Soc 124:12380–12388

65. Ortiz RP, Casado J, Hernández V, López Navarrete JT, Ortí E, Viruela PM, Milián B, Hotta S, Zotti G, Zecchin S, Vercelli B (2006) Magnetic properties of quinoidal oligothiophenes: more than good candidates for ambipolar organic semiconductors? Adv Funct Mater 16:531–536

66. Ribierre JC, Fujihara T, Watanabe S, Matsumoto M, Muto T, Nakao A, Aoyama T (2010) Direct laser writing of complementary logic gates and lateral p–n diodes in a solution-processible monolithic organic semiconductor. Adv Mater 22:1722–1726

67. Pappenfus TM, Chesterfield RJ, Frisbie CD, Mann KR, Casado J, Raff JD, Miller LL (2002) A π-stacking terthiophene-based quinodimethane is an n-channel conductor in a thin film transistor. J Am Chem Soc 124:4184–4185

68. Chesterfield RJ, Newman CR, Pappenfus TM, Ewbank PC, Haukaas MH, Mann KR, Miller LL, Frisbie CD (2003) High electron mobility and ambipolar transport in organic thin-film transistors based on a π-stacking quinoidal terthiophene. Adv Mater 15:1278–1282

69. Casado J, Zgierski MZ, Ewbank PC, Burand MW, Janzen DE, Mann KR, Pappenfus TM, Berlin A, Pérez-Inestrosa E, Ortiz RP, López Navarrete JT (2006) Exploration of ground and excited electronic states of aromatic and quinoid *S,S*-dioxide terthiophenes. Complementary systems for enhanced electronic organic materials. J Am Chem Soc 128:10134–10144

70. Handa S, Miyazaki E, Takimiya K, Kunugi Y (2007) Solution-processible n-channel organic field-effect transistors based on dicyanomethylene-substituted terthienoquinoid derivative. J Am Chem Soc 129:11684–11685

71. Yui K, Ishida H, Aso Y, Otsubo T, Ogura F, Kawamoto A, Tanaka J (1989) Novel electron acceptors bearing a heteroquinonoid system. Synthesis and conductive complexes of 2,5-bis(dicyanomethylene)-2,5-dihydrothieno[3,2-b]thiophene, 2,6-bis(dicyanomethylene)-2,6-dihydrodithieno[3,2-b:2′,3′-d]thiophene, and their derivatives. Bull Chem Soc Jpn 62:1547–1555

72. Wu Q, Li R, Hong W, Li H, Gao X, Zhu D (2011) Dicyanomethylene-substituted fused tetrathienoquinoid for high-performance, ambient-stable, solution-processable n-channel organic thin-film transistors. Chem Mater 23:3138–3140

73. Yoshida S, Fujii M, Aso Y, Otsubo T, Ogura F (1994) Novel electron acceptors bearing a heteroquinonoid system: syntheses, properties, and charge-transfer complexes of 2,7-bis (dicyanomethylene) -2,7-di hydrobenzo[2, 1-b:3,4-b']dithiophene, 2,7-Bis(dicyanomethylene)-2,7-dihydrobenzo[1,2-b:4,3-b']dithiophene, and 2,6-Bis(dicyanomethylene)-2,6-dihydrobenzo[1,2–h4,5-b]dithiophene. J Org Chem 59:3077–3081

74. Kashiki T, Miyazaki E, Takimiya K (2009) Alkylated 2,6-Bis(dicyanomethylene)-2,6-dihydrobenzo[1,2-b:4,5-b0]dithiophenes: new soluble n-channel organic semiconductors for air-stable OFETs. Chem Lett 38:568–569

75. Mori T, Yanai N, Osaka I, Takimiya K (2014) Quinoidal naphtho[1,2-b:5,6-b']dithiophenes for solution-processed n-channel organic field-effect transistors. Org Lett 16:1334–1337

 205 

76. Pappenfus TM, Hermanson BJ, Helland TJ, Lee GGW, Drew SM, Mann KR, McGee KA, Rasmussen SC (2008) Reduced band gap dithieno[3,2-b:2′,3′-d]pyrroles: new n-type organic materials via unexpected reactivity. Org Lett 10:1553–1556

77. Qiao Y, Guo Y, Yu C, Zhang F, Xu W, Liu Y, Zhu D (2012) Diketopyrrolopyrrole-containing quinoidal small molecules for high-performance, air-stable, and solution-processable n-channel organic field-effect transistors. J Am Chem Soc 134:4084–4087

78. Xiong Y, Tao J, Wang R, Qiao X, Yang X, Wang D, Wu H, Li H (2016) A furan–thiophene-based quinoidal compound: a new class of solution-processable high-performance n-type organic semiconductor. Adv Mater 28:5949–5953

79. Zhang C, Zang Y, Gann E, McNeill CR, Zhu X, Di C, Zhu D (2014) Two-dimensional π-expanded quinoidal terthiophenes terminated with dicyanomethylenes as n-type semiconductors for high-performance organic thin-film transistors. J Am Chem Soc 136:16176–16184

80. Suzuki Y, Miyazaki E, Takimiya K (2010) ((Alkyloxy)carbonyl)cyanomethylene-substituted thienoquinoidal compounds: a new class of soluble n-channel organic semiconductors for air-stable organic field-effect transistors. J Am Chem Soc 132:10453–10466

81. Zafra JL, Qiu L, Yanai N, Mori T, Nakano M, Peña Alvarez M, López Navarrete JT, Gómez-García CJ, Kertesz M, Takimiya K, Casado J (2016) Reversible dimerization and polymerization of a janus diradical to produce labile C-C bonds and large chromic effects. Angew Chem Int Ed 55:14563–14568

82. Takahashi K, Suzuki T (1989) p-Diphenoquinone analogues extended by dihydrothiophenediylidene insertion: a novel amphoteric multistage redox system. J Am Chem Soc 111:5483–5485

83. Lanata M, Bertarelli C, Gallazzi MC, Bianco A, Del Zoppo M, Zerbi G (2003) Molecules with quinoid ground state: a new class of large molecular optical nonlinearities? Synth Met 138:357–362

84. D'Amore F, Lanata M, Gallazzi MC, Zerbi G (2003) Optical properties of a quinoid molecule. Chem Phys Lett 377:243–248

85. Canesi EV, Fazzi D, Colella L, Bertarelli C, Castiglioni C (2012) Tuning the quinoid versus biradicaloid character of thiophene-based heteroquaterphenoquinones by means of functional groups. J Am Chem Soc 134:19070–19083

86. Osaka I, Abe T, Mori H, Saito M, Takemura N, Koganezawa T, Takimiya K (2014) Small band gap polymers incorporating a strong acceptor, thieno[3,2-*b*]thiophene-2,5-dione, with p-channel and ambipolar charge transport characteristics. J Mater Chem C 2:2307–2312

87. Kawabata K, Osaka I, Nakano M, Takemura N, Koganezawa T, Takimiya K (2015) Thienothiophene-2,5-dione-based donor–acceptor polymers: improved synthesis and influence of the donor units on ambipolar charge transport properties. Adv Electron Mater 1:1500039

88. Kawabata K, Saito M, Osaka I, Takimiya K (2016) Very small bandgap π-conjugated polymers with extended thienoquinoids. J Am Chem Soc 138:7725–7732

89. Kawababa K, Osaka I, Sawamoto M, Zafra JL, Burrezo PM, Casado J, Takimiya K (2017) Dithienyl acenedithiophenediones as new and #x03C0;-extended quinoidal cores: synthesis and properties. Chem Eur J 23:4579–4589

90. Shi X, Burrezo PM, Lee S, Zhang W, Zheng B, Dai G, Chang J, López Navarrete JT, Huang KW, Kim D, Casado J, Chi C (2014) Antiaromatic bisindeno-[n]thienoacenes with small singlet biradical characters: syntheses, structures and chain length dependent physical properties. Chem Sci 5:4490–4503

91. Rudebusch GE, Fix AG, Henthorn HA, Vonnegut CL, Zakharov LN, Haley MM (2014) Quinoidal diindenothienoacenes: synthesis and properties of new functional organic materials. Chem Sci 5:3627–3633

92. Shi X, Quintero E, Lee S, Jing L, Herng TS, Zheng B, Huang KW, López Navarrete JT, Ding J, Kim D, Casado J, Chi C (2016) Benzo-thia-fused [*n*]thienoacenequinodimethanes with small to moderate diradical characters: the role of pro-aromaticity versus anti-aromaticity. Chem Sci 7:3036–3046

93. Anthony J (2008) The larger acenes: versatile organic semiconductors. Angew Chem Int Ed 47:452–483

94. Weil T, Vosch T, Hofkens J, Peneva K, Müllen K (2010) The rylene colorant family-tailored nanoemitters for photonics research and applications. Angew Chem Int Ed 49:9068–9093

95. Shi X, Lee S, Son M, Zheng B, Chang J, Jing L, Huang KW, Kim D, Chi C (2015) Pro-aromatic bisphenaleno-thieno[3,2-*b*]thiophene versus anti-aromatic bisindenothieno[3,2-*b*]thiophene: different ground-state properties and applications in field-effect transistors. Chem Commun 51:13178–13180

96. Clar E (1939) Vorschläge zur nomenklatur kondensierter ringsysteme (aromatische kohlenwasserstoffe, XXVI. Mitteil.). Chem Ber 72:2137–2139
97. Kaur I, Jia W, Kopreski RP, Selvarasah S, Dokmeci R, Pramanik C, Mcgruer NE, Miller GP (2008) Substituent effects in pentacenes: gaining control over HOMO-LUMO gaps and photooxidative resistances. J Am Chem Soc 130:16274–16286
98. Purushothaman B, Bruzek M, Parkin SR, Miller AF, Anthony JE (2011) Synthesis and structural characterization of crystalline nonacenes. Angew Chem Int Ed 50:7013–7017
99. Berg O, Chronister EL, Yamashita T, Scott GW, Sweet RM, Calabrese J (1999) s-Dipentacene: structure, spectroscopy, and temperature- and pressure-dependent photochemistry. J Phys Chem A 103:2451–2459
100. Purushothaman B, Parkin SR, Anthony JE (2010) Synthesis and stability of soluble hexacenes. Org Lett 12:2060–2063
101. Ye Q, Chang J, Shi X, Dai G, Zhang W, Huang KW, Chi C (2014) Stable 7,14-disubstituted-5,12-dithiapentacenes with quinoidal conjugation. Org Lett 16:3966–3969
102. Dong S, Herng TS, Gopalakrishna TY, Phan H, Lim ZL, Hu P, Webster RD, Ding J, Chi C (2016) Extended bis(benzothia)-quinodimethanes and their dications: from singlet diradicaloids to isoelectronic structures of long acenes. Angew Chem Int Ed 55:9316–9320
103. Zeng Z, Lee S, Son M, Fukuda K, Burrezo PM, Zhu X, Qi Q, Li RW, López Navarrete JT, Ding J, Casado J, Nakano M, Kim D, Wu J (2015) Push-pull type oligo(N-annulated perylene)quinodimethanes: chain length and solvent-dependent ground states and physical properties. J Am Chem Soc 137:8572–8583
104. Hu P, Lee S, Herng TS, Aratani N, Gonçalves TP, Qi Q, Shi X, Yamada H, Huang KW, Ding J, Kim D, Wu J (2016) Toward tetraradicaloid: the effect of fusion mode on radical character and chemical reactivity. J Am Chem Soc 138:1065–1077
105. Shi X, Kueh W, Zheng B, Huang KW, Chi C (2015) Dipolar quinoidal acene analogues as stable isoelectronic structures of pentacene and nonacene. Angew Chem Int Ed 54:14412–14416
106. Yamaguchi S, Xu C, Tamao K (2003) Bis-silicon-bridged stilbene homologues synthesized by new intramolecular reductive double cyclization. J Am Chem Soc 125:13662–13663
107. Xu C, Yamada H, Wakamiya A, Yamaguchi S, Tamao K (2004) Ladder bis-silicon-bridged stilbenes as a new building unit for fluorescent π-conjugated polymers. Macromolecules 37:8978–8983
108. Toulokhonova IS, Stringfellow TC, Ivanov SA, Masunov A, West R (2003) A disilapentalene and a stable diradical from the reaction of a dilithiosilole with a dichlorocyclopropene. J Am Chem Soc 125:5767–5773
109. Shintani R, Iino R, Nozaki K (2016) Rhodium-catalyzed stitching reaction: convergent synthesis of quinoidal fused oligosiloles. J Am Chem Soc 138:3635–3638

Top Curr Chem (Z) (2017) 375:73
DOI 10.1007/s41061-017-0163-2

REVIEW

Para-Quinodimethanes: A Unified Review of the Quinoidal-Versus-Aromatic Competition and its Implications

Juan Casado[1]

Received: 3 February 2017 / Accepted: 17 July 2017 / Published online: 31 July 2017
© Springer International Publishing AG 2017

Abstract In this article, some quinoidal *p*-quinodimethanes compounds that convert partially or completely to diradicals or biradicaloids are analyzed. The aromatic/quinoidal balance is revisited with the objective of providing a common interpretation for most of them. For that purpose, important structural and energetic parameters such as the bond length alternation pattern and the singlet–triplet gaps are analyzed and interpreted in the framework of double spin polarization and π-conjugation. *p*-Quinodimethanes based in oligothiophenes, polycyclic aromatic hydrocarbons, oligophenylenes, thienothiophenes, charged dications and cyclic conjugated molecules are discussed. There are excellent reviews in the field of singlet diradicals; however, a revision similar to that proposed here can help the reader to have another perspective on these promising new functional materials. The focus has been put on molecules which are well known by the author and another of relevance in the field. In this regard, the article finishes with a discussion of some important applications of these diradicals in organic electronics. New chemical systems based on the *p*-quinodimethane building blocks are waiting us around the corner, bringing us new and challenging structures and fascinating novel properties, which describe a very rich field of research in chemistry and in physics with an excellent present and a bright future.

Keywords Diradicals · Quinoidal-to-aromatic transition · Bond length alternation · Singlet–triplet gap · Double spin polarization · Applications of diradicals

---

Chapter 5 was originally published as Casado, J. Top Curr Chem (Z) (2017) 375: 73. DOI 10.1007/s41061-017-0163-2.

✉ Juan Casado
   casado@uma.es

[1] Department of Physical Chemistry, Faculty of Science, University of Málaga, Campus de Teatinos s/n, 29071 Málaga, Spain

1 Introduction and electronic structure concepts of *p*-quinodimethanes

The pyrolysis of *p*-xylene [1–4] produces the reactive *p*-xylylene intermediate in Scheme 1, an unstable molecular species that represents the first member of the benzene-based *p*-quinodimethane family, or, in general, *p*-quinodimethanes (i.e., abbreviated in this article as *p*-QDM). Such as formulated in its closed-shell configuration, *p*-QDM in Scheme 1 has a typical quinoidal structure for the benzenoid ring and, therefore, a resonant structure with a net gain of aromaticity in the benzene at the expense of the rupture of one double bond can be formulated, a "broken" structure with diradical character also drawn in Scheme 1 [5]. This general chemical formula in which a central quinoidal ring (i.e., one or more rings, either of benzoaromatic or heteroaromatic nature) is flanked in the para positions by exocyclic double bonds (i.e., often substituted) is the focus of this review article. *p*-QDM is highly reactive and has been stabilized only at very low −78 °C temperature using particular solvent conditions [6, 7]. Upon reaction, this reactive diradical intermediate gives rise to a series of very interesting chemical products such as [2] paracyclophanes by dimerization [8–10], or poly-*p*-xylylene by polymerization. The attention of the reader is brought into the competition between the quinoid and diradical aromatic resonant forms in brackets in Scheme 1 as this highlights the main point of this paper.

The most famous *p*-QDM derivatives, and fathers of the *p*-QDM chemistry, are the Thiele and Chichibabin's hydrocarbons in Scheme 2 first prepared at the beginning of the last century [11, 12]. Nowadays, more than hundred years since their discoveries, a full understanding of the electronic structure of these molecules has not been completely achieved and a persistent debate of some particular aspects of their spin configurations still persists. The underlying diradical character of these molecules and their inherent reactivity is the main cause that hinders the full elucidation of the unique characteristics of the electronic structure of the Thiele and Chichibabin hydrocarbons.

Herewith, some of the most relevant and recent *p*-QDM derivatives are reviewed with especial emphasis on those molecular systems, where the participation and consequences of the diradical structure in their electronic and functional properties has been described. Today, a number of excellent reviews of quinoidal *p*-QDM molecules are available to the reader [13–21]. Most of these molecular systems will be presented here again, however, the distinctive point now will be the attempt to rationalize and unify the common aspects of their quinoidal/diradical fingerprints trying to explain the origin of the tuning, from molecule to molecule, of properties such as the singlet–triplet gap and the bond length alternation parameter.

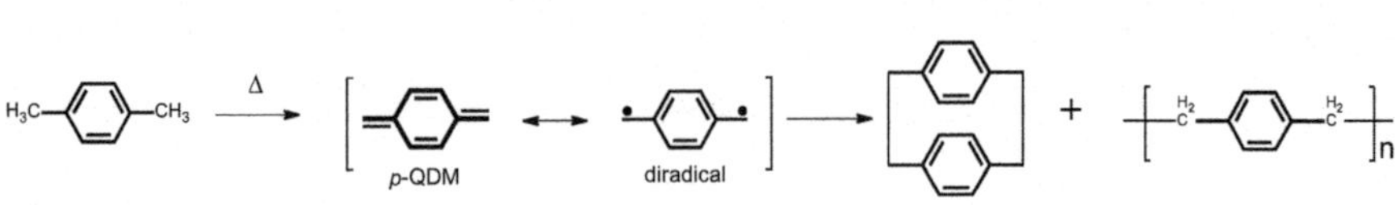

Scheme 1 Pyrolysis of *p*-xylene giving the intermediate *p*-QDM

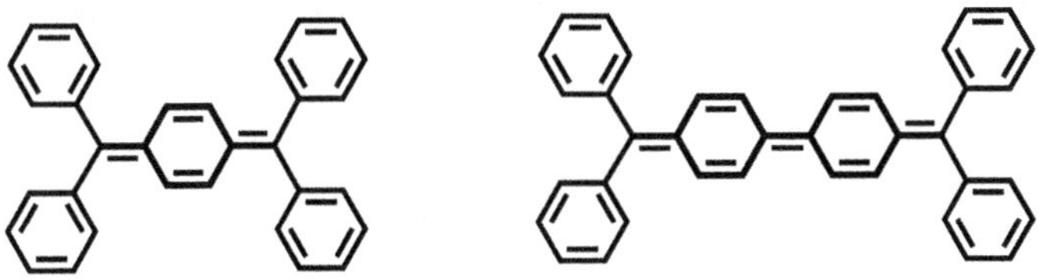

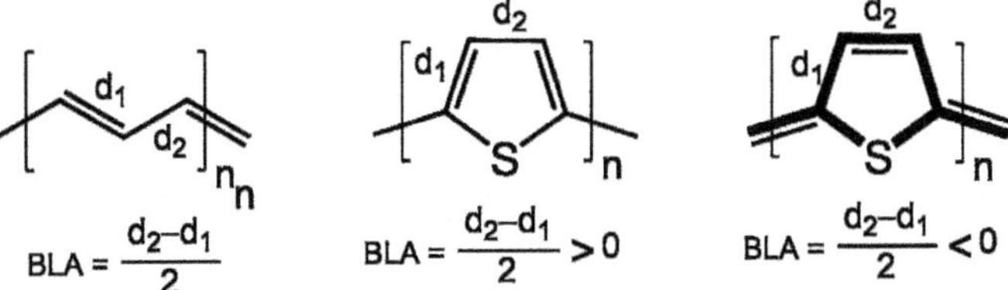

Scheme 2 Thiele (*left*) and Chichibabin (*right*) hydrocarbon molecules

1.1 Bond Length Alternation Parameter

In the history of conducting or conjugated polymers, their structural and electronic features have been understood, among others, in terms of the bond length alternation structural parameter, or BLA in short [22]. The BLA definition is reminiscent of the studies of aromaticity as a structural index to account for the degree of aromatic character [23]. The application of this concept to π-conjugated molecules is straightforward as larger π-conjugation tends to equalize the distances of alternant bonds. The extension of the BLA definition to polyconjugated conducting polymers took place with the pioneering conducting polymer, polyacetylene, where its BLA responds to the average value of the bond distances of alternant single and double bonds (Scheme 3 for the mathematical definition).

For polyacetylene, Fig. 1 shows the potential energy curve of the ground electronic state along the BLA reaction coordinate (identical to the BLA parameter). It is by representing the potential energy curve versus the BLA that emerges the typical features of polyacetylene associated with the appearance of two minima of the same energy, which describe the structure of the polymer under reversal of the BLA (points A and B in the curve in Fig. 1) [24]. This discussion exemplifies the pertinence of the BLA parameter to characterize the structure of polyacetylene, and by extrapolation of other π-conjugated polymers and π-conjugated oligomers. Given that the two A/B structures of polyacetylene are equally probable (same energy), they are dynamically interconverting through the rupture of one double bond giving way to the formation of a diradical structure which corresponds to the singular point at BLA $= 0$ (point C) in the potential energy curve of polyacetylene. In other words, there is a fraction of the chain of polyacetylene with bond length equalization giving rise to segments in the polymer with full π-electron delocalization and radicaloid character, the so-called soliton states in polyacetylene [25].

Scheme 3 Definition of BLA for polyacetylene and for a conjugated polymer (polythiophene) in its aromatic (BLA > 0) and quinoidal (BLA < 0) structures; d_1 and d_2 correspond to the distances of the two representative consecutive bonds

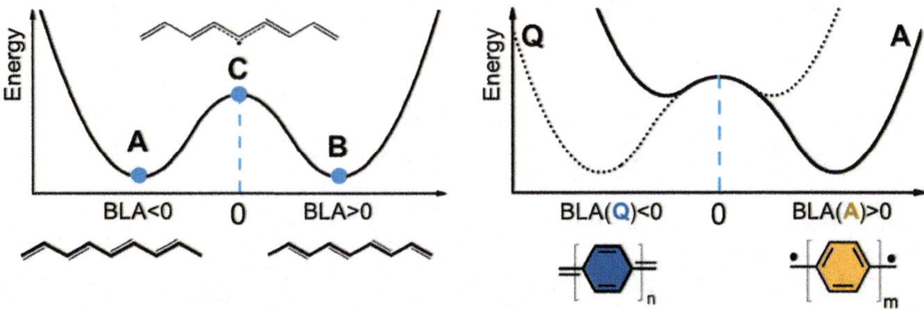

Fig. 1 Two minima representation of the ground electronic state potential energy curve as a function of the BLA parameter for polyacetylene (*left*) and for aromatic or hetero-aromatic polymers (*right*). For the latter the aromatic curve is represented by *A* (*solid line*, right minimum, BLA > 0), whereas the quinoidal curve is represented by *Q* (*dotted line*, left minimum, BLA < 0). The two representations are valid for the corresponding oligomers, and the case of the evolution of the electronic structure of phenylene-based *p*-QDM oligomers is included, where $n < m$ exemplifies the quinoidal to aromatic conversion with the increment of the rings

In the case of the ground electronic state of aromatic and heteroaromatic oligomers and polymers, also in Fig. 1, the two minima of the potential energy curve no longer have the same energy and both appear separated by a gap [26]. Going from BLA = 0 (maximum of the curve), the lowest energy minimum is the result of the aromatic stabilization in the repeating monomer structure and, thus, corresponds to the *aromatic* form. The other minimum from BLA = 0 pertains to the *quinoidal* mode of inter-ring π-electron delocalization (BLA < 0). The separation energy gap is the difference between the aromatic (i.e., BLA > 0) and the quinoidal stabilizations that are connected through the BLA = 0 in which both stabilization modes disappear. From the point of view of the BLA, the aromatic structure has a positive value whereas the quinoidal structure is characterized by a negative one, thus, from one to another the BLA is reversed (see Scheme 3). Contrarily to polyacetylene, the absolute values of these BLAs in the minima of the energy curve of the heteroaromatic polymers are not the same, since they are the result of two different stabilization modes. Hence, the use of the BLA parameter is a very convenient and simple way to characterize π-electron conjugated polymers and oligomers in terms of participation of aromatic and quinoidal forms and, therefore, to address the quinoidal→aromatic conversion.

In the particular case of the formation of aromatic diradicals from quinoidal *p*-QDM closed-shell molecules, as a result of the increasing number of quinoidal rings, the smaller oligomers (Fig. 1, *n* small) are mostly closed-shell quinoidals. For larger values, $n \rightarrow m$ (with $n < m$) the molecules progressively transform into diradicals featuring aromatic positive BLAs. At intermediate values, the systems go through the vicinity of bond length equalization. In this way, it is rather straightforward to visualize the oligomer size at which the transformation takes place. Furthermore, it is also possible to represent how the molecules undergo the quinoidal→aromatic conversion as a function of the functionalization, or depending of the nature of the rings for different families of compounds, etc. Finally, we will also try to establish correlations between these data and the singlet–triplet energy gaps [27–29].

1.2 Singlet–triplet energy gap and double spin polarization

In the scale of Fig. 2, large negative values of BLAs mean that we are far from bond length equalization in the quinoidal side, and residual signs of diradical character are expected for such π-conjugated systems. These type of molecules will have all their π-electrons paired (i.e., bonded) mostly featuring defined closed-shell singlet configurations in their ground electronic states (Fig. 2, BLA ≪ 0).

When we approach BLA→0 from the quinoidal side, and pass it (BLA > 0), the two highest energy electrons (which occupy the HOMO orbital in the closed-shell configuration) now pass to reside in singly occupied molecular orbitals (i.e., SOMOs) which are of a disjointed character highlighting no direct spatial overlap. However, the electronic configuration of the ground electronic state (triplet or singlet) along the whole range of the BLA reaction coordinate does not change (see the S_0 reference line in Fig. 2) indicating that the low spin singlet configuration is always the ground electronic state. This can be understood by assuming that the unpaired electrons of the diradical occupy disjoint orbitals, placed in different molecular regions, but keeping an antiparallel arrangement of their spins. In other words, in the bond length equalization border, and surpassed, the chemical bond between the highest energy electrons of the diradical weakens, given that their

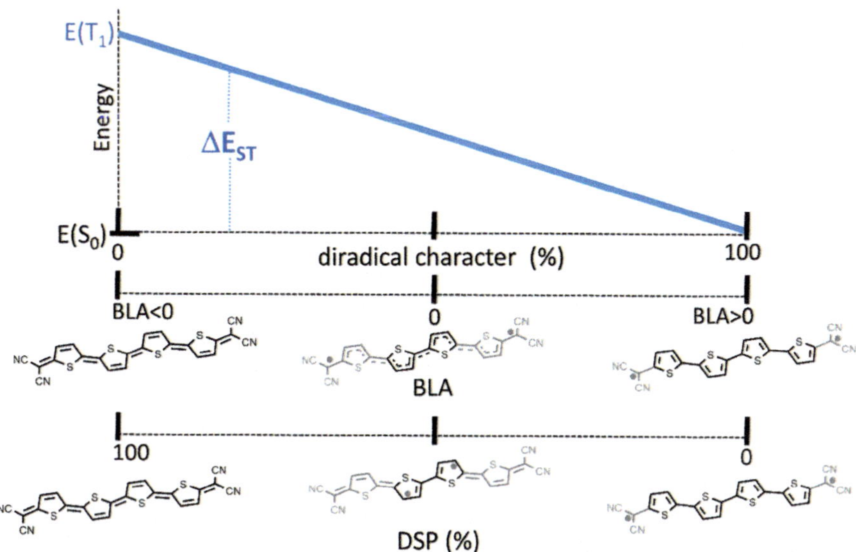

Fig. 2 Variation of the diradical character and of the bond length alternation parameter (BLA) along the conversion from a closed-shell molecule (quinoidal, BLA < 0) to a diradical species (aromatic, BLA > 0) and variation of the singlet–triplet gap (ΔE_{ST}). A prototypical quinoidal tetracyanoquaterthiophenequinodimethane that converts to diradical is taken as an example. The criteria of signs for the BLA are those in Fig. 1

SOMO wavefunctions progressively reduce their overlap, but they keep an effective through bond interaction, even at rather large distances in the biradicaloid form, which is responsible for the stabilization of the singlet.

For these type of diradicals, it is important to evaluate the energy of the first triplet excited state. The placement of the electrons in disjoint orbitals imparts a progressive mitigation of their electron repulsion. Since the first triplet excited electronic state (T_1) differs in energy from the singlet ground electronic state (S_0) in the electron repulsion integral extended over the overlap region, then the energy difference between the S_0 and T_1 is the largest in closed-shell molecules and continuously decreases as the overlap tends to nullify. In the strictest sense, when the ground electronic state configuration is a singlet, it means that some (indirect, through the bridge) bonding interaction exists, and; therefore, the triplet is always slightly higher in energy. When this interaction collapses to zero, the S_0 and T_1 are degenerate and the ground electronic state should be better referred to as a double doublet (biradicals or fully independent electrons, 100% biradicaloid character in Fig. 2). Therefore, the singlet–triplet energy gap (i.e., ΔE_{ST}) is another intrinsic property of the singlet diradical species, which is inversely proportional to the degree of diradical character (larger gaps indicate preferential closed-shell structures and small values indicate large diradical character. ΔE_{ST} is directly correlated with other diradical indexes.

The stabilization in these kind of diradicals of the singlet configuration regarding the triplet violates the Hund's rule of highest multiplicity for the ground electronic state of a distribution of unpaired electrons. In order to explain this apparent contradiction, and to quantitatively evaluate ΔE_{ST}, the phenomenon of double spin polarization (i.e., abbreviated as DSP, also called dynamic spin polarization) enters into play [30–34]. DSP is in essence a conjugation or delocalization stabilizing effect that takes place between the π-electrons of the doubly occupied molecular orbitals (HOMO$'$, HOMO-1$'$, etc.) and the SOMOs containing the unpaired π-electrons (see Fig. 3).

For the singlet diradical configuration in Fig. 3, it turns out that the electrons of the HOMO$'$ can delocalize and occupy the SOMO vacancies according to the Pauli exclusion principle; hence, two possible π-delocalization effects can be drawn, one of the β-electron of the HOMO$'$ on the SOMO occupied by an α-electron and another by the α-electron of the HOMO$'$ on the SOMO containing the β-electron. However, for the triplet configuration, only one stabilizing effect is accessible consisting of the delocalization of the α-electron of the HOMO$'$ on one of the two SOMOs containing β-electrons. One quickly realizes of the double stabilization of the singlet compared with only one in the triplet and of the reason why the singlet is always the configuration of smaller energy (and thus the ground electronic state) of most of these p-QDM diradicals. The association of DSP with a π-conjugation effect is pertinent, since it provides us with well-known arguments to evaluate the extent and efficacy of DSP, and by extension of ΔE_{ST}, in terms of the nature of the conjugating bridge, conformation and length of the bridge, etc.

It is interesting to consider DSP from the point of view of the resulting molecular structures. This is represented in Fig. 4 where four hypothetical situations are schematically depicted for a prototypical p-QDM oligothiophene: (1) the closed-

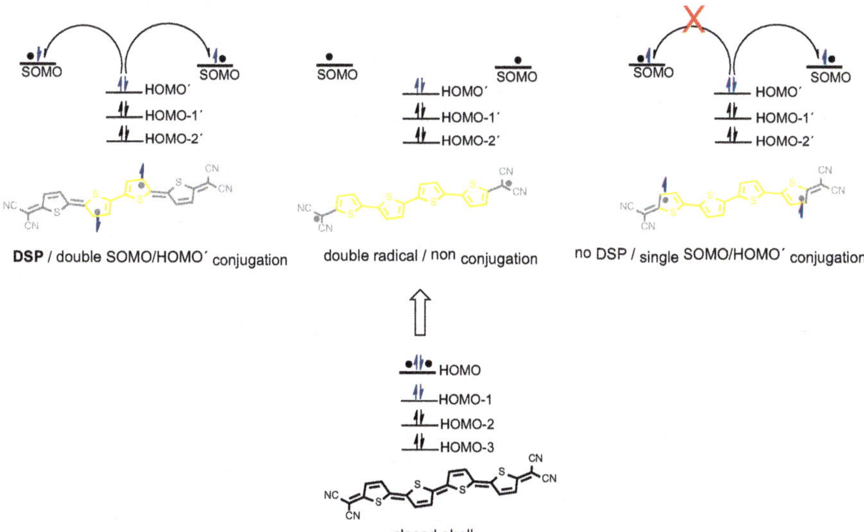

Fig. 3 Molecular orbital view of DSP where the α and β electrons in the SOMO are represented by arrows oriented either with up or down headings. A hypothetical quinoidal tetracyanoquaterthiophenequinodimethane (*closed-shell*) that converts to diradical is taken as an example. Note the differentiation between the HOMO and HOMO' orbitals, which do not correspond to each other, in fact, HOMO' in the open-shell relates to the HOMO-1 in the closed-shell configuration

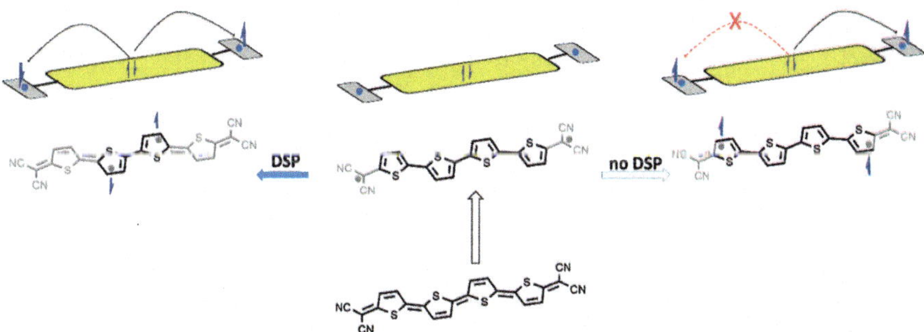

Fig. 4 DSP and chemical structures. A prototypical quinoidal tetracyanoquaterthiophenequinodimethane that converts to diradical is taken as an example

shell situation, which corresponds to the quinoidal tetracyanoquinodimethane quaterthiophene (full π-bonding), (2) the full biradical state, with noninteracting radicals (nonconjugation), (3) the structure of the singlet diradical with DSP π-conjugation, and (4) the corresponding triplet without DSP π-conjugation. Since this oligothiophene diradical in Fig. 4 is formed by aromatization of the thiophene rings, the larger the DSP, the larger the quinoidization and the smaller the diradical character.

This article, thus, commences with a brief introduction to quinoidal *p*-QDMs, its presentation and main discoveries, followed by the definition and description of the important structural parameter BLA in connection with the ΔE_{ST} and the phenomenon of DSP. In what follows, a unified analysis of the variation of their BLA (when available) and of their ΔE_{ST} in terms of the factors that affect and modulate DSP will be discussed. Such factors are the length of the spacers, the conformation and chemical nature of the bridge between the radicals and the substitution of the radical centers with other stabilizing groups, to cite just a few. This will allow the reader to have a unified understanding of the electronic structure of a set of important polyaromatic diradicals. Hence, it is not the central objective of the present article to review all existing cases formed from quinoidal p-QDMs but some comparative studies of BLA and ΔE_{ST} on those stable diradicals. Along the whole manuscript, the term diradical, biradicaloid and biradical can be found, the criterion for which is the value of the singlet–triplet gap (i.e., J): diradicals or biradicaloids are used for those with small J, while biradical is preferred for those with vanishing values of J and for triplets. The article will close with some promising applications and prospects of the use of these molecules in organic electronics.

1.3 Substituted p-quinodimethane

The basic quinoidal structure of p-QDMs can be modified in several ways: (1) by replacing the hydrogen atoms in the exocyclic bonds; (2) by extending the central pro-aromatic spacer with more than one ring; and (3) by changing the nature of the rings. The most popular and known derivatization of p-QDMs is by replacing the exocyclic H atoms by cyano groups giving way to tetracyanoquinodimethane abbreviated as TCNQ and shown in Scheme 4 [35–37]. TCNQ became very popular in the seventies of the last century when combined with the electron-donor tetrathiafulvalene allowed the formation of organic ionic solids and charge transfer salts, which resulted in the first organic-based electrical conductor [38–41]. This discovery took place in the preludes of the appearance of the electrically conducting polymers, with polyacetylene as the pioneering actor [42]. At the origin of the ability to form ionic salts of TCNQ with electron donor is the great propensity to stabilize anions given the recovery of aromaticity by the quinoidal benzene upon reduction. Indeed, the modulation by chemical design of the aromatic character of these rings and of the electron affinity in order to create improved conductors was

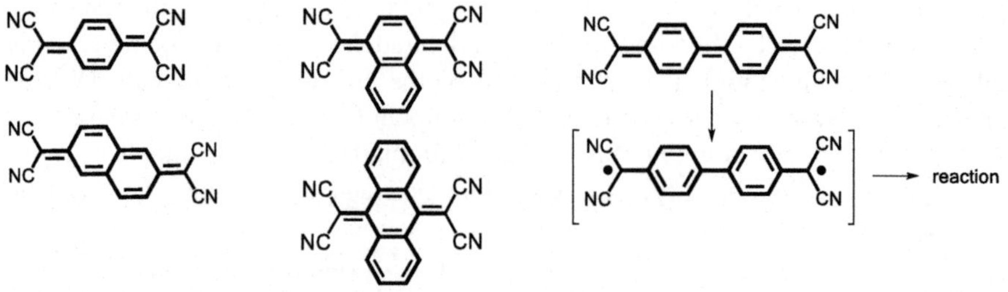

Scheme 4 Tetracyanoquinodimethane and derivatives [43–47]

the driving force for the further development of the chemistry of TCNQ derivatives during the next decades of the 1980s and the 1990s. [43–47].

If one interrogates about the differential stability of *p*-QDM and TCNQ, one should argue in terms of the presence of more electron-withdrawing terminal groups, which are able to largely stabilize the closed-shell configuration versus the reactive *p*-QDM. However, passing from TCNQ to the derivative with a central quinoidal biphenyl group, the stabilization by aromatization of two benzenes surpasses the energy required to break a double bond by which a diradical is formed (Scheme 4). Nonetheless, this biphenyl diradical derivative, in analogy with the Chichibabin hydrocarbon, is also highly unstable and reactive [43]. Alternatively, derivatives of TCNQ with naphthalene and anthracene central groups, either linearly or perpendicularly (i.e., tetracyanonaphtha- and anthra-*p*-quinodimethanes in Scheme 4), have been prepared as stable dyads [48]. The smaller aromaticity of naphthalene is much smaller than that of biphenyl (aromatization of two benzenes) and is the cause of the chemical robustness and absence of diradical character on these *p*-QDMs. These TCNQ acene derivatives are all well-defined quinoidals molecules with singlet preferential closed-shell electronic configurations, pronounced negative BLA values and, therefore, featuring large ΔE_{ST} gaps.

2 Extended *p*-quinodimethanes

The initial development of *p*-QDM molecules was made by replacing the central benzene ring by other aromatic heterocycles, standing out the inclusion of thiophene, or tetracyanothienoquinodimethanes (i.e., *p*-QDMTn in Scheme 5). Thiophene derivatives provided the suitable platform for the development of the first consistent series of extended *p*-QDMs; in fact, up to today, the tetracyanothienoquinodimethane series is one of those with the largest number of members, although it has taken almost 20 years for these to see the light. The first element

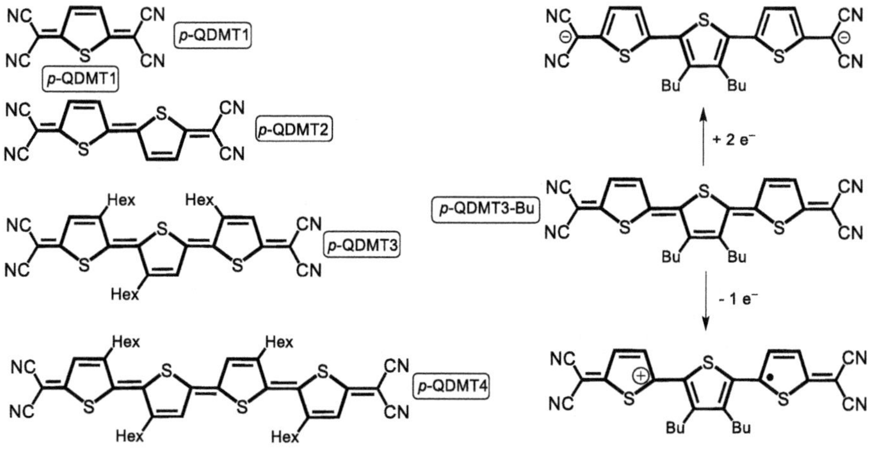

Scheme 5 Tetracyanothienoquinodimethanes, *p*-QDMTn (*n* = number of thiophenes)

Springer

with one thiophene (i.e., *p*-QDMT1) was prepared by Gronowitz and Uppström in 1974 [49]. Conversely, in the case of the biphenyl-based quinoidals, the tetracyanobithienoquinodimethane derivative (i.e., *p*-QDMT2), prepared by Otsubo and Ogura in the late 80s, displayed great chemical stability owing to its robust closed-shell quinoidal structure, much more stable than the diradical canonical form owing to the much smaller aromaticity of the thiophene [50, 51]. Quickly after, the tetracyanoterthieno and quaterthienoquinodimethane derivatives (i.e., *p*-QDMT3 and *p*-QDMT4, see Scheme 5) were synthesized [52, 53]. The turning point on the research of these TCNQ thiophene derivatives came across with the discovery of amphoteric redox behavior (i.e., simultaneous presence of stable/reversible oxidation and reduction electrochemical processes) in *p*-QDMT4 in 2002 [54]. The manifestation of this type of amphoteric redox behavior is realized when the molecule contains a balanced composition of electron donors and electron acceptors, for instance, *p*-QDMT4 with four quinoidal thiophene rings and four cyanos. This finding triggered the preparation of new tetracyanoterthienoquin-odimethane molecules (i.e., *p*-QDMT3-Bu in Scheme 5) by the Minnesota's group (Larry L. Miller and Kent R. Mann) which also showed amphoteric redox behavior [55]. *p*-QDMT3-Bu was implemented in a thin film organic field-effect transistor and showed the best electron mobility (*n*-type behavior) at that time, an effect correlated with the presence of the tetracyano electron accepting groups and with the recovery of aromaticity in the thiophenes upon reduction (Scheme 5). Both effects confer great electron affinity to *p*-QDMT3-Bu at the origin of the *n*-type or electron transport.

The same group, one year latter, reported *p*-QDMT3-Bu as the first case of ambipolar charge transport (both *p* and *n*-conduction) in a purely organic thin film field-effect transistor [56], an electrical performance based on the ability to stabilize charges of different sign, in both cases by aromaticity recovery on the thiophene backbone (Scheme 5). The discovery of ambipolar conduction greatly urged on the research of thiophene-based tetracyano substituted quinodimethanes and, since then, a significant increase in publications of advanced material applications based on these compounds has been realized (some of these applications will be reviewed in Sect. 9) [57–67].

In 2005, the tetracyanoquinquethieno and hexathienoquinodimethanes (i.e., *p*-QDMT5-CP and *p*-QDMT6-CP) were successfully prepared by the group of Tetsuo Otsubo, Kazuo Takimiya and Yoshio Aso [68]. These two compounds are the most challenging members of the first and largest series of homogeneous (same substitution pattern, see Scheme 6) of tetracyanothienoquinodimethane compounds, existing from a monomer to an hexamer [69]. *p*-QDMTn-CP's represents a set of molecules with very interesting structural, electronic and spectroscopic properties. If these molecules are depicted in their quinoidal forms in Scheme 6, one would expect a progressive increment of π-conjugation with the increment of the number of rings (i.e., π-electrons), an electronic property that translates into a progressive reduction of the optical gap (wavelength red-shift of the absorption bands) accompanied by strong coloration of the samples. In the whole series, a significant red-shift of the absorption maximum (λ_{max}) of approximately 100 nm per ring is observed (412 nm in *p*-QDMT1-CP, 548 nm in *p*-QDMT2-CP, 657 nm in

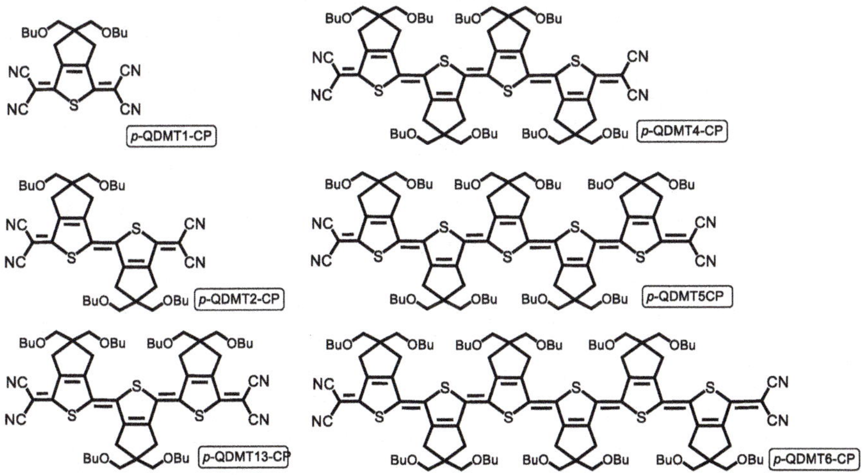

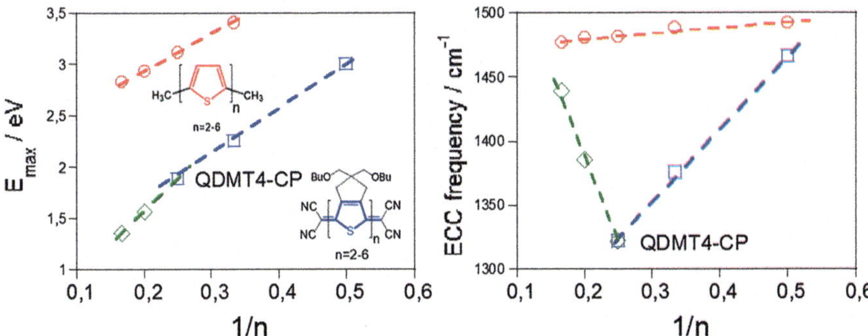

Scheme 6 The largest series of tetracyanothienoquinodimethanes, *p*-QDMTn-CP (*n* = number of thiophenes) from a monomer to a hexamer

p-QDMT3-CP, 788 nm in *p*-QDMT4-CP, 913 nm in *p*-QDMT5-CP and 1012 nm in *p*-QDMT6-CP) [69]. The chain length evolution of the energies of the band maxima (E_{max}) is represented in Fig. 5 together with those E_{max}'s of the same lowest energy lying electronic absorption bands, corresponding to the dimethyl oligothiophenes of the same sizes (from a trimer to an hexamer) which are characterized by an aromatic structure in their thiophene rings [70]. For the *p*-QDMTn-CP quinoidal compounds, the evolution of E_{max} with size displays two linear tendencies, one from the dimer to the tetramer, and another up to the hexamer; with the tetramer placed at the tuning point of the two linear fits.

Fig. 5 *Left* representation of the energy maxima (in eV) of the strongest electronic absorption bands of the aromatic oligothiophenes (dimethyl oligothiophenes, see insert, from a dimer, *n* = 2, to a hexamer, *n* = 6, red circles) and of the tetracyanoquinodimethane oligothiophenes (QDMTn-CP, from a monomer, *n* = 1, to a tetramer, *n* = 4, *blue squares* and from the tetramer, *n* = 4, to the hexamer, *n* = 6, *green squares*) versus the reciprocal of the number of repeating units (1/n). *Right* the same for the wavenumber (in cm^{-1}) of the maxima of the strongest Raman bands in both aromatic and quinoidal series

Trying to explain this, Fig. 5 also shows the behavior of the frequency of the most intense bands of the Raman spectra of both series of quinoidal [71] and aromatic oligthiophenes [72]. These strongest Raman bands are assigned to C=C/C–C skeletal bond stretching vibrations of the conjugated path, which correspond to the so-called ECC modes in the Effective Conjugation Coordinate theory, or ECC theory [73]. The ECC vibrational mode is defined as a fully in phase stretching vibration of the successive C=C–C bonds of the delocalized over most of the conjugated structure. Due to the nature of these vibrations, the associated frequencies are markers of the degree of bond length alternation among the successive single and double bonds of the π-conjugated path resulting in a spectroscopic observable of the degree of π-conjugation in the ground electronic state. Because in these quinoidal and aromatic molecules, π-conjugation is the result of the competition between intra-ring aromaticity and inter-ring π-electron delocalization, the frequencies of these Raman bands are the outcome of the "fight" between these two opposite electronic forces. For the aromatic dimethyl oligothiophenes, from the dimer to the hexamer, the Raman frequencies progressively downshift with the enlargement of the molecular length [72]. For the p-QDMTn-CP quinoidal compounds, two differentiated trends are observed, one from the dimer to the tetramer, with a clear frequency downshift, such as that shown by the aromatic oligothiophenes and that denotes the progressive increment of π-conjugation with the increment of the molecular length in both series of compounds. Conversely, from the quinodial tetramer to the hexamer, the frequencies of the main Raman bands up-shift, with the value in p-QDMT6-CP approaching that of the aromatic dimethyl hexathiophene. This behavior gives an account of the formation in p-QDMT4-CP of a structure (Scheme 7) where a double bond of the full quinoidal canonical form is broken allowing the aromatization of the thiophene rings and the formation of a diradical species. Once the diradical is formed in the tetramer, up to the hexamer, aromatization of more rings originates the upshift of the Raman ECC frequencies (Scheme 7).

The X-ray molecular structures of the p-QDMTn-CP compounds are not available due to their amorphous character, and; thus, their BLAs cannot be experimentally obtained. Alternatively, the close correlation between the behavior of the BLA parameter and that of the ECC Raman frequencies can be used to represent the p-QDMTn-CP oligomers in the BLA energy curve already discussed in the Introduction [74]. This is represented in Fig. 6, as follows: p-QDMT2-CP is well assumed to display a mostly quinoidal structure given the insufficiency of the aromaticity gaining by two thiophenes to break a double bond and, therefore, p-QDMT2-CP is placed in the quinoidal side of the curve (BLA < 0). The frequency downshift of the Raman ECC mode from p-QDMT2-CP to p-QDMT3-CP and to p-QDMT4-CP reveals the increment of π-conjugation, which means that the BLA of these compounds in absolute values is decreasing with the enlargement of the chain. Then, this trimer and tetramer can be placed in the BLA energy curve closer to bond length equalization in the quinoidal part of the diagram of Fig. 6. On passing to p-QDMT5-CP and p-QDMT6-CP, the Raman frequency upshifts as a result of the aromatization of the thiophenes, and; therefore, these compounds should be located in the aromatic branch of the BLA energy curve (BLA > 0) with p-QDMT6-CP

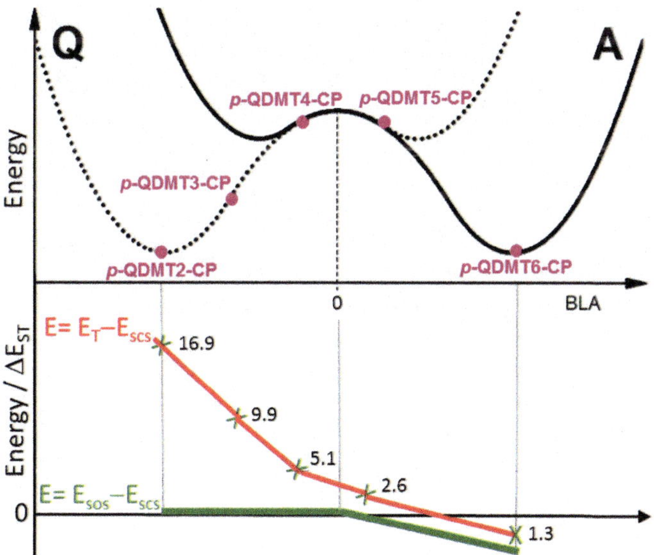

Scheme 7 Quinoidal to aromatic conversion and formation of a singlet diradical species in *p*-QDMT4-CP and enlargement of the aromatic structure of the diradicals in the longer compounds according to the Raman frequency behavior

Fig. 6 *Top* two wells generalized BLA energy curve with *purple circles* showing the values of the BLA according to the Raman shifts. This *curve* represents a generalized curve since it connects all minima of all compounds in comparison (each compound resides in its own absolute potential energy minimum and the generalized curve connects of all of them). *Bottom* evolution of the ΔE_{ST} gap in kcal/mol calculated at the DFT/(U)B3LYP/6-31G** (exact values are shown) for the *p*-QDMTn-CP series

bearing the largest BLA values. In this representation we see that, either in the quinoidal or aromatic wells, when π-conjugation increases, the BLA (as deduced on the basis of the Raman frequency) moves towards BLA $= 0$. Hence, on p-QDMT2-CP$\rightarrow p$-QDMT4-CP and on p-QDMT6-CP$\rightarrow p$-QDMT4-CP π-conjugation increases, in nice correspondence with the interpretation of the formation of a diradical species in p-QDMT4-CP that strengthens its aromatic character passing to p-QDMT5-CP and p-QDMT6-CP.

We are in position now to comparatively evaluate the evolution of the ΔE_{ST} gap in this p-QDMTn-CP series such as represented in Fig. 6. For the mostly closed-shell quinoidal p-QDMT2-CP, the ΔE_{ST} is the largest on the series, given the full bonding situation of all their π-electrons [71]. Passing to p-QDMT3-CP and p-QDMT4-CP, the ΔE_{ST} decreases since the progressive reduction of the BLA (in absolute terms) and appearance of appreciable diradical character in the tetramer. In the cases of p-QDMT5-CP and p-QDMT6-CP, the ΔE_{ST} is further reduced since the accentuation of the diradical structure. In this case, the larger aromatization of the thiophene backbone, the smaller the ΔE_{ST}. In all these molecules, due to the pattern of conjugation through the para-positions of the p-QDM units, the ground electronic state always has a singlet configuration as all of them have a portion of bonding (large in the closed-shell forms and smaller in the others) which is in essence the π-conjugation represented by DSP. As a result, the ΔE_{ST} is reduced in the series given the "dilution" of the π-conjugation (shared by more rings) of the central bridge towards the radical sites.

The case of p-QDMT6-CP is particularly interesting, given that it shows the smaller ΔE_{ST} value in the series by which a certain population of triplets should be present in the thermally equilibrated sample at room temperature, such as confirmed by EPR spectroscopy [68]. Owing to this small population of triplets, its Raman spectrum was obtained by resonance Raman spectroscopy at room temperature and compared with that of the singlet diradical ground electronic state. The Raman frequency of the ECC mode of the triplet of p-QDMT6-CP is at 1487 cm^{-1}, a frequency value typical of aromatic oligothiophenes (see Fig. 5 for dimethyl oligothiophenes) further reaffirming the aromatization effect in the diradical/biradicaloid. On the other hand, the corresponding Raman frequency of the singlet diradical of p-QDMT6-CP is at 1439 cm^{-1}, which is significantly smaller than that of the triplet. This 1487/1439 cm^{-1} difference in the triplet/singlet biradicaloid of p-QDMT6-CP can be accounted for by DSP as summarized in Scheme 8. If we consider vanishing conjugation (null DSP effect) of the radical sites with the bridge, the situation could be described by a canonical form with the radicals placed on the carbon atoms of the dicyanomethylenes. Now, if DSP is active, therefore delocalization of the bridge electrons (HOMO$'$, HOMO-1$'$, etc.,) over the radical SOMOs will occur, provoking a partial quinoidization of the external thiophenes and the localization of the radicals inside the hexathiophene bridge. This quinoidization effect would be larger as stronger is DSP and the larger the bonding character of the unpaired electrons is, reaching the ultimate case of full pairing and full quinoidization (closed-shell). The frequency downshift (upshift) from $1487 \rightarrow 1439$ cm^{-1} ($1439 \rightarrow 1487$ cm^{-1}) is revealing the larger quinoidization

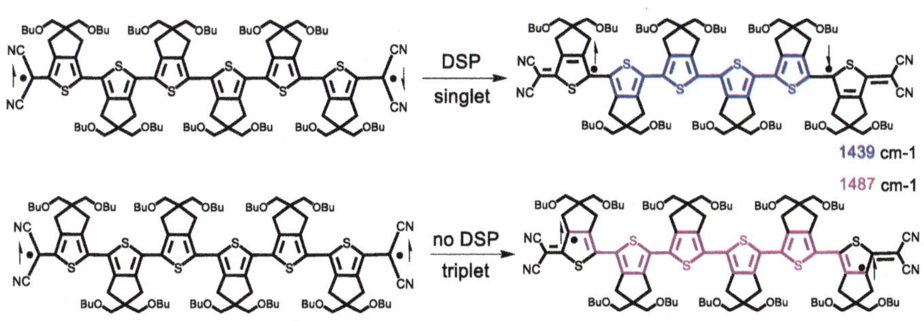

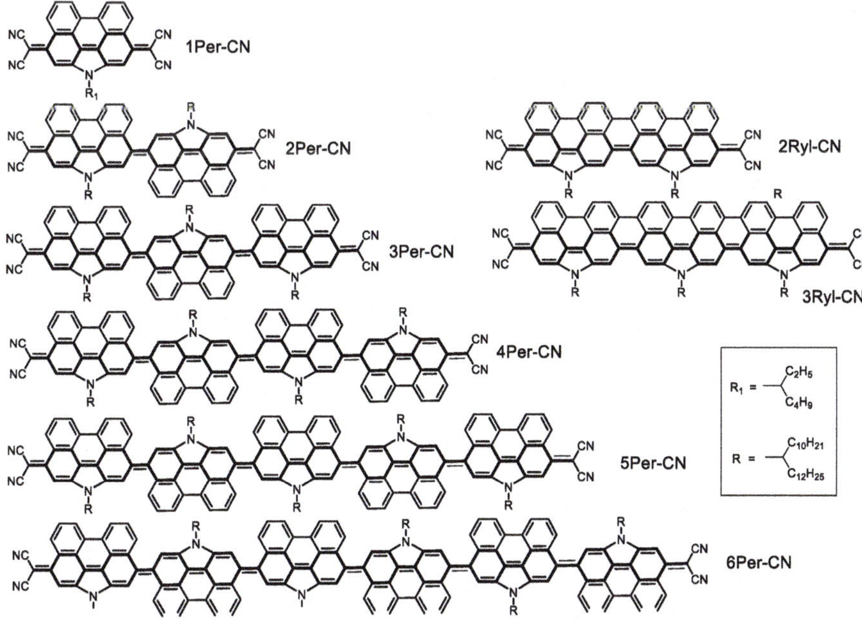

Scheme 8 Qualitative representations of the chemical structures of the singlet diradicals (*blue*) and triplets (*purple*) of *p*-QDMT6-CP according to DSP and from the Raman spectra

(aromatization) of the singlet diradical (triplet) relative to the triplet (singlet) of *p*-QDMT6-CP by DSP.

Another interesting series of extended *p*-QDM quinoidal molecules in which DSP dictates the nature and structure of the ground electronic state is represented by the tetracyanooligoperylenequinodimethanes (nPer-CN with $n = 1$–6) and the tetracyanooligorylenequinodimethanes (nRyl-CN, with $n = 1$–3) in Scheme 9 [75–78]. The shortest member of the series, 1Per-CN, can be identified by mostly having a closed-shell electronic shape, where the recovery of aromaticity in the perylene unit is unable to surpass the C=C rupture energy. The case of 2Per-CN is transitory between quinoidal closed-shell and aromatic diradical, whereas for $n > 2$

Scheme 9 Chemical structures of the tetracyanooligoperylenequinodimethanes (nPer-CN, $n = 1$–6) and tetracyanooligorylenequinodimethanes (nRyl-CN, $n = 2$–3) the number of benzenes is large enough to break the quinoidal structure and to generate a full diradical species. Nonetheless, not all oligoperylenes with $n > 2$ are identical: for $n = 2$–4, π-conjugation of the terminal unpaired electrons with the central oligoperylene bridge is significant so that the singlet ground electronic state for these diradicals is clearly preferred by DSP. As a result, their bridges disclose increasing aromatic character in the perylenes from 2Per-CN to 4Per-CN. However, for 5Per-CN and 6Per-CN, the recovery of aromaticity in the rings between dicyanomethylenes is extensive and liberates inter-perylene conformational rotation provoking distortions and overall lack of planarity (1-4Per-CN show rather flat structures, Scheme 10). In this scenario, DSP is only effective in the terminal rings and the bonding effect between the unpaired electrons is residual or zero.

The collapse of DSP by conformational distortions in these longer oligomers makes the Hind's rule to be valid, and these molecules show triplet ground electronic states, which is the case of 5Per-CN and 6Per-CN. The comparison with the p-QDMTn-CPs allows us to see that the number of pro-aromatic rings in the largest perylene molecule, 6Per-CN, is twice that of QDMT6-CP by which the two radical centers are largely separated causing decoupling of the radical π-conjugation and null bonding. p-QDMT6-CP is comparable to 3-4Per-CN, and the three display singlet open-shell character. It is interesting to compare the ΔE_{ST} values of the two series in Fig. 7: at the B3LYP/6-31-G**, for 2Per-CN is 0.755 kcal/mol (experimental, 0.342 kcal/mol), for 3Per-CN, 0.042 kcal/mol (experimental 0.107 kcal/mol) and for 4Per-CN, 0.002 kcal/mol. The B3LYP/6-31-G** values of ΔE_{ST} are: 5.070 kcal/mol for p-QDMT4-CP, 2.600 kcal/mol for p-QDMT5-CP and 1.350 kcal/mol for p-QDMT6-CP, revealing that they are larger in the thiophene series due to the smaller aromaticity of thiophene, which facilitates π-conjugation and DSP and increased ΔE_{ST}. This confirms how benzenoid spacers isolate the radicals by restricting inter-radical π-conjugation.

Another follow-up case is that of tetracyanooligorylenequinodimethane compounds in Scheme 9, which are the fully fused analogues of the oligoperylenes. In this case the quaterrylene (2Ryl-CN) and hexarylene (3Ryl-CN) are compared with the 2Per-CN and 3Per-CN, respectively. A priori, since the planar format favour π-conjugation one would expect more effective DSP in these rylenes and, therefore, larger singlet–triplet gaps: this is confirmed since $\Delta E_{ST} = 4.21$ kcal/mol in 3Ryl-CN is one order of magnitude larger than in 3Per-CN [78].

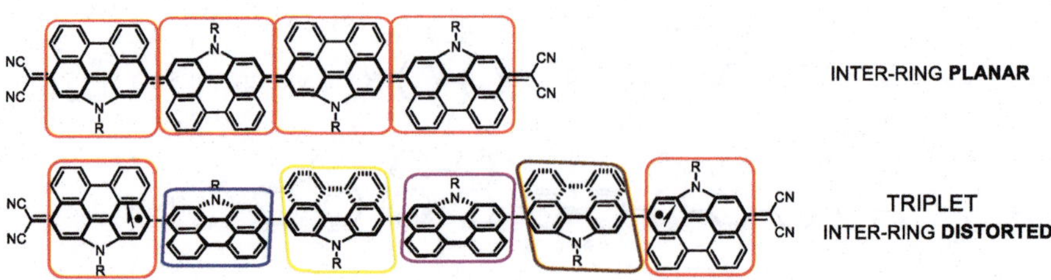

Scheme 10 Tentative chemical structure for the triplet ground electronic state of 6Per-CN

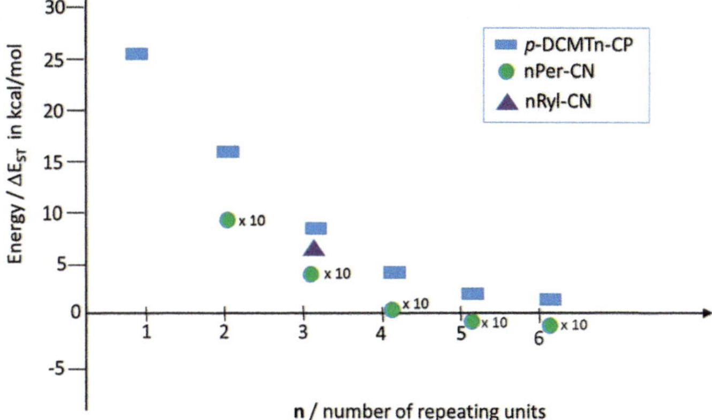

Fig. 7 Comparative ΔE_{ST} obtained at the DFT/B3LYP/6-31G** level of calculation for the three series of quinoidal tetracyanoquinodimethanes (for the nPer-CN values are multiplied by 10)

3 Polycyclic *p*-quinodimethanes

The *p*-QDM quinoidal unit has been also exploited inserted in polycyclic aromatic hydrocarbons (i.e., PAH) in several forms, such as benzo-, naphto- and anthra-quinodimethanes. One of the most interesting systems of this class is the family of zethrenes in Scheme 11. Zethrene is a PAH with six benzenoid rings resulting from the condensation of two phenalenyl groups through the generation of a stable *trans*-diene unit in the rings of connection [79–84]. This diene moiety can be converted into a *p*-QDM by inserting a quinoidal ring between the double bonds of the *trans*-diene giving way to a system with seven benzenoid rings or heptazethrene [85, 86].

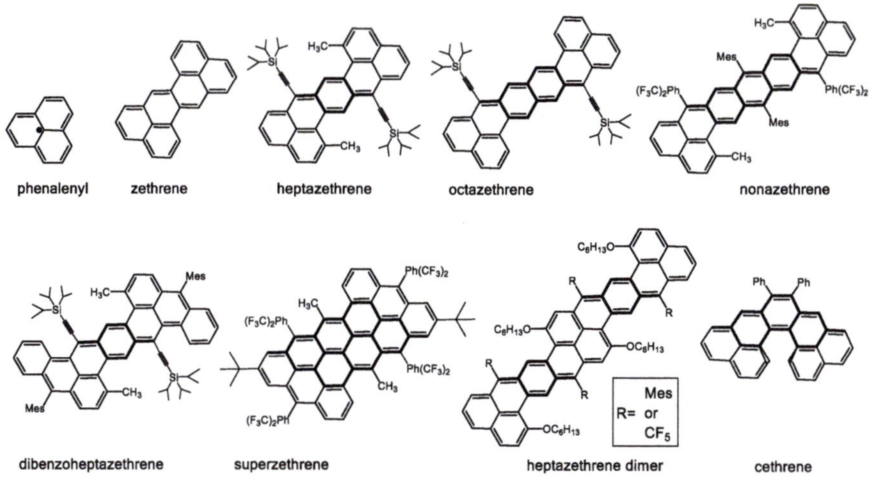

Scheme 11 Chemical structures of the zethrene structures. *Top*, from the left, phenalenyl, zethrene, heptazethrene, octazethrene and nonazethrene. *Bottom*, from the left, dibenzoheptazethrene, superheptazethrene, heptazethrene dimer and cethrene

Such as in many other embedded or protected *p*-QDM, heptazethrene is stable, and the electronic configuration of the ground electronic state is a predominantly closed-shell singlet characterized by a moderate ΔE_{ST} which preludes the conversion to a singlet diradical in the next extended zethrene. This is octazethrene, where the central unit is now a *p*-naphtoquinodimethane, which is already able to recover the aromaticity, hence, stabilizing a diradical as its ground electronic state [86]. In most of the cases, the aromatization of a naphthalene is insufficient to produce a persistent diradical. The unexpected stabilization of the naphto-aromatic form in octazethrene might be related to the additional stabilization of the radical centers by π-conjugation in the external rings. The DFT/(U)B3LYP/6-31G** ΔE_{ST} for octazethrene is 4.4 kcal/mol (experimental, 3.87 kcal/mol).

The situation in octazethrene shares with previous tetracyanoquinodimethanes the fact that the singlet diradical configuration is preferred regarding the triplet owing to DSP and π-conjugation with the central core again (see blue arrows in case A of Scheme 12). However, octazethrene is different because the two radicals might be additionally stabilized by single spin polarization with the external rings (see red arrows in case A of Scheme 12). Delocalization of the radicals in the outer rings gives rise to the diradical canonical form B of Fig. 8. In the case of B with the radical centers more separated from the *p*-QDMs, DSP (blue arrows) occurs through within a larger central core by which resonance is locally confined to the terminal sites of the bridge (i.e., such as in *p*-QDMT6-CP) translating into a decreased ΔE_{ST} compared with the hypothetical case of A with DSP being extensive through the central naphthalene.

Recently, the anthra *p*-QDM zethrene derivative, nonazethrene in Scheme 11, has been prepared, which also shows a diradical ground electronic state owing to the recovery of the aromaticity in the central anthracene [87]. The configuration of the ground electronic state is a singlet again due to DSP within the anthracene. However, the experimental $\Delta E_{ST} = 5.2$ kcal/mol, is slightly higher than in the smaller octazethrene. In this case, octazethrene and nonazathrene have rather

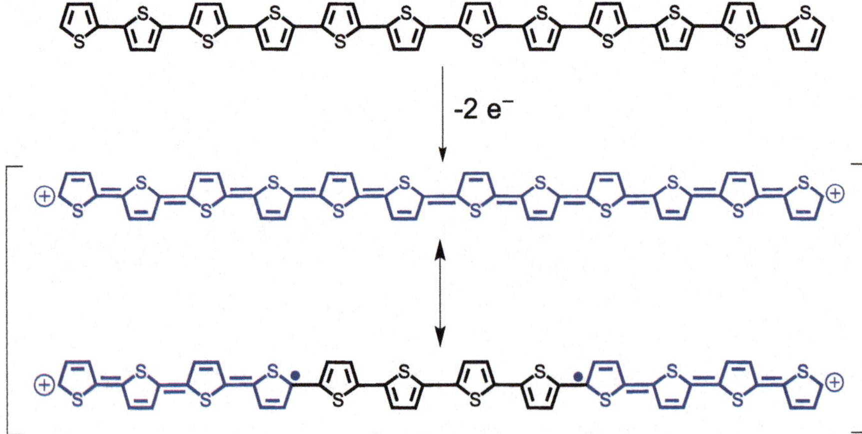

Scheme 12 Structures of neutral 12T and of the dication upon oxidation

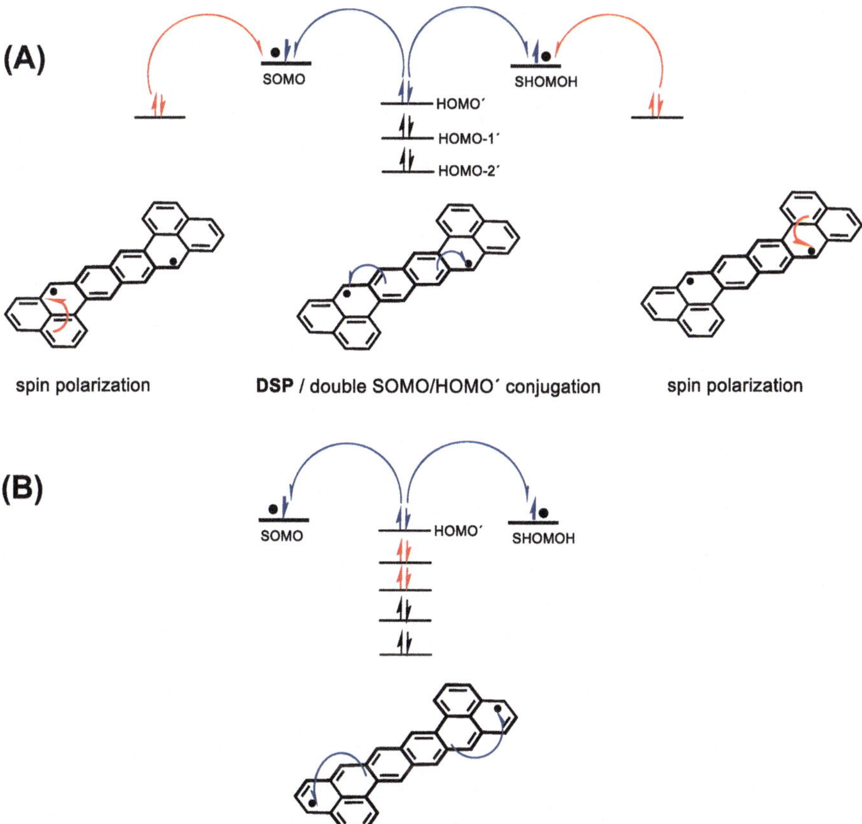

Fig. 8 DSP and spin polarization in octazethrene. Note the differentiation between the HOMO and HOMO′ orbitals, which do not correspond each other, in fact, HOMO′ in the open-shell relates to the HOMO-1 in the closed-shell configuration

different substitution patterns, which could influence the ΔE_{ST}. The derivative of heptazethrene in which the external rings are extended with more benzenoid rings, or dibenzoheptazethrene in Scheme 11, shows a smaller ΔE_{ST} of 3.7 kcal/mol regarding heptazethrene (closed-shell) further indicating better π-electron delocalization in the external rings that reduces the gap by competitive spin polarization [88].

The condensadion of two heptazethrenes gives rise to a new member or superheptazethrene (Scheme 11) which resulted in a net diradical species with a very small $\Delta E_{ST} = 0.94$ kcal/mol, disclosing a ground electronic state of singlet character but with the triplet state very close in energy [89]. In terms of π-conjugation and DSP, the presence of pseudo-cross-conjugated paths perpendicularly disposed in relation to the inter-radical conjugated core describes a situation of interference in the inter-radical π-conjugated core, which perturbs DSP and reduces the gap. Furthermore, such as in dibenzoheptazethrene, the presence of extended π-platforms at the molecular periphery increases spin delocalization that also reduces

Springer

the gap. Finally, the isomer of heptazethrene, cethrene in Scheme 11, has been prepared, which shows a diradical ground electronic state, with a singlet triplet gap of 5.6 kcal/mol in accordance with the impeded π-conjugation along a *cis* disposition of the double bonds in the core [90].

4 Dicationic *p*-quinodimethanes

In analogy with the tetracyano *p*-QDM oligomers, aromatic oligomers upon oxidation to the dication state (and partially in the radical cation) also generate quinoidal structures (Scheme 12) [91]. Dicationic species of oligothiophenes [92–95] and oligoparaphenylene-vinylenes [96–101] are typical examples, which have been broadly characterized. When a given number of quinoidal rings is accumulated in the dicationic structure, these can eventually recover the aromaticity allowing the development of biradicaloid structures with a ground electronic state configuration and singlet–triplet balance very similar to that previously discussed for neutral tetracyano *p*-QDM's [94].

It is interesting to compare the dications of oligothiophenes with the neutral tetracyanooligothiophenes. Ones and others can be viewed as quinoidal structures stabilized at the terminal positions by dicyanomethylene or carbocation groups, respectively. In the particular case of oligothiophenes, it has been reported that the diradical species is more stabilized by the dicyanomethylne substitution than by the carbocation as the latter needs many more pro-aromatic rings to break the closed-shell quinoidal structure [102]. In fact, $12T^{+2}$ in Scheme 12 is the first oligothiophene dication with a well-defined diradical structure. This can be indicating by the greater electron-withdrawing character of the cyano moieties that better stabilizes the radical center and its inherent charge. As already mentioned, whereas tetracyanothienoquinodimethanes turn out to diradicals at the level between the tetramer and the pentamer, in the case of the dicationic oligothiophenes, the quinoidal-to-aromatic transformation occurs at some point between the octamer and the decamer (the oligothiophene dodecamer dication, $12T^{+2}$, already is extensively biradicaloid). The values of DFT/(U)B3LYP/6-31G** for the ΔE_{ST} are 7.39 kcal/mol for the dication of the sexithiophene (this sexithiophene dication has a predominately closed-shell structure) which compares with the 1.35 kcal/mol of *p*-DCMT6-CP. Further, in the case of the dication of 12T, the theoretically estimated ΔE_{ST} is 0.24 kcal/mol, revealing the open-shell diradical/biradicaloid structure of the singlet ground electronic state.

Terminally functionalized dications have been also reported to be good examples of competition between the quinodimethane and the diradical structures. This has been the case of bis(amino) functionalized π-conjugated oligomers which, upon two-electron oxidation, generated the quinoidal structures, similarly to the case of Scheme 12, that might eventually break in favour of the aromatic diradical species. It must be mentioned the nice examples of thiophene-vinylenes reported by Barlow [103–105], the dications of disubstituted bis(arylamino) oligophenylenes (i.e., phenylene, bisphenyl and terphenyl derivatives) by Wang [106–109] and those with phenylene-vinylenes bridges [113, 114]. The discussion is now focused on the

Wang's dicationic in Scheme 13 as the experimental values of ΔE_{ST} and of BLAs are available.

In many of these cases, the formation of diradical species is reported in all of them having singlet ground electronic states resulting from DSP of the unpaired electrons centred in the nitrogens (Fig. 9). Interestingly, there is a sizeable dependence of the singlet diradical character and of ΔE_{ST} as a function of: (1) the increase in the number of quinoidal or pro-aromatic units in the central bridge between the nitrogens, such as on $\mathbf{1}^{2+}$ (predominantly closed-shell) $\rightarrow \mathbf{2}^{2+}$ (open-shell, $\Delta E_{ST} = 2.8$ kcal/mol) $\rightarrow \mathbf{3(Me)}^{++}$ (open-shell, $\Delta E_{ST} = 2.5/1.1$ kcal/mol); and (2) the conformation of the central bridge relative to the nitrogen atoms highlighted by the insertion of steric crowding in the central units, such as on $\mathbf{1}^{2+}$ (predominantly closed-shell) $\rightarrow \mathbf{1Naph}^{2+}$ (predominantly closed-shell) $\rightarrow \mathbf{1Anth}^{2+}$ (open-shell, $\Delta E_{ST} = 1.4$ kcal/mol) $\rightarrow \mathbf{1Me_4}^{++}$ (open-shell, $\Delta E_{ST} = 1.1/0.54$ kcal/mol). Interestingly, in $\mathbf{2}^{2+}$ (open-shell, $\Delta E_{ST} = 2.8$ kcal/mol) $\rightarrow \mathbf{2(Py)}^{2+}$ (open-shell, $\Delta E_{ST} = 2.8$ kcal/mol) the planarization of the biphenyl group in the pyrene derivative and the enrichment of the π-electron structure with lateral double bonds seems to cancel each other [110]. In addition, the analogue of $\mathbf{2(Py)}^{2+}$ replacing the electron donor nitrogen by boron in the corresponding dianion state, $\mathbf{2(BPy)}^-$, has been shown to bear a predominantly quinoidal structure in the bridge, which can be attributed to the preferred DSP and conjugation through the bridge, given the impossibility of delocalizing the negative charges in the terminal electron donor mesithyls [111, 112]. In these latter cases,

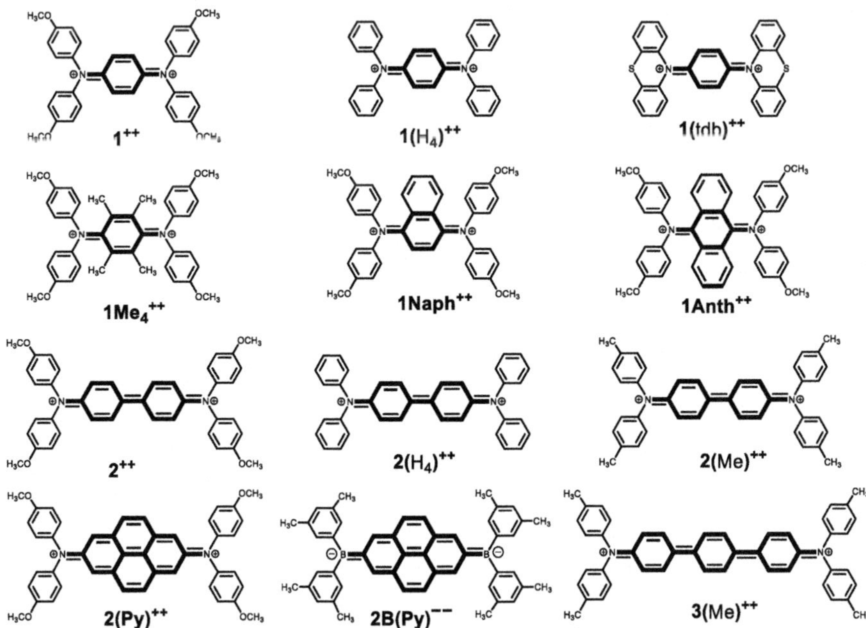

Scheme 13 Chemical structures of the bis(arylamino) oligophenylene dications reported by Wang and resonance between the *p*-quinodiimine and the diradical aromatic triphenyl amine. Compound $\mathbf{2B(Py)}^{2-}$ is the boron based homologue dianion

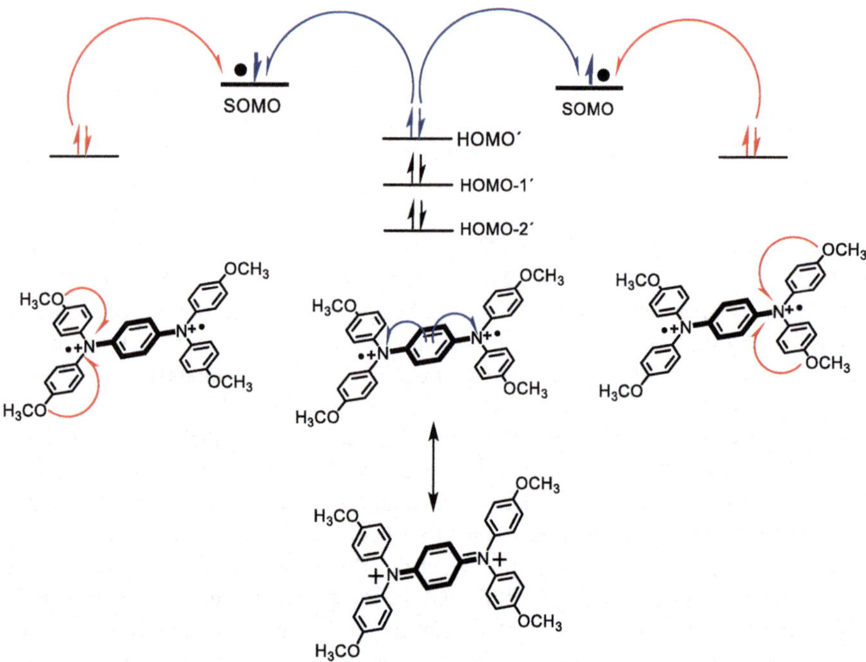

Fig. 9 Chemical structures of the bis(arylamino) oligophenylene dications reported by Wang and resonance between the *p*-quinodiimine and the diradical aromatic triphenyl amine. Note the differentiation between the HOMO and HOMO′ orbitals, which do not correspond each other, in fact, HOMO′ in the open-shell relates with the HOMO-1 in the closed-shell configuration

steric strain forces the twisting around the carbon–nitrogen double quinodiimine bond with the subsequent weakening, facilitating the formation of the open-shell structures (i.e., the energy required to break the bond is smaller). Simultaneously, from the perspective of the diradical structure, this distorted conformation impedes π-conjugation of the unpaired electrons on the nitrogens with the central site. As a result, for the derivative with an unsubstituted benzene in the bridge, a predominantly closed-shell structure is reported. However, for the tetramethyl and anthracene derivatives open-shell diradical species are stabilized. The case of the derivative with a napththalene bridge shows a mostly closed-shell structure as well as due to intermediate steric crowding. On the other hand, these small values of ΔE_{ST} reveal a diminishing of the DSP in these systems, despite the proximity between the radicals as a result of the large sterically induced distortions.

In addition, interestingly, the biradical character and ΔE_{ST} are further modulated by functionalization on the external aryl groups, for instance, on $2(H_4)^{2+}$ (predominantly closed-shell) → $2(Me_4)^{2+}$ (predominantly closed-shell) → 2^{2+} (open-shell, $\Delta E_{ST} = 1.4$ kcal/mol), as well as on $1(H_4)^{++}$ (predominantly closed-shell) → 1^{2+} (predominantly closed-shell) → $1(tbd)^{2+}$ (open-shell, $\Delta E_{ST} = 0.02$ - kcal/mol). Therefore, substitution with π-mesomeric electron donor groups $[2(Me_4)^{2+} → 2^{2+}]$, such as methoxy instead of methyls, in the para-positions of the outermost phenyl rings, reduces significantly the ΔE_{ST}. Furthermore on

$1^{2+} \rightarrow 1(tbd)^{2+}$, the external groups are planarized in such a way that also favour π-conjugation of these moieties with the nitrogen radical centers. This has been explained in terms of favourable spin delocalization on these p-methoxy or tdb groups, which further separate the spin active centers and reduces their antiferromagnetic coupling and ΔE_{ST}. However, in the context of DSP, this delocalization effect might be explained by the consideration again of two competitive double and single spin polarization π-delocalization/π-conjugation paths (see Fig. 9). In this case, spin polarization of the doubly occupied orbitals of the external groups into the SOMOs would compete with DSP in the center, thus reducing the ΔE_{ST}. With the same argument, if the external rings are planarized regarding the nitrogen radical centers, better external π-conjugation and smaller ΔE_{ST} are expected when comparing diradicals with the same bridge.

The case of the open-shell biradical $1Me_4^{2+}$ is particularly interesting because it shows magnetic bistability [105], a rare and highly desirable property for applications of these functional organic materials. According to the authors, the bistability is associated with the diradical character of the dicationic species, which is modulated upon lowering the temperature. In fact, two different ΔE_{ST} were measured at high (i.e., 0.54 kcal/mol) and at low (i.e., 1.06 kcal/mol) temperatures. The authors discuss this in terms of a phase transition associated with two conformations of the two bis(amino) terminal groups regarding the central tetramethyl benzene unit. That at high temperature, corresponds with the external groups more distorted regarding the central unit, a conformation that facilitates π-delocalization of the external groups on the SOMO of the nitrogens restricting or decoupling DSP from the central benzene and reinforcing the diradical character and smaller ΔE_{ST}. Conversely, on cooling, conformational freedom is restricted, the distortion angles decreases and the nitrogen radical more delocalizes on the central ring favouring DSP and enlarging the ΔE_{ST}. Similarly Wang et al. [110] also report the existence on $3(Me)^{2+}$ (open-shell) of two different values of the ΔE_{ST} as a function of the temperature. In this case, the situation is reversed and at high temperatures the gap is larger than at low temperatures, 2.5 and 1.1 kcal/mol, respectively.

The BLA values from the X-ray structures for some of these structures are reported that allows us to correlate the singlet diradical character by means of ΔE_{ST} with the structure of the conjugated bridge, such as shown in Fig. 10. For the predominantly closed-shell compounds, these display negative values of the BLAs and correspond to quinoidal structures with large ΔE_{ST} values (not available experimentally). On the other hand, all open-shell diradical structures are placed on the aromatic well due to the recovery of aromaticity as the driving force for the diradical formation. In addition, they all disclose certainly small ΔE_{ST} as the rings get more and more aromatic as a result of the functionalization.

5 Antiaromatic p-quinodimethanes

The s-indacene chemical unit in Scheme 14 is a paradigmatic antiaromatic compound with 4n (12) Hückel π-electrons [115]. Many derivatives of s-indacene have been prepared exploring more stable alternative structures such as those of

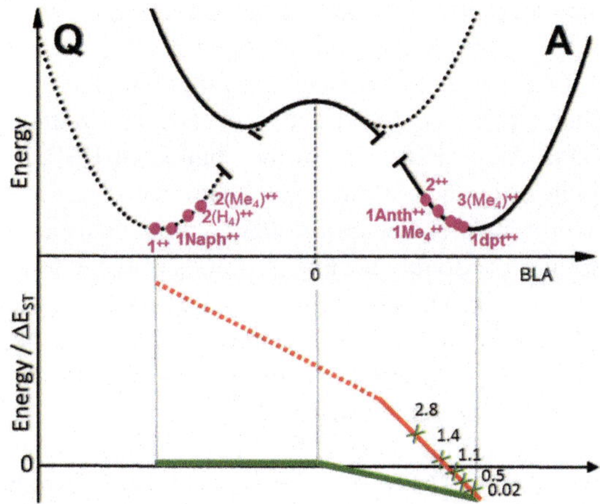

Fig. 10 *Top* two wells BLA energy curve with *purple circles* showing the values of the BLA from the X-ray. Note that the BLAs have been recalculated from the data of bond distances reported, given that our definition in Scheme 3 is different. This *curve* represents a generalized curve, since it connects all minima of all compounds in comparison (each compound resides in its own absolute potential energy minimum and the generalized curve connects all of them). In addition, quinoidal and aromatic BLAs are not to scale. Bottom: evolution of the experimentally reported ΔE_{ST} gaps in kcal/mol for the bis(arylamino) oligophenylene dications in Scheme 13

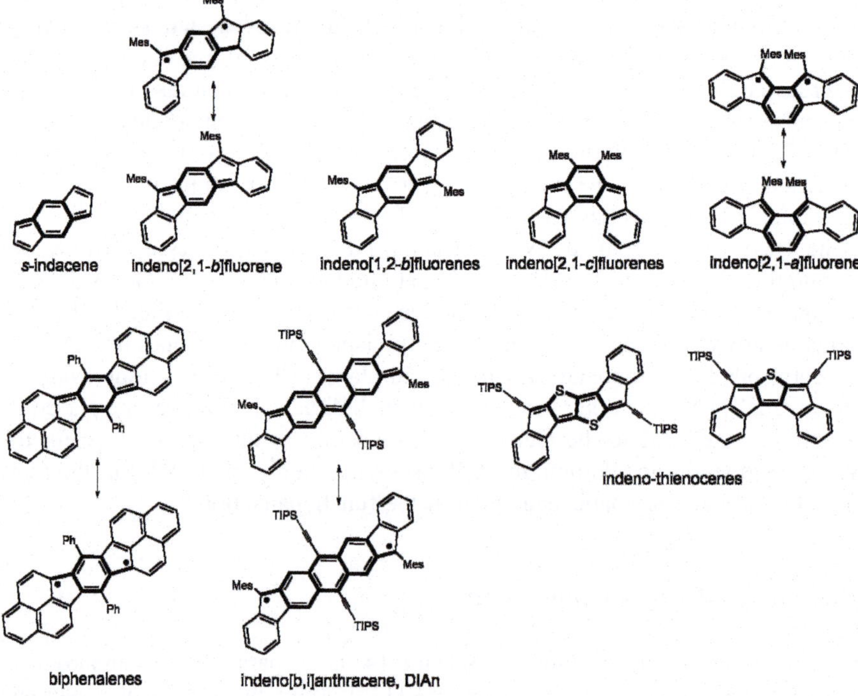

Scheme 14 Chemical structures of the *s*-indacene derivatives

Tobe and Haley. A central p-QDM quinoidal or pro-aromatic unit can be drawn in s-indacene, where the exocyclic double bonds are embedded in the five member rings. This is relevant in order to account for the electronic properties of these molecules given that they can be viewed either as pro-aromatic or as antiaromatic blocks, both of which can yield diradical structures.

Most indenofluorenes correspond to an s-indacene unit with the two external ciclopentadiene rings stabilized with benzene rings such as displayed in Scheme 14. For instance, for the indeno[2,1-b]fluorene compound, the ΔE_{ST} amounts to 4.2 kcal/mol with the singlet as the ground electronic state revealing efficient DSP π-conjugation of the unpaired electrons either through the central benzene ring or through the whole 12 electrons π-belt depending on the origin of the formation of the diradical [116]. An interesting aspect of this molecule is that it is formulated in the diradical canonical form, and it is a $meta$-substituted xylylene which, if the relevant inter-radical conjugation path should act through the central aromatic benzene, then it should display a triplet ground electronic state. Since the ground electronic state is a singlet, this might indicate that the antiaromatic contribution is dominant. In line with this discussion is the detection of singlet diradical character in the indeno[2,1-a]fluorene [117] which contains a $ortho$-QDM and that would stabilize triplet biradicals under the assumption that inter-radical π-conjugation is effective through the aromatic benzene (see next section for further discussion). Since the ground electronic state is again a singlet, this would indicate antiaromatic character as well.

A conveniently substituted indeno[1,2-b]fluorene [118, 119] has also been prepared, which responds to a predominant closed-shell system without any sign of diradical character highlighting maximal bonding through π-conjugation and delineating a molecule better described in its quinoidal canonical form. This is also the case of the analogue indeno[2,1-c]fluorenes with a strained p-QDM unit [120] and derivatives [121].

The π-extended derivative of indeno[1,2-b]fluorene replacing the central p-QDM by a p-anthra-QDM (abbreviated as DIAn in Scheme 14) has been recently reported [122]. Such as in the cases of its homologues, the electronic structure of DIAn can be described as the superposition of an antiaromatic path, as one can count 4n (20) electrons in the periphery (excluding the outermost benzenes), and a quinoidal or pro-aromatic anthracene in the center. From our previous comparison, and extrapolating from the smaller indacene-fluorenes, one should argue that the origin of the diradical stabilizations is the antiaromatic character, however, in the case of the extended DIAn analogue, the overall antiaromatic contribution has been found to be negligible, and the molecule displays an open-shell ground electronic state owing to the recovery of the aromaticity in the central anthracene, with a $\Delta E_{ST} = 4.18$ kcal/mol. The diradical configuration of DIAn converts the molecule in a pseudo-heptacene derivative, from it which takes its excellent performance in OFETs (DIAn behaves as an ambipolar organic semiconductor, see Sect. 9). Heptacene, after hexacene, represents the first member of the acene series in which the singlet diradical state is the ground electronic state [123–125]. However, the intrinsically different origin of the diradical structure in DIAn and in heptacene makes the difference in stability. In fact, heptacene and longer acenes are highly reactive species only characterized at low temperature in isolated phases. DSP by inter-radical π-conjugation over the anthracene moiety in DIAn entitles great stability and allows ambient characterization and implementation in devices.

Now the electronic structure of the indeno[2,1-*b*]fluorene, which is a singlet ground electronic state, can be compared with that of a isomer of heptazethrene, in Scheme 15, which, conversely, has been reported to disclose a triplet ground electronic state [126]. The absence of effective conjugation and DSP between the radicals through the meta connection in the isomer of heptazethrene makes the triplet the ground electronic state, which is in analogy with *meta*-substituted xylylenes. Such as mentioned above, the antiaromatic nature of the diradical species in the indeno[2,1-*b*]fluorene makes the singlet the ground electronic state configuration. Furthermore, in Scheme 15, it is shown the structures of two isomers of octazethrene with the central naphthalene (in the diradical forms) either substituted at the 2,6 positions (*p*-naphthoquinodimethane) or at the 1,5 positions [127]. According to DFT/(U)B3LYP/6-31G**calculations, in the 2,6 derivative, a theoretical E_{ST} of 5.80 kcal/mol (experimental $E_{ST} = 3.81$ kcal/mol) is predicted, whereas for the 1,5 analogue the equivalent theoretical value is 2.10 kcal/mol. Since π-conjugation of the radicals through the central naphthalene is more efficient for the *p*-naphthoquinodimethane derivatives, DSP is also more efficient, and, therefore, the E_{ST} is larger compared to the situation of perturbed π-conjugation through the 1,5 positions.

The case of indeno oligothienoacenequinodimethanes in Scheme 14 is also of particular interest [128]. These are derivatives of indeno[2,1-*c*]fluorene and indeno[1,2-*b*]fluorene where the central *p*-QDM unit has been replaced by *p*-thienoquinodimethane. A similar 4n electron count, such as in the fluorene derivatives, is feasible (for this, one should consider the replacement of the two electrons of a double bond by the two electrons of the lone pair of sulphur). By assuming the role of the antiaromatic path in these indeno-oligothienoacenes, it was possible to explain the evolution of the structural properties (bond length alternation from X-ray analysis) and the vibrational Raman spectra, and also the evolution of the UV–Vis absorption properties as a function of the number of thienoquinoidal units. The Raman spectra of these thienoacene compounds are particularly

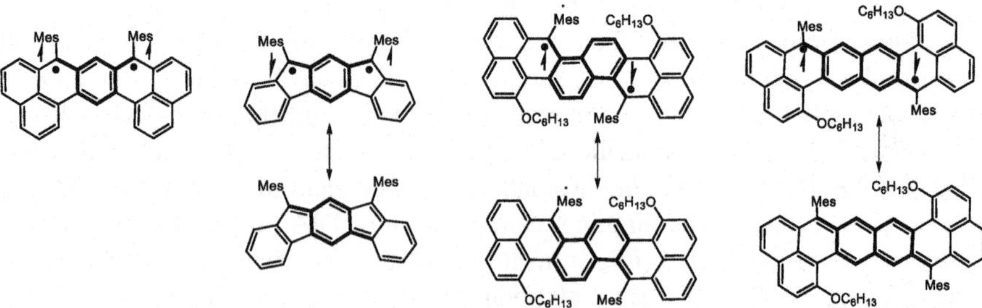

Scheme 15 Chemical structures of: (1) indeno[2,1-*b*]fluorene and a related compound of heptazethrene with triplet ground electronic state both having the same central meta-substituted biradical, and (2) mesithyl octazethrrene with a 2,6 substitution path in the central naphthalene and the derivative of octazethrrene with the substitution path through the 1,5 positions

informative of the evolution of the electronic structure as a function of size. It is particularly noticeable that the large shift of the strong Raman band from the monomer to the trimer contrasts with the much smaller shifts of the same strong Raman bands of the spectra of the aromatic thienoacenes of the same length. This behaviour of the Raman frequencies of the strongest bands in indeno oligothienoacene is a clear indication of the quinoidal π-conjugation, similarly to that described for the tetracyano quinoidal oligothiophenes [128]. In line with this, no diradical character has been detected in these series of indenothienothiophenes up to the element with three fused thiophenes.

Recently, we have reported a tetracyanooligothiophene, which combines the quinoidal p-QDM structure and a 1,6 methane [17] annulene group in the middle (i.e., TMTQ, Scheme 16). A $\Delta E_{ST} = 4.87$ kcal/mol has been measured in TMTQ, which reveals that it corresponds to a singlet diradical ground electronic state [129]. Given the annulene nature of the central group, it is weakly aromatic and the presence of two thiophene rings seems to be insufficient to provide enough driving force to break a double bond forming a diradical. We thus wonder about the origin of the rather small ΔE_{ST}. The dicyano groups are well known as strong electron acceptor moieties and, therefore, an electron-withdrawing effect on the vicinal thiophenes and on the 1,6 methane [17] annulene unit is feasible (Scheme 16). This results in a canonical structure with a dication in the middle part, and the external dicyanos bearing a negative charge (the aromatization of the central part will help to stabilize this contribution). In this zwitterionic canonical form, the planar [17] annulene dication corresponds to an antiaromatic structure (4n π-electrons, with $n = 2$) in which a particular type of aromatic stabilization takes place in the first triplet excited state, or Baird aromaticity [130, 131]. Baird aromaticity is based in the fact that the antibonding nature of the triplet removes two electrons from coupling, thus leaving 6 π-electronically active electrons, which are of a $4n - 2$ $[4(n - 1) + 2]$ Hückel aromatic species, resulting in the fact that the triplet is energetically stabilized. As a result, TMTQ would show a quinoidal singlet closed-

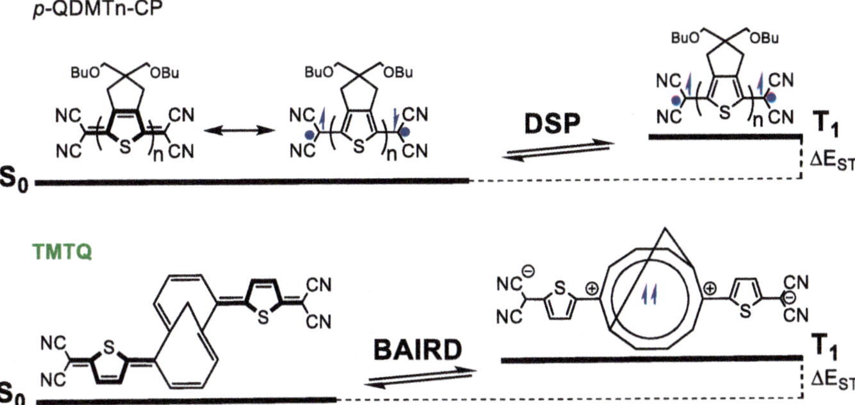

Scheme 16 Chemical structure and canonical forms of TMTQ and comparison of the stabilizing effect of Baird and Hückel aromaticities

shell ground electronic state with a Baird stabilized triplet, which might justify the small ΔE_{ST}. This is an interesting counterpoint example with the cases above in which the singlet diradical ground electronic state is formed by Hückel aromaticity recovery and stabilized by DSP, whereas in TMTQ the singlet is rather unaltered and the triplet stabilized by Baird aromaticity. Recent high level quantum chemical calculations by Ottosson et al. have shown that TMTQ actually behaves as a Hückel–Baird hybrid which receives stabilization by the two types of aromatic stabilization, the singlet by Hückel and the triplet by Baird [132]. These new types of mix aromatic/antiaromatic co-oligomers represent a new avenue for stable molecules with tuned ΔE_{ST}.

6 Cyclic *p*-quinodimethanes

The tetracyclopenta[*def,jkl,pqr,vwx*]tetraphenylene compound (i.e., TCPTP) in Scheme 17 is, to the best of our knowledge, the first cyclically embedded arrays of *p*-QDM's. This has been recently prepared by Tobe et al. and consists of the condensation of two angle-shaped indeno[2,1-*c*]fluorenes in such a way that the outermost benzenes of one unit stabilize the two consecutive *p*-QDMs [133].

Such as mentioned in the previous section, the indeno[2,1-*c*]fluorene building block has a predominant closed-shell structure by which, by extrapolation, the cyclic structure of TCPTP is expected to show negligible open-shell character based on the assumption of the insufficient pro-aromatic driving force of this indeno[2,1-*c*]fluorene. Nonetheless, Scheme 17 shows the possible resonant forms that would give way to the polyradical character based on the aromaticity recovery. Furthermore, the ground electronic state would be singlet in all cases given the

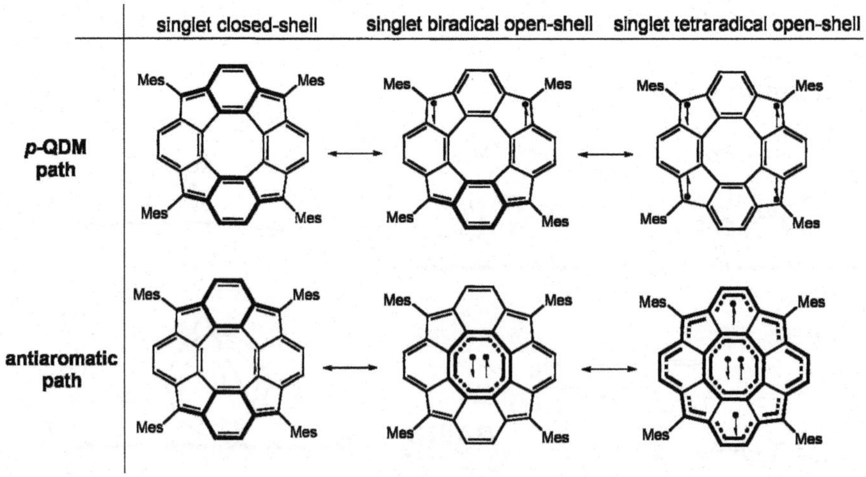

Scheme 17 Chemical structure and canonical forms of cyclic compound TCPTP. Note that in the tetraradical resonant form of the antiaromatic path, the outermost diradical is delocalized over the entire belt, not confined in any particular ring

DSP effect. However, by taking into consideration the central antiaromatic 8 π-electrons cyclooctatetraene ($4n$, with $n = 2$), in its planar conformation, the system could result in a new source of diradical contribution (innermost diradical). In addition, another outermost π-belt containing 28 π-electrons ($4n$, with $n = 7$) can be formulated, which could stabilize another open-shell diradical (outermost diradical). Overall, the complete structure only based on antiaromatic arguments would result in a singlet tetraradical form. This is similar to the case of the singlet open-shell character in acenes, where each zig–zag ribbon accommodates one radical and the two are coupled antiferromagnetically whereas in TCPTP there are two diradicals coupled antiferromagnetically as well [123–125]. These effects are nicely described computationally by the authors according to the progressive conversion along a internal reaction coordinate (typical intramolecular conversion of antiaromatic systems).

Given the small diradical character developed by the indeno[2,1-c]fluorene unit in TCPTP, Wu et al. [134] designed new cyclic structures incorporating two and three quinoidal p-QDMs carbazoles alternating with aromatic carbazoles as well (4MC and 6MC in Schemes 18, 19). Compared to the Tobe's TCPTP compound, the driving force for the development of diradical and polyradical character (up to tetraradical in 4MC and hexaradical in 6MC) was significantly higher owing to the biphenyl quinoidal core of the carbazole. In contrast with TCPTP, in 4MC and 6MC the fusion or condensation mode between successive rings of the macrocycle is broken in the carbazole unit since this is a biphenyl rigidified with an electronically

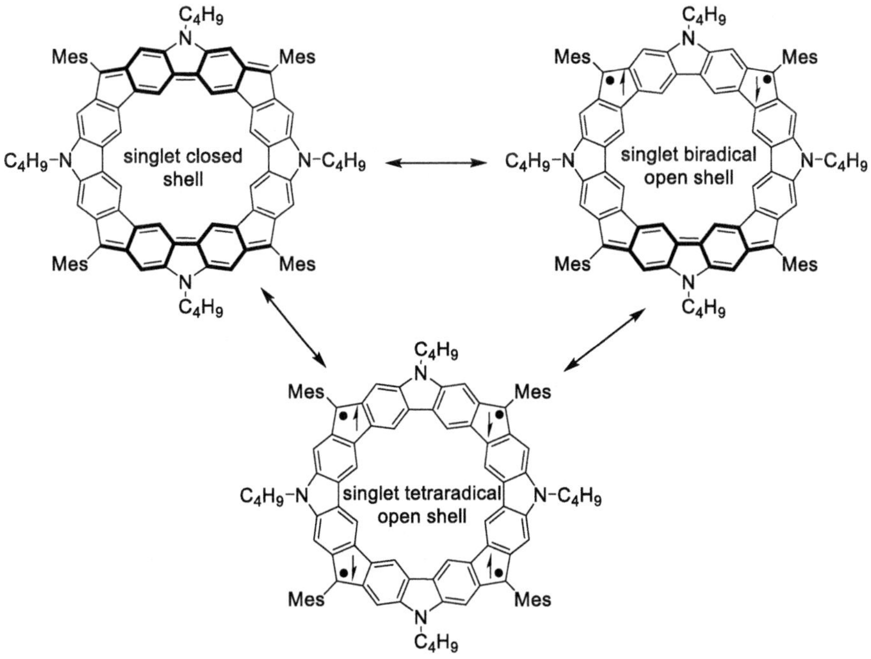

Scheme 18 Chemical structure and canonical forms of 4MC

inactive nitrogen. As a result, it is not possible to write, neither inside nor outside, antiaromatic frameworks, and the only source of radicaloid character comes from the aromaticity recovery in the *p*-QDM units.

Once the multiradical centers are formed (tetraradical and hexaradical in 4MC and 6MC, respectively) maximal conjugation among them would be achieved in the case of the full planar structures in which these systems would display their larger singlet multiradical expression. Acording to quantum chemical calculations, both 4MC and 6MC display bowl-shaped forms in their minimum energy conformations whereas their planar isomeric structures (along the internal reaction coordinate) appear, for example in 4MC, at impassable energies of 40–50 kcal/mol. This energy penalty to get planar is largely decreased in the largest macrocycle, and this barrier in 6MC becomes much smaller (3–4 kcal/mol), revealing the mitigation of strain and allowing developing a significant diradical/tetraradical/hexaradical character. This is important since the presence of tetraradical and hexaradical characters allows the stabilization of the corresponding quintet and heptuplet high spin excited electronic states in cohabitation with the triplet, that is, a cascade of thermal transitions between the singlet/triplet/quintet/heptuplet might be present in 6MC. This is relevant, since if the polyradical character is large, small energy separation

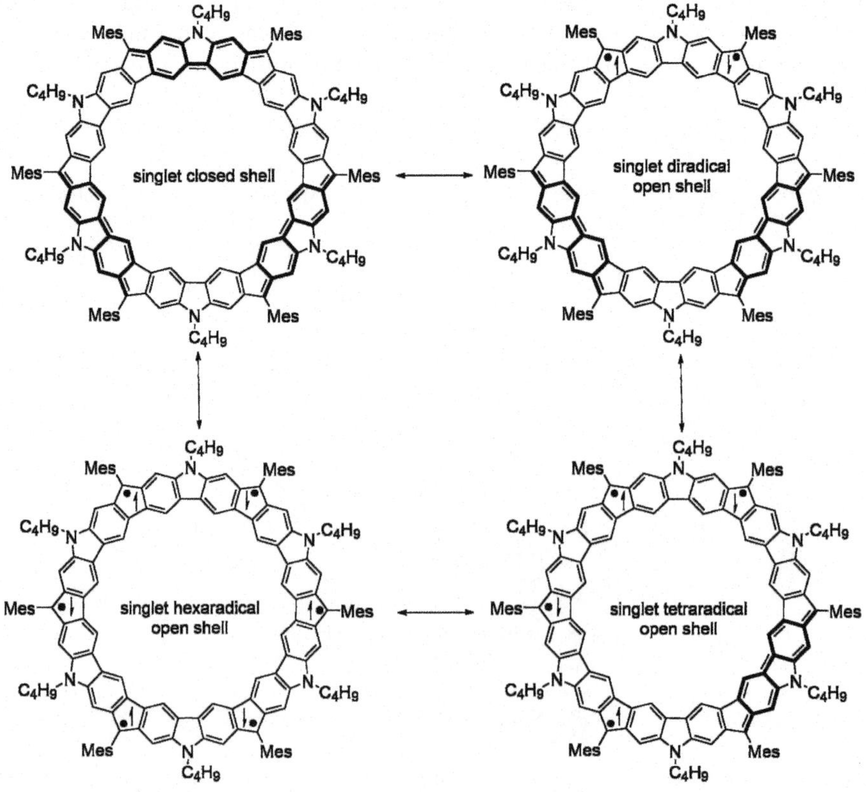

Scheme 19 Chemical structure and canonical forms of 6MC

are expected for the low-to-high spin transitions, thus enriching the magnetic response to external stimuli and permitting envisaging new phenomena and materials. Nonetheless, rather similar ΔE_{ST}'s have been measured for these two macrocycles, 0.30 and 0.25 kcal/mol for 6MC and 4MC, highlighting singlet ground electronic states for both compounds due to the effective DSP in the diradical format of 4MC and also in 6MC (in this latter more favoured because of the larger planarity of the structure).

For these three *p*-QDM cyclic compounds, the π-electronic structure more or less lies perpendicular to the plane containing the diameters. To finish this section, *p*-QDMs in which the π-electronic structure is in the same plane of the diameters, and therefore inherently distorted by the curvature, or distorted *p*-QDMs, is reviewed. These are cyclo oligoparaphenylenes (i.e., [*n*]CPP where *n* denotes the number of phenylenes, see Scheme 20) that have been intensively developed in the last few years due to the enormous interest as perfectly chemically defined molecular models of arm-chair single wall carbon nanotubes. In reality, [*n*]CPPs represent slices of these [135–140]. In their neutral forms, the longer [*n*]CPP members ($n > 8$) preserve the aromaticity of the benzenes, whereas in the smaller analogues ($n = 4$–7), ring curvature is pronounced and the planarity of the benzenes broken as well as their aromaticity. In these short [*n*]CPPs the rupture of aromaticity, at different extents depending on the curvature/size, is the way to accommodate the molecular strain imparted by the molecular cyclation. The rupture of the aromaticity is at the expense of the formation of a pseudo *p*-QDM and quinoidal structure in the six-member rings.

These [*n*]CPPs, with $n = 6$ to $n = 12$, can be oxidized by chemical treatments up to the radical cations and dications species [141–143]. Particularly interesting are the dications of [*n*]CPP in Scheme 20 represented for [6]CPP. In fact, the dications of the smaller members of the series show strong ring quinoidization and full π-conjugation of the positive charges through the whole molecular perimeter. This is indeed a nice example of cyclic π-conjugation, which acts here stabilizing these divalent oxidation states. The case of the dication of [6]CPP represents a "real" example of the existence of cyclic π-conjugation, which has been intensively searched in the past due to the interest in the analysis and properties of a full conjugated molecule without any boundary or terminal effects. Indeed, up until today, cyclacenes have been often proposed as models of cyclic π-conjugation, but

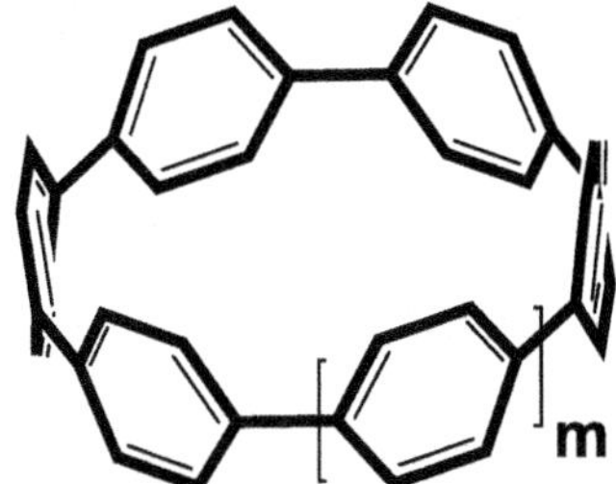 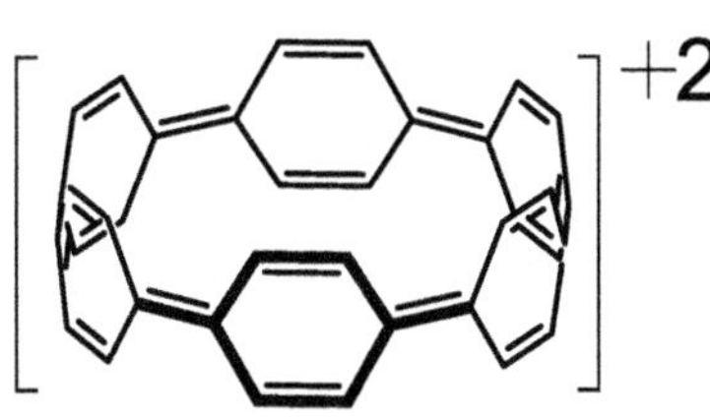

Scheme 20 Chemical structures of the neutral [*n*]CPPs ($n = 6 + m$, with $m = 0$–6) on the left and chemical structure of the dication of [6]CPP

these are only at the stage of theoretical models since they have not been synthesized yet [23, 144, 145]. Electron confinement and localization in extended linear π-conjugated molecules as a result of the molecular termini has burdened the maximization of the electronic properties of π-conjugated molecules. The especial and unique curvature of the [*n*]CPP has permitted the realization of π-conjugation unperturbed by electronic confinement. The existence of a well-developed quinoidal structure in the [6]CPP dication, such as shown in Scheme 20, reveals the formation of *p*-QDM, which are not planar anymore and appear severely distorted as the only way to accommodate and facilitate cyclic π-conjugation [146].

7 Softened Chemical Reactivity of Extended *p*-QDM

Let me finish the review of the different forms of *p*-QDM by revisiting their chemical properties as it was outlined in the Introduction. The high reactivity of *p*-QDMs was the first and most noticeable characteristic since their discovery. Inspired in this reactivity, the syntheses of some very important conductive or conjugated polymers, such as polyparaphenylenevinylene, have been accomplished based on the formation of *p*-QDMs intermediates. Owing to their diradical content, unstable *p*-QDM's polymerize by means of fast and irreversible reactions such as that represented for polyparaphenylenevinylene in Scheme 21. The incorporation of a larger number of aromatic rings "between" the radical centers, or extended *p*-QDMs in previous sections, has made possible the stabilization of these diradical states and, consequently, the softening of their reactivity permitting a number of new properties to emerge, such as reversible chromism and new polymeric soft phases [147–149].

The case of di(cyanoacyl)dithienonapthoquinodimethane in Scheme 22 is a good example of this softened reactivity, which gives rise to the reversible formation of oligomeric and polymeric structures that are present in dynamic equilibrium with the reactant monomers [150]. This reaction is in clear analogy with that of the pyrolisis of *p*-xylene having both in common the formation of *p*-QDM intermediates. In the case of our compound, we were able to characterize the generation of paracyclophane dimers and polymers by the recovery of the aromaticity in the monomer building blocks such as shown in Scheme 22. The existence of these reactions in extended *p*-QDMs seems to have some particular requirements. For instance, it does not take place in compounds with aromatic units that are flexible

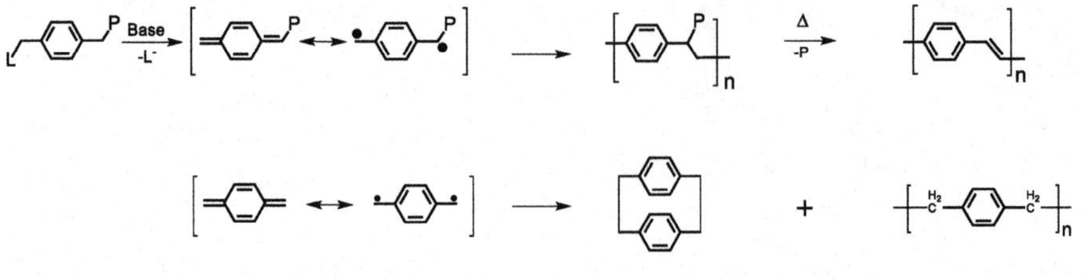

Scheme 21 Irreversible reaction of polymerization of p-QDM diradical intermediates

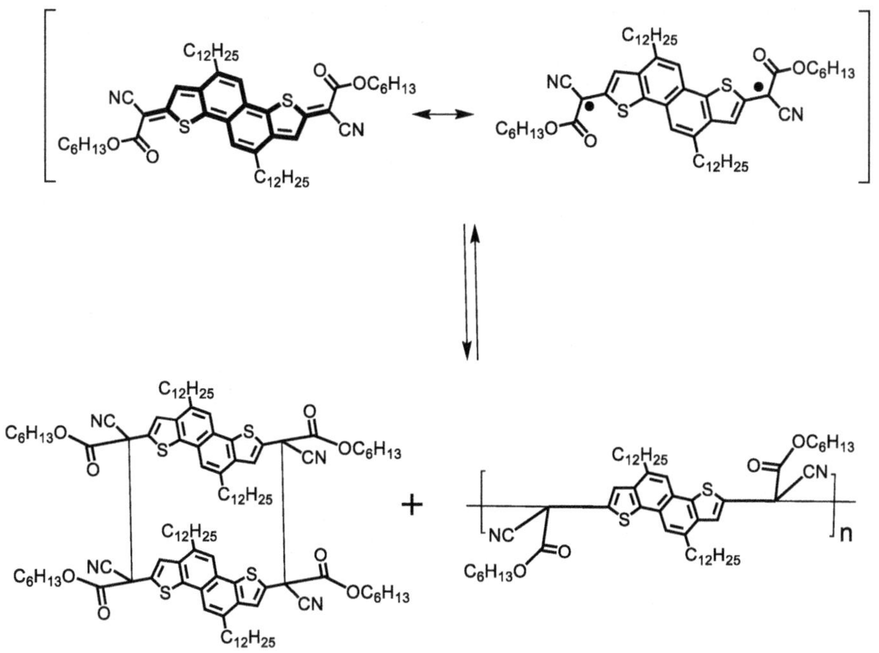

Scheme 22 Reaction in equilibrium between the diradical form of the di(cyanoacyl) dithienonapthoquinodimethane and its cyclophane and polymer products

given the impossibility of making them bulge them out in order to mitigate steric strain. But it is also related with the electron-withdrawing character of the external groups on the p-QDM units since strong acceptors impede sharing electron density needed in order to form the key CC inter-monomer bonds. In fact, the tetracyano analogue of the dicyanoacyl compound in Scheme 22 (i.e., one carbonyl group at each end is replaced by one cyano) does not form any of these products.

8 Applications of p-quinodimethanes

Modern applications of extended p-QDMs have significantly evolved in the last few years, but still need a "real" take off in the field of organic electronics. At the origin of this situation of impasse is, unavoidably, the instability inherent to such highly π-conjugated molecules with diradical character and small HOMO–LUMO gaps. In spite of this, a number of applications have been developed and reported, some of which are summarized in continuation.

8.1 Organic Field-Effect Transistor (OFET) Applications

One of the pioneering applications of tetracyanooligothiophenequinodimethanes was in organic field-effect transistors (OFET) by the group of the University of Minnesota, at that time headed by Larry L. Miller who firstly implemented *p*-QDMT3-Bu (see Scheme 5) as the organic semiconductor in the OFET device [55, 56]. In this first study, the authors reported a large electron *n*-type mobility of 0.005 cm2/Vs, which was certainly unique in a field clearly dominated by the p-type semiconductors. This discovery in 2002 closely followed the first report on *n*-type mobility in a perfluoroalkyl sexithiophene by Facchetti et al. [151] with an electron mobility of 0.02 cm2/Vs. However, remarkably, the smaller molecular weight and better processability of *p*-QDMT3-Bu (three vs six thiophenes) made the *n*-channel mobility in the p-QDM terthiophene an important achievement. The same Minnesota group, the next (2003) year, reported the ambipolar charge transport behavior of the same *p*-QDMT3-Bu compound, a distinctive aspect not featured in the Fachetti perfluorinated sexithiophene.

Many other *p*-QDM quinoidal closed-shell oligothiophenes have been implemented in OFET devices by the group of Takimiya [57–59] (see for instance p-QDM-T3-Bo in Scheme 23) with certainly impressive *n*-type mobilities. Also, molecules designed by Ribière et al. (see for instance *p*-QDMT4-Hx in Scheme 23) reported good performance as semiconductor devices with ambipolar behaviour in the largest quinoidal oligothiophene implemented in logic circuits devices, a tetramer with similar substitution profile of *p*-QDMT4 [60–64]. In this line of application of diradicals as organic semiconductors, it must be mentioned the work of Kubo in the implementation of biphenalenyls [152] and of DIAn very recently [122]. Both types of semiconductors behave as ambipolar materials entrenched in the existence of semi-empty high energy levels due to the diradical structures, which can be either fulfilled or emptied for *n*- or *p*-type conduction.

8.2 Organic Photovoltaic Applications

Whereas OFETs applications have been developed continuously from 2002 and more will come due to the intrinsic interest in *n*-type semiconductors for which *p*-QDM derivatives are ideally suited, now it is interesting to go with the more recent applications of *p*-QDM, which mostly have occurred in the field of organic

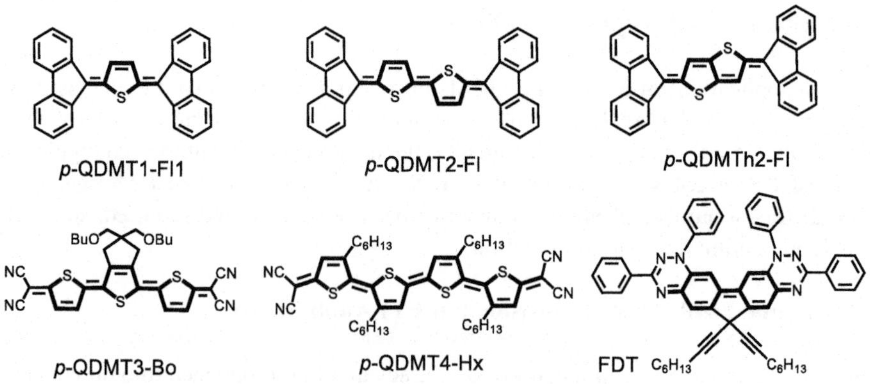

Scheme 23 Some representative quinoidal *p*-QDMs implemented in modern organic electronic applications

photovoltaics, and in particular in singlet exciton fission (SEF). The main condition for a given molecular system to work in SEF is based in the energy conservation of the process, which should verify that $E(S_1) \approx 2\,E(T_1)$ [153] a condition not fulfilled for most of the cases of π-conjugated closed-shell molecules in which $E(S_1) \ll 2\,E(T_1)$. The particularity of p-QDM substituted molecules for SEF is that they are impregnated of diradical character which largely stabilizes the first triplet excited state regarding the optically active singlet excited state [154, 155]. In Fig. 11 is represented the variation of the energy of the first excited states of two types of π-conjugated molecules, aromatic predominantly closed-shell systems, where the energies of the first singlet excited state and first triplet are rather similar, and therefore, $E(S_1) \ll 2E(T_1)$ and the condition for SEF is not attained. Conversely to the aromatic case, for the quinoidal analogues, the energy of T_1 is significantly lowered regarding that of S_1 owing to the gaining of diradical character and, therefore, there is a given point at intermediate diradical contents in which the $E(S_1) \approx 2\,E(T_1)$ condition is matched. It is also interesting to highlight that although SEF could happen also under the condition of $E(S_1) > 2\,E(T_1)$, this would imply systems with small ΔE_{ST} for which other competitive mechanisms appear diminishing the efficiency of SEF. Along these lines, quinoidal oligothiophenes in Scheme 23 (p-QDMT1-Fl, p-QDMT2-Fl and p-QDMTh1-Fl) have been prepared and implemented in organic photovoltaics yielding promising results for SEF [156]. The size or number of thiophenes seems to be small to confer intermediate diradical character needed to $E(S_1) \approx 2\,E(T_1)$. However, as it has been explained in previous sections, the presence of external fluorenic groups help to stabilize the triplet state in

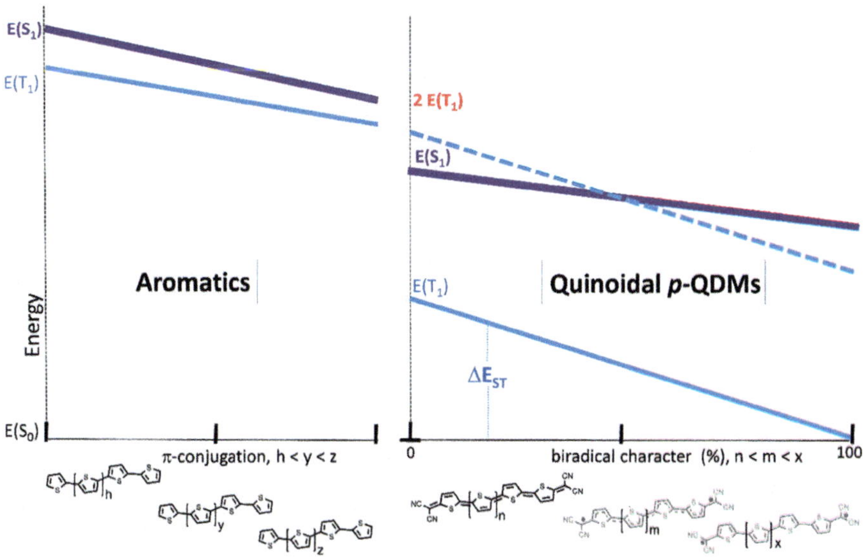

Fig. 11 Variation of the energy of the S_1 and T_1 states (relative to S_0) in typical aromatic and tetracyano quinoidal p-QDMs oligothiophenes with the enlargement of the molecular size and of the diradical character

a similar way as that described to explain the small ΔE_{ST} in zethrenes and in the bis(arylamino) phenylene dications.

The above statements in SEF refer to the case of intermolecular SEF (note that the term $2\,E(T_1)$ indicate the energy required to form two triplets in a dimer, one triplet in each molecule). However, the intramolecular case of SEF is also very attractive in which the two triplets are embedded in the same molecule, that is, the initial singlet excited state should produce a quintet excited state. Recently, we have shown that compound p-QDMT2-CP in Scheme 6 yields intramolecular SEF with a quantum yield close to 180% (≈ 1.8 triplets for each singlet formed) [157]. In this regard, the ability of the dicyano external bonds to rotate around the exocyclic double bonds of the p-QDMs (which convert in pseudo single bonds in the optically active singlet excited state) is the key structural aspect to stabilize a quintet excited state in intramolecular SEF.

Finally, the diradical source of extended p-QDMs makes open-shell structures to disclose certainly small (HOMO→SOMO and SOMO→LUMO) energy gaps that further imparts small optical gaps for the optical excitations, which have been exploited in photodetectors such as compound FDT in Scheme 23 [158]. In this FDT molecule and derivatives, the phenomenon of self-doping (based on the amphoteric redox behaviour described in previous sections) has been demonstrated, which opens the way for new applications of these open-shell diradicals formed from p-QDMs [159].

9 Conclusions and perspectives

The preparation and applications of p-QDM-based molecules is now 50 years old since TCNQ. In the last 15 years, they have revived with the applications in organic electronic as promising n-type semiconductors. In the last 10 years, the boost in their preparation has been by the formation of diradicals, which promises a lot of new applications of these systems in new and fancy devices, SEF for instance. In this review article, the main p-QDM compounds have been presented, in particular, those that are able to produce diradical structures. In fact, the preparation and analysis of the structure of diradical molecules is another area of great importance in chemistry since the discovery of the Chichibabin and Thiele molecules more than 100 years ago. p-QDMs and diradicals "marry" in this way. There are excellent reviews of diradicals and p-QDMs; however, here, I wished to revise their main structural and electronic properties and how these can be addressed in a more or less unified way. This has been done under the well-known concept of BLA in the area of π-conjugated molecules and of ΔE_{ST} and double spin polarization in the field of diradical systems. The focus has been put on those molecules which are well known by the author and another of relevance in the field owing to their novelty as well as for their interest in applications. New chemical systems based on the p-QDM building blocks are waiting us around the corner with new and challenging structures and fascinating properties, which overall describes a very rich field of research in chemistry and physics with an excellent present and a bright future.

Acknowledgements The author wishes to acknowledge many of his collaborators at the University of Málaga and many other bright organic chemists who have provided beautiful p-QDM molecules in the last 20 years. A number of institutions are also acknowledged, such as the University of Málaga, The Junta de Andalucía and the Government of Spain for continuous support of my research (CTQ2015-69391-P).

References

1. Szwarc M (1947) Discuss Faraday Soc 2:46
2. Szwarc M (1947) Nature 160:403
3. Szwarc M (1951) J Polym Sci 6:319
4. Szwarc M (1948) J Chem Phys 16:128
5. Coulson CA, Craig DP, Maccoll A, Pullman MA (1947) Discuss Faraday Soc 2:36
6. Errede LA, Landrum BF (1957) J Am Chem Soc 79:4952
7. Pearson JM, Six HA, Williams DJ, Levy M (1971) J Am Chem Soc 93:5034
8. Brown CJ, Farthing AC (1949) Nature 164:915
9. Auspos LA, Burnam CW, Hall LAR, Hubbard JK, Kirk W Jr, Schaefgen JR, Speck SB (1955) J Polym Sci 15:19
10. Schaefgen JR (1955) J Polym Sci 15:203
11. Tschitschibabin AE (1907) Chem Ber 40:1810
12. Thiele J, Balhorn H (1904) Chem Ber 37:1463
13. Salem L, Rowland C- (1972) Angew Chem Int Ed 11:92
14. Rajca A (1994) Chem Rev 94:871
15. Breher F (2007) Coor Chem Rev 251:1007
16. Sun Z, Ye Q, Chi C, Wu J (2012) Chem Soc Rev 41:7857
17. Casado J, Ortiz RP, Navarrete JTL (2012) Chem Soc Rev 41:5672
18. Abe M (2013) Chem Rev 113:7011
19. Sun Z, Zeng Z, Wu J (2014) Acc Chem Res 47:2582
20. Kubo T (2015) Chem Rec 15:218
21. Zeng Z, Shi X, Chi C, Navarrete JTL, Casado J, Wu J (2015) Chem Soc Rev 44:6578
22. Kertesz M, Choi CH, Yang S (2005) Chem Rev 105:3448
23. Krygowski TM, Cyranski MK (2001) Chem Rev 101:1385
24. Peierls R (1991) In: Treiman SB (ed) More surprises in theoretical physics. Princeton University Press, Princeton
25. Heeger AJ, Schrieffer JR (1983) Solid State Commun 48:207
26. Chance RR, Boudreaux DS, Bredas JL, Silbey R (1986) In: Skotheim TA (ed) Handbook of conducting polymers, vol 2. Marcel Dekker, New York, pp 825–857
27. Karpfen A, Kertesz M (1991) J Phys Chem 95:7680
28. Kertesz M (1995) Synth Met 69:641
29. Kertesz M, Lee YS (1987) J Phys Chem 91:2690
30. Borden WT, Iwamura H, Berson JA (1994) Acc Chem Res 27:109
31. Borden WT (ed) (1982) Diradicals. Wiley, New York
32. Borden WT, Davidson ER (1977) J Am Chem Soc 99:4587
33. Dougherty D (1991) Acc Chem Res 24:88
34. Karafiloglou P (1989) J Chem Educ 66:816
35. Melby LR, Harder RJ, Hertler WR, Mahler W, Benson RE, Mochel WE (1962) J Am Chem Soc 84:3374
36. Acker D-S-, Harder RJ, Hertler WR, Mahler W, Melby LR, Benson RE, Mochel WE (1960) J Am Chem Soc 82:6408
37. Kepler RG, Bierstedt PF, Merrifield RE (1960) Phys Rev Lett 5:503
38. Wudl F, Wobschall D, Hufnagel EJ (1972) J Am Chem Soc 94:670
39. Ferraris J, Cowan DO, Walatka VV, Perlstein JH (1973) J Am Chem Soc 95:948
40. Coleman LB, Cohen MJ, Sandman DJ, Yamagishi FG, Garito AF, Heeger AJ (1973) Solid State Commun 12:1125
41. Torrance JB (1979) Acc Chem Res 12:79
42. Shirakawa H (2001) Angew Chem Int Ed 40:2574

43. Maxfield M, Bloch AN, Cowan DO (1985) J Org Chem 50:1789
44. Addison AW (1977) Can J Chem 55:4191
45. Yamashita Y, Suzuki T, Saito G, Mukai T (1986) Chem Lett 5:715
46. Martin N, Seoane C, Segura JL, Marco JL, Hanack M (1991) Synth Met 41–43:1873
47. Martin N, Perez I, Sánchez L, Seoane C (1997) J Org Chem 62:870
48. Kini AM, Cowan DO, Gerson F, Mockel R (1985) J Am Chem Soc 107:556
49. Gronowitz S, Uppström B (1974) Acta Chem Scand B 28:981
50. Yui K, Aso Y, Otsubo T, Ogura F (1989) Bull Chem Soc Jpn 62:1539
51. Yui K, Ishida H, Aso Y, Otsubo T, Ogura F, Kawamoto A, Tanaka J (1989) Bull Chem Soc Jpn 62:1547
52. Higuchi H, Nakayama T, Koyama H, Ojima J, Wada T, Sasabe H (1995) Bull Chem Soc Jpn 68:2363
53. Higuchi H, Yoshida S, Uraki Y, Ojima J (1998) Bull Chem Soc Jpn 71:2229
54. Casado J, Miller LL, Mann KR, Pappenfus TM, Higuchi H, Orti E, Milian B, Pou-Amerigo R, Hernandez V, Navarrete JTL (2002) J Am Chem Soc 124:12380
55. Pappenfus TM, Chesterfield RJ, Frisbie CD, Mann KR, Casado J, Raff JD, Miller LL (2002) J Am Chem Soc 124:4184
56. Chesterfield RJ, Newman CR, Pappenfus TM, Ewbank PC, Haukaas MH, Mann KR, Miller LL, Frisbie CD (2003) Adv Mater 15:1278
57. Suzuki Y, Miyazaki E, Takimiya K (2010) J Am Chem Soc 132:10453
58. Handa S, Miyazaki E, Takimiya K, Kunugi Y (2007) J Am Chem Soc 129:11684
59. Suzuki Y, Shimawaki M, Miyazaki E, Osaka I, Takimiya K (2011) Chem Mater 23:795
60. Ribierre JC, Watanabe S, Matsumoto M, Muto T, Nakao A, Aoyama T (2010) Adv Mater 22:4044
61. Ribierre JC, Fujihara T, Watanabe S, Matsumoto M, Muto T, Nakao A, Aoyama T (2010) Adv Mater 22:1722
62. Ribierre JC, Takaishi K, Muto T, Aoyama T (2011) Opt Mater 33:1415
63. Ribierre JC, Watanabe S, Matsumoto M, Muto T, Aoyama T (2010) Appl Phys Lett 96:34
64. Ribierre JC, Zhao L, Furukawa S, Kikitsu T, Inoue D, Muranaka A, Takaishi K, Muto T, Matsumoto S, Hashizume D, Uchiyama M, Andre P, Adachi C, Aoyama T (2015) Chem Commun 51:5836
65. Wu Q, Li R, Hong W, Li H, Gao X, Zhu D (2011) Chem Mater 23:3138
66. Zhang C, Zang Y, Gann E, McNeill CR, Zhu X, Di C, Zhu D (2014) J Am Chem Soc 136:16176
67. Li J, Qiao X, Xiong Y, Li H, Zhu D (2014) Chem Mater 26:5782
68. Takahashi T, Matsuoka K-I, Takimiya K, Otsubo T, Aso Y (2005) J Am Chem Soc 127:8928
69. Casado J, Navarrete JTL (2011) Chem Rec 11:45
70. Hotta S, Waragai K (1993) J Phys Chem 97:7427
71. Ortiz RP, Casado J, Hernandez V, Navarrete JTL, Viruela PM, Orti E, Takimiya K, Otsubo T (2007) Angew Chem Int Ed Engl 46:9057
72. Hernández V, Casado J, Ramírez FJ, Zotti G, Hotta S, Navarrete JTL (1996) J Chem Phys 104:9271
73. Castiglioni C, Tommasini M, Zerbi G (2004) Philos Trans R Soc Lond Ser A Math Phys Eng Sci 362:2425
74. Burrezo P-M, Zafra JL, Navarrete JTL, Casado J (2017) Angew Chem Int Ed 56:2250
75. Zeng Z, Ishida M, Zafra JL, Zhu X, Sung YM, Bao N, Webster RD, Lee BS, Li R-W, Zeng W, Li Y, Chi C, Navarrete JTL, Ding J, Casado J, Kim D, Wu J (2013) J Am Chem Soc 135:6363
76. Zeng Z, Lee S, Son M, Fukuda K, Burrezo PM, Zhu X, Qi Q, Li R-W, Navarrete JTL, Ding J, Casado J, Nakano M, Kim D, Wu J (2015) J Am Chem Soc 137:8572
77. Zeng Z, Lee S, Zafra JL, Ishida M, Bao N, Webster RD, Navarrete JTL, Ding J, Casado J, Kim D, Wu J (2014) Chem Sci 5:3072
78. Zeng Z, Lee S, Zafra JL, Ishida M, Zhu X, Sun Z, Ni Y, Webster RD, Li R-W, Navarrete JTL, Chi C, Ding J, Casado J, Kim D, Wu J (2013) Angew Chem Int Ed 52:8561
79. Clar E, Lang KF, Sehulz-Kiesow H (1955) Chem Ber 88:1520
80. Umeda R, Hibi D, Miki K, Tobe Y (2009) Org Lett 11:4104
81. Umeda R, Hibi D, Miki K, Tobe Y (2010) Pure Appl Chem 82:871
82. Wu T, Chen C, Hibi D, Shimizu A, Tobe Y, Wu Y (2010) Angew Chem Int Ed 49:7059
83. Shan L, Liang Z, Xu X, Tang Q, Miao Q (2013) Chem Sci 4:3294
84. Hsieh Y-C, Fang H-Y, Chen Y-T, Yang R, Yang C-I, Chou P-T, Kuo MY, Wu Y-T (2015) Angew Chem Int Ed 54:3069
85. Sun Z, Huang K, Wu J (2011) J Am Chem Soc 133:11896

86. Li Y, Heng W, Lee BS, Aratani N, Zafra JL, Bao N, Lee R, Sung YM, Sun Z, Huang K, Webster RD, Navarrete JTL, Kim D, Osuka A, Casado J, Ding J, Wu J (2012) J Am Chem Soc 134:14913
87. Huang R, Phan H, Seng Herng T, Hu P, Zeng W, Dong S-Q, Das S, Shen Y, Ding J, Casanova D, Wu J (2016) J Am Chem Soc 138:103231
88. Sun Z, Lee S, Park KH, Zhu X, Zhang W, Zheng B, Hu P, Zeng Z, Das S, Li Y, Chi C, Li R-W, Huang K-W, Ding J, Kim D, Wu J (2013) J Am Chem Soc 135:18229
89. Zeng W, Sun Z, Herng TS, Gonzalves TP, Gopalakrishna TY, Huang KW, Ding J, Wu J (2016) Angew Chem Int Ed 55:8615
90. Ravat P, Solomek T, Rickhaus M, Hussinger D, Neuburger M, Baumgarten M, Juricek M (2016) Angew Chem Int Ed 55:1183
91. Bredas JL, Street GB (1985) Acc Chem Res 18:309
92. Hill MG, Penneau JF, Zinger B, Mann KR, Miller LL (1992) Chem Mater 4:1106
93. Bauerle P, Segebacher U, Maier A, Mehring M (1993) J Am Chem Soc 115:10217
94. van Haare J, Havinga EE, van Dongen JLJ, Janssen RAJ, Cornil J, Bredas JL (1998) Chem Eur J 4:1509
95. Furukawa Y (1996) J Phys Chem 100:15644
96. Bredas JL, Chance RR, Silbey R (1981) Mol Cryst Liq Cryst 77:319
97. Khanna RK, Jiang YM, Srinivas B, Smithhart CB, Wertz DL (1993) Chem Mater 5:1792
98. Sakamoto A, Furukawa Y, Tasumi M (1992) J Phys Chem 96:3870
99. Sakamoto A, Furukawa Y, Tasumi M (1994) J Phys Chem 98:4635
100. Banerjee M, Shukla R, Rathore R (2009) J Am Chem Soc 131:1780
101. Wang D, Talipov MR, Ivanov MV, Rathore R (2016) J Am Chem Soc 138:16337
102. González SR, Ie Y, Aso Y, Navarrete JTL, Casado J (2011) J Am Chem Soc 133:16350
103. Barlow S, Risko C, Odom SA, Zheng S, Coropceanu V, Beverina L, Brédas JL, Marder SR (2012) J Am Chem Soc 134:10146
104. Zheng S, Barlow S, Risko C, Kinnibriugh TL, Khrustalev VN, Jones SC, Yu Antipin M, Tucker NM, Timofeeva TV, Coropceanu V, Brédas JL, Marder SR (2016) J Am Chem Soc 126:1817
105. Su Y, Wang X, Li Y, Song Y, Sui Y, Wang X (2015) Angew Chem Int Ed 54:1634
106. Li T, Tan G, Shao D, Li J, Zhang Z, Song Y, Song Y, Sui Y, Chen S, Fang Y, Wang X (2016) J Am Chem Soc 138:10092
107. Wang X, Zhang Z, Song Y, Su Y, Wang X (2015) Chem Commun 51:11822
108. Su Y, Wang X, Zheng X, Zhang Z, Song Y, Sui Y, Li Y, Wang X (2014) Angew Chem Int Ed 53:2857
109. Su Y, Wang X, Wang L, Zhang Z, Wang X, Song Y, Power PP (2016) Chem Sci 7:6514–6518
110. Kurata R, Tanaka K, Ito A (2016) J Org Chem 81:37–145
111. Zheng Y, Xiong J, Sun Y, Pan X, Wu J (2015) Angew Chem Int Ed 54:12933–12936
112. Ji L, Edkins RM, Lorbach A, Krummenacher I, Bruckner C, Eichhorn A, Braunschweig H, Engels B, Low PJ, Marder TB (2015) J Am Chem Soc 137:6750–6753
113. Burrezo PM, Zhu X, Zhu S-F, Yan Q, Navarrete JTL, Tsuji H, Nakamura E, Casado J (2015) J Am Chem Soc 136:3834
114. Zhu X, Tsuji H, Nakabayashi K, Ohkoshi S-I, Nakamura E (2011) J Am Chem Soc 133:16342
115. Tobe Y (2015) Chem Rec 15:86
116. Shimizu A, Kishi R, Nakano M, Shiomi D, Sato K, Takui T, Hisaki I, Miyata M, Tobe Y (2013) Angew Chem Int Ed 52:6076
117. Shimizu A, Tobe Y (2011) Angew Chem Int Ed 50:6906
118. Chase DT, Rose BD, McClintock SP, Zakharov LN, Haley MM (2011) Angew Chem Int Ed 50:1127
119. Chase DT, Fix AG, Kang SJ, Rose BD, Weber CD, Zhong Y, Zakharov LN, Lonergan MC, Nuckolls C, Haley MM (2012) J Am Chem Soc 134:10349
120. Fix AG, Deal PE, Vonnegut CL, Rose BD, Zakharov LN, Haley MM (2013) Org Lett 15:1362
121. Rudebusch GE, Fix AG, Henthorn HA, Vonnegut CL, Zakharov LN, Haley MM (2014) Chem Sci 5:3627
122. Rudebusch GE, Zafra JL, Jorner K, Fukuda K, Marshall JL, Gómez-García CJ, Zakharov LN, Nakano M, Ottosson H, Casado J, Haley MM (2016) Nat Chem 8:753
123. Zade SS, Bendikov M (2010) Angew Chem Int Ed 49:4012
124. Tonshoff C, Bettinger HF (2010) Angew Chem Int Ed 49:4125
125. Bendikov M, Duong HM, Starkey K, Houk KN, Carter EA, Wudl F (2004) J Am Chem Soc 126:7416

126. Luo D, Lee S, Zheng B, Sun Z, Zeng W, Huang K-W, Furukawa K, Kim D, Webster RD, Wu J (2014) Chem Sci 5:4944
127. Lee S, Park KH, Das S, Herng T-S, Goncalves TP, Huang K-W, Ding J, Kim D, Wu J (2016) J Org Chem 81:2911
128. Shi X, Burrezo PM, Lee S, Zhang W, Zheng B, Dai G, Chang J, López Navarrete JT, Huang K-W, Kim D, Casado J, Chi C (2014) Chem Sci 5:4490
129. Streifel BC, Zafra JL, Espejo GL, Gómez-García CJ, Casado J, Tovar JD (2015) Angew Chem Int Ed 54:5888
130. Baird NC (1972) J Am Chem Soc 94:4941
131. Jorner K, Emanuelsson R, Dahlstrand C, Tong H, Denisova AV, Ottosson H (2014) Chem Eur J 20:9295
132. Jorner K, Feixas F, Ayub R, Lindh R, Sola M, Ottosson H (2016) Chem Eur J 22:2793
133. Nobusue S, Miyoshi H, Shimizu A, Hisaki I, Fukuda K, Nakano M, Tobe Y (2015) Angew Chem Int Ed 54:2090
134. Das S, Seng Herng T, Zafra JL, Burrezo PM, Kitano M, Ishida M, Hu P, Osuka A, Casado J, Ding J, Casanova D, Wu J (2016) J Am Chem Soc 138:7782
135. Jasti R, Bhattacharjee J, Neaton JB, Bertozzi CR (2008) J Am Chem Soc 130:17646
136. Xia J, Jasti R (2012) Angew Chem Int Ed 51:2474
137. Sisto TJ, Golder MR, Hirst ES, Jasti R (2011) J Am Chem Soc 133:15800
138. Iwamoto T, Watanabe Y, Sakamoto Y, Suzuki T, Yamago S (2011) J Am Chem Soc 133:8354
139. Yamago S, Watanabe Y, Iwamoto T (2010) Angew Chem Int Ed 49:757
140. Iwamoto T, Watanabe Y, Sadahiro T, Haino T, Yamago S (2011) Angew Chem Int Ed 50:8342
141. Golder MR, Wong BM, Jasti R (2013) Chem Sci 4:4285
142. Kayahara E, Kouyama T, Kato T, Takaya H, Yasuda N, Yamago S (2013) Angew Chem Int Ed 52:13722
143. Toriumi N, Muranaka A, Kayahara E, Yamago S, Uchiyama M (2015) J Am Chem Soc 137:82
144. Chandrasekhar J, Jemmis ED, Schleyer PVR (1979) Tetrahedron Lett 39:3707
145. Kawase T, Oda M (2004) Angew Chem Int Ed 43:4396
146. Alvarez MP, Delgado MCR, Taravillo M, Baonza VG, Navarrete JTL, Evans P, Jasti R, Yamago S, Kertesz M, Casado J (2016) Chem Sci 7:3494
147. Iwatsuki S, Itoh T, Iwai T, Sawada H (1985) Macromolecules 18:2726
148. Iwatsuki S, Itoh T, Sato T, Higuchi T (1987) Macromolecules 20:2651
149. Kubo T, Shimizu A, Sakamoto M, Uruichi M, Yakushi K, Nakano M, Shiomi D, Sato K, Takui T, Morita Y, Nakasuji K (2005) Angew Chem Int Ed 44:6564
150. Zafra JL, Qiu L, Yanai N, Mori T, Nakano M, Alvarez MP, Navarrete JTL, Gómez-García CJ, Kertesz M, Takimiya K, Casado J (2016) Angew Chem Int Ed 55:14563
151. Facchetti A, Yoon M-H, Stern CL, Katz HE, Marks TJ (2003) Angew Chem Int Ed 42:3900
152. Koike H, Chikamatsu M, Azumi R, Tsutsumi J, Ogawa K, Yamane W, Nishiuchi T, Kubo T, Hasegawa T, Kanai K (2016) Adv Func Mater 26:277
153. Smith MB, Michl J (2010) Chem Rev 110:6891
154. Minami T, Nakano M (2012) J Phys Chem Lett 3:145
155. Ito S, Nagami T, Nakano M (2016) J Phys Chem A 120:6236
156. Kawata S, Pu Y-J, Saito A, Kurashige Y, Beppu T, Katagiri H, Hada M, Kido J (2016) Adv Mater 28:1585
157. Varnavski O, Abeyasinghe N, Aragó J, Serrrano-Pérez J, Ortí E, Navarrete JTL, Takimiya K, Casanova D, Casado J, Goodson T III (2015) J Phys Chem Lett 6:1375
158. Zhang Y, Zheng Y, Zhou H, Miao M-S, Wudl F, Nguyen T-Q (2015) Adv Mater 27:7412
159. Zheng Y, Miao M-S, Dantelle G, Eisenmenger ND, Wu G, Yavuz I, Chabinyc ML, Houk KN, Wudl F (2015) Adv Mater 27:1718

Top Curr Chem (Z) (2017) 375:80
DOI 10.1007/s41061-017-0162-3

1,3-Diphenylisobenzofuran: a Model Chromophore for Singlet Fission

Justin C. Johnson[1] · Josef Michl[2,3]

Received: 17 February 2017 / Accepted: 12 July 2017 / Published online: 11 September 2017
© Springer International Publishing AG 2017

Abstract In this review we first provide an introductory description of the singlet fission phenomenon and then describe the ground and electronically excited states of the parent 1,3-diphenylisobenzofuran chromophore (**1**) and about a dozen of its derivatives. A discussion of singlet fission in thin polycrystalline layers of these materials follows. The highest quantum yield of triplet formation by singlet fission, 200% at 80 K, is found in one of the two known crystal modification of the parent. In the other modification and in many derivatives, excimer formation competes successfully and triplet yields are low. A description of solution photophysics of covalent dimers is described in the next section. Triplet yields are very low, but interesting phenomena are uncovered. One is an observation of a separated-charges (charge-transfer) intermediate in highly polar solvents. The other is an observation of excitation isomerism in both singlet and triplet states, where in one isomer the excitation is delocalized over both halves of the covalent dimer, whereas in the other it is localized on one of the halves. In the last section we present the operation of a simple device illustrating the use of triplets generated by singlet fission for charge separation.

Chapter 6 was originally published as Johnson, J. C. & Michl, J. Top Curr Chem (Z) (2017) 375: 80. DOI 10.1007/s41061-017-0162-3.

✉ Josef Michl
    Josef.Michl@colorado.edu

[1] National Renewable Energy Laboratory, 15013 Denver West Pkwy, Golden, CO 80401, USA

[2] Department of Chemistry and Biochemistry, University of Colorado, 215 UCB, Boulder, CO 80309-0215, USA

[3] Institute of Organic Chemistry and Biochemistry, Academy of Sciences of the Czech Republic, Flemingovo nám. 2, 16610 Prague 6, Czech Republic

Keywords 1,3-Diphenylisobenzofuran · Photophysics · Solar energy · Singlet fission · Covalent dimers

1 Introduction

Singlet fission (SF) is of interest for the conversion of solar energy into electricity or fuel, since a suitable combination of an ordinary sensitizer with an SF sensitizer should increase the efficiency of a single-junction solar cell. For the production of electricity, the theoretical efficiency rises from the Shockley–Queisser limit of $\sim$32% [1] to $\sim$46% [2]. Before describing SF in solids containing 1,3-diphenylisobenzofuran (**1**) and its derivatives, analogs, and covalent dimers, we provide a brief introduction to the phenomenon. Additional detail can be found in review articles [3–5], and in an earlier review dealing specifically with **1** [6].

SF is a process in which a singlet excited chromophore transfers some of its energy to a neighboring chromophore and both end up as triplets. The triplets are initially coupled into an overall singlet, making the process spin-allowed and possibly very fast. Figure 1 shows the energy levels of two chromophores, one on the left and the other on the right, and the initial singlet excitation of the left chromophore, normally by absorption of a photon (process 1). The SF event, i.e., the production of a triplet state on each chromophore, is shown as process 2. The description is oversimplified in that the initial excitation may be shared by more than one molecule.

Properly speaking, SF is not over until the two resulting triplets have separated and become independent, and this is not shown in Fig. 1. SF requires overcoming any binding energy between the triplets, which typically appears to be small if the chromophores are located in two distinct molecules. The separation often happens rapidly when the chromophores occur in the form of crystals, aggregates, or polymers, where triplet hopping from molecule to molecule is generally facile, or even in solution, where solute molecules can diffuse apart. If the binding energy of the two triplets is sufficiently large, the bound triplet pair (biexciton) might be separately observable and kinetically significant, in which case the total SF process would be a two-step event. In isolated covalent dimers, the separation cannot occur and the process stops after step 2 in Fig. 1, making such dimers a somewhat special case.

Fig. 1 Singlet excitation (1) followed by singlet fission (SF) (2). Reprinted with permission from [3]. Copyright 2010 American Chemical Society

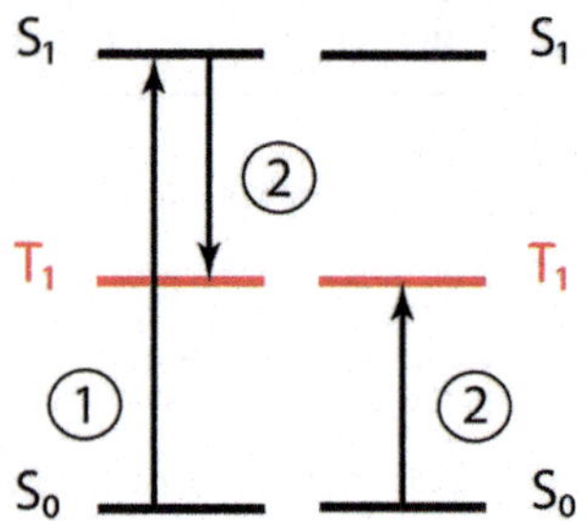

A simplified schematic representation of the critical step 2 of Fig. 1 in the diabatic representation is provided in Fig. 2, and we shall see below that in principle it itself also could occur in a single step, or in two steps. The final separation of the two triplets will not be discussed here.

Figure 2 is only symbolic in that it shows the dominant electron configurations, but in reality other electron configurations contribute as well. The left-hand side of Fig. 2 shows frontier orbital occupancies and spins in the initial singlet state, with the left chromophore singly excited and the right chromophore in its ground state. This initial state is normally prepared by the absorption of a photon (in reality, the singlet excitation is often shared by several molecules). The right-hand side of Fig. 2 shows frontier orbital occupancies and spins after SF has taken place. In the resulting doubly excited state of the dimer (biexciton) each chromophore is in its triplet state, but the two triplets are at first coupled into an overall singlet. Again, the real situation is more complicated since a combination of two triplets gives rise to nine substates and spin-dependent parts of the interaction Hamiltonian cause the initially reached singlet to develop in time, giving rise to magnetic field effects that are one of the hallmarks of SF [7, 8] (but will not be discussed here).

There are two ways (mechanisms) in which step 2 can proceed. The usual mechanism is a single-step process with no intermediate. Its rate is proportional to the square of a Hamiltonian matrix element that in the present approximation contains only three contributions. A small one is due to a direct interaction of the initial configuration with the final configuration and is indicated by a red arrow in Fig. 2. Two large contributions, possibly of mutually opposite signs, are due to the mediated interactions indicated by the cyan and blue arrows in Fig. 2. The two mediating states that provide the interaction (superexchange) paths are the charge-transfer configurations shown at the top and bottom of Fig. 2. In these, an electron has been removed from one or the other chromophore and added to its partner. Although there still is some debate as to whether the direct or the mediated contribution to the matrix element is more important, to our knowledge in all cases

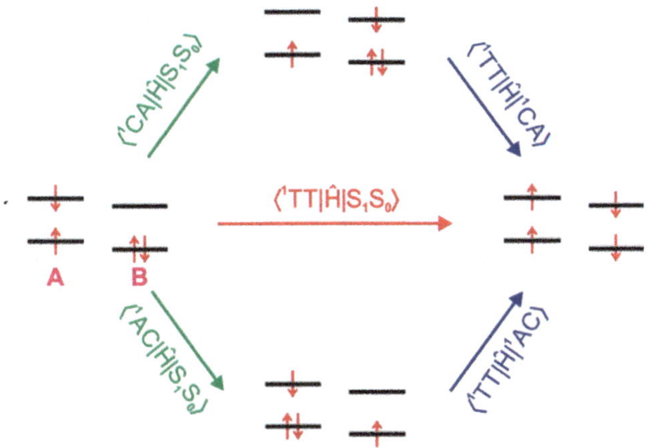

Fig. 2 A schematic representation of SF paths (see text)

in which the two have been carefully separated in a computation the direct contribution was found to be negligible relative to the algebraic sum of the mediated ones. The possibly destructive interference between the two mediated paths has interesting implications for the dependence of the matrix element on the choice of the mutual disposition (distance and orientation) of the two chromophores if they are identical; we return to this topic in subsequent text. Destructive interference can be largely avoided if the two chromophores are different, and one acts as an electron donor and the other as an electron acceptor.

In rare instances the energy of one or both charge-transfer configurations can be so favorable that they no longer represent virtual states but become observable real states with finite lifetimes (minima in the lowest excited singlet hypersurface). Then, process 2 in Fig. 1 can proceed by a two-step mechanism with a charge-transfer species as an observable intermediate. We shall see below that certain covalent dimers of **1** dissolved in highly polar solvents are good candidates for this two-step mechanism.

SF was initially observed in crystals, with the first report being for anthracene [9]. It was subsequently invoked to explain thermal quenching of photoexcited fluorescence in tetracene [10]. The phenomenon was confirmed beyond doubt by further studies on tetracene and pentacene [11–19] and was thoroughly reviewed [20]. Additional studies of polyacene crystals followed [21–29]. Once the process was considered to be well established, interest in it abated. SF was also found in a few other compounds, including non-crystalline ones (e.g., carotenoids and conjugated polymers), and in solution for a molecule composed of two tetracenes linked via a covalent bridge [30, 31], but the yields of triplets formed in these additional structures were mostly just a few per cent and always below 30–35%. SF thus acquired the reputation of being a rare and generally inefficient process. The exceptions were tetracene and pentacene, which were suspected to give very high triplet yields, although the actual numbers were not known.

When it comes to potential practical utility, triplet yields that lie substantially below the theoretical upper limit of 200%, and certainly those below 100%, are of no interest. SF materials need to meet a large number of additional conditions, among which stability in the sun under small amounts of atmospheric oxygen is the most obvious. This requirement eliminates both parent tetracene and pentacene as candidates for industrial applications, but leaves the hope that some of their analogs and derivatives might be acceptable.

The pioneering studies firmly established the existence of the SF process and recognized that it is favored in materials in which the first singlet excitation energy $E(S_1)$ is roughly equal to twice the first triplet excitation energy 2 $E(T_1)$, as in tetracene and pentacene. However, it was only the realization by Nozik and his collaborators [2] that SF could be extremely helpful in boosting solar cell efficiency that launched the present wave of interest in the phenomenon. Until then, there was little motivation to design and then synthesize sturdy and light-fast chromophores whose dimers, oligomers, or crystals would be optimized for high triplet yields from SF. In 2009–2010, several reports of highly efficient SF materials appeared [32–35], which led to a thorough reinvestigation of the phenomenon. After a decade of effort, there now is general consensus that efficient SF indeed is rare, but that under the right circumstances, the triplet yield can actually be very close to 200%, as is the case in the two champions, tetracene and pentacene. We believe now, as we did then, that finding a similar yield in a material that is actually practically useful needs to be the top priority in SF research.

The first study that addressed the issue of chromophore design for optimal SF was published in 2006 and used simple theoretical considerations to identify two partially overlapping classes of likely candidates among the vast number of possibilities: large alternant hydrocarbons and biradicaloids [36]. One of the biradicaloid structures identified as promising was 1,3-diphenylisobenzofuran (**1**) (Fig. 3), a compound of previously known [37, 38] triplet energy that had already been in use for some time for the detection of singlet oxygen [39, 40]. This is the species to which the present review is dedicated. It is the first and to date the only highly efficient structure entirely different from the traditional polyacenes chosen rationally for SF investigations, and it is even less useful in the practical sense than tetracene and pentacene. Not only is it sensitive to the combination of light and air, it also does not absorb sufficiently into the visible light spectrum. However, its properties prove that a rational design is possible, and it represents an interesting model chromophore. Numerous derivatives of **1**, shown in Fig. 3, are synthesized

1

cmpd	A	B	C	D	E	F	G	H	I	J	K	L	M	N
1	H	H	H	H	H	H	H	H	H	H	H	H	H	H
1a	H	H	F	H	H	H	H	H	H	H	H	H	H	H
1b	H	H	F	H	H	H	H	H	H	H	H	F	H	H
1c	F	H	F	H	F	H	H	H	H	H	H	H	H	H
1d	F	F	F	F	F	F	H	H	H	H	H	H	H	H
1e	F	F	F	F	F	F	H	H	F	F	F	F	F	F
1f	H	H	CH$_3$	H	H	H	H	H	H	H	H	H	H	H
1g	H	H	t-Bu	H	H	H	H	H	H	H	H	H	H	H
1h	H	t-Bu	H	t-Bu	H	H	H	H	H	H	H	H	H	H
1i	H	H	C$_2$H$_3$< (polymer)	H	H	H	H	H	H	H	H	H	H	H
1j	N	H	H	H	H	H	H	H	H	H	H	H	H	H
1k	MeN$^+$OTf$^-$	H	H	H	H	H	H	H	H	H	H	H	H	H
1l	H	MeN$^+$PF$_6^-$	H	H	H	H	H	H	H	H	H	H	H	H
1m	H	H	MeN$^+$OTf$^-$	H	H	H	H	H	H	H	H	H	H	H
1n	H	H	(CH$_2$)$_5$COOH	H	H	H	H	H	H	H	H	H	H	H

Fig. 3 Formulas of 1,3-diphenylisobenzofuran (**1**) and its derivatives

much more easily than those of tetracene and pentacene, and so are its covalent dimers.

Subsequent work elaborated the role of the biradicaloid character in the chromophore still further [41]. Various specific biradicaloid structures have been proposed [42–44], but only a very few have been tested, of which none have shown to possess a high efficiency to date. The requirement of biradicaloid character is somewhat unfortunate from a practical point of view, since most known biradicaloids are fairly unstable. By far most of the hundreds of SF studies published in the last decade have been devoted to acquiring a detailed understanding of the SF process in the higher polyacenes and to examining substituted and otherwise modified tetracenes and pentacenes. Although interesting and important, these lie outside the scope of the present review.

The other urgent and difficult issue to address in the search for SF materials is the optimal mutual disposition of the chromophores in a crystal, aggregate, oligomer, or dimer [45]. This problem is difficult because the space of mutual disposition of a pair of rigid bodies is already six-dimensional, and the dimensionality gets out of hand rapidly for larger sets of chromophores. The challenge was first addressed tentatively in 2010 [3]. A simple old model that only considers the highest occupied (HOMO) and the lowest unoccupied (LUMO) molecular orbital on each partner in a pair was selected (Fig. 2), simple formulas for the SF matrix element were worked out, and it was suggested that a slip-stacked geometry of two planar π-electron systems was especially favorable, whereas the exactly stacked geometry was not. In calculations that are not limited to the HOMO and LUMO orbitals, the contrast is less stark, but the slip-stacked geometry is still favored [46]. The size of this matrix element is critical, since in the Fermi golden rule approximation the rate of SF is proportional to its square.

As noted above, in the HOMO/LUMO model, the matrix element for SF consists of the 'direct' term and the 'mediated' terms, which involve virtual charge transfer between the two members of a pair [3]. Although it was not clear at first which of these contributions to the matrix element is dominant, so far, in all computations in which the terms were carefully separated, the mediated one dominates overwhelmingly.[1] We believe that in most cases it is the only one that needs to be considered seriously.

Further simplification of the HOMO/LUMO model has yielded an extremely simple approximate formula for the SF matrix element [48] that requires only the knowledge of coefficients of the frontier molecular orbitals and the overlaps of the atomic orbitals (AOs) on one and the other partner. This formula permits a computer search for local maxima in the six-dimensional space at hundreds of millions of dimer geometries in a few hours or days, depending on chromophore size.

For a pair of ethylene molecules taken as the simplest model, a few dozen best geometries were identified in this manner, but only a few of these offer really large

---

[1] The only exception that we are aware of is a claim that for two stacked tetracenes slipped strongly along the long axis and only slightly along the short axis the direct term dominates completely over the mediated term: Tamura et al. [47]. We have been unable to reproduce this result within the HOMO/LUMO model and find that this case is unexceptional in that the mediated term exceeds the direct term by more than two orders of magnitude.

matrix elements [5]. The simplicity of the formula allows the formulation of a design rule by inspection, and the rule is indeed satisfied in the favorable structures found by the computer: one of the π-symmetry atomic orbitals (AOs) on the first ethylene needs to overlap with both π-symmetry AOs of the second ethylene, while the second AO on the first ethylene should not overlap with the AOs of the second ethylene at all, or at least as little as possible. A generalization of this simple rule applies to more complicated chromophores [5].

In the Fermi golden rule approximation, both the electronic matrix element just discussed and a density of states term are important. In the Marcus theory approximation, the latter can be expressed through reorganization energy and reaction exoergicity, requiring consideration of the possible differences between excitation energies in an isolated chromophore and those in a dimer (exciton splitting) or a crystal (Davydov splitting). Simple approximate formulas have been developed for the dimer case and added to the computer program that automatically finds optimal geometries in the six-dimensional space of dimer geometries, thereby permitting a removal of some geometries that appear to be favorable on the basis of the electronic matrix element alone [5].

Maximizing the rate of SF represents only one-half of the task at hand; the other half is minimizing the rates of competing processes. Intramolecular competing processes, such as intersystem crossing, internal conversion, and unimolecular photochemical reactions (e.g., cis–trans isomerization), are relatively easy to avoid by a suitable choice of the chromophore used. It is more difficult to avoid intermolecular processes that need to be considered as soon as a crystal, aggregate, or a dimer are used. The most important of these is the formation of excimers and charge-separated states; intermolecular photochemical reactions can usually be avoided by a suitable choice of chromophores. Both excimer formation and charge separation are detrimental even though their formation preserves most of the electronic excitation energy and does not necessarily preclude subsequent SF (compare the two-step mechanism described above). In practice, however, these processes represent the most common threat to high efficiency in SF. First, they lower the energy of the excited singlet state, making it less likely that the formation of two triplets will be exoergic. Second, in their lowest energy electronically excited state, excimers may have a smaller matrix element for SF. Third, both excitons and charge-separated states often have efficient channels for internal conversion to the ground or the lowest triplet state (e.g., intersystem crossing in the charge-separated species). We shall see in the following sections that in the case of **1** and its derivatives, formation of excimers is often competitive and lowers the yield of triplets from SF. Since it may require motion within the crystal, its occurrence is not easy to predict.

At present, efforts are underway to identify favorable geometries in pairs of chromophores other than the simple ethylene model [5], including **1**. Although the project has not yet been completed, it has become clear that the results are quite sensitive to minor changes in dimer geometry, and we shall see below that indeed the two known crystal structures of **1**, although quite similar, provide very different triplet yields from SF [49]. Thus, even though the structure of **1** was selected for investigation in a rational way based on first principles, it was only sheer luck that in

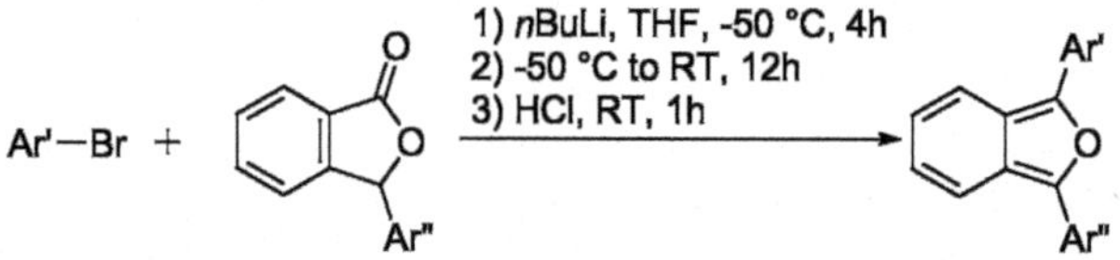

Fig. 4 The standard synthetic route to **1** and its derivatives

the initially investigated polycrystalline layers [33] **1** was packed in the crystal form that provides triplet yields close to 200%.

It needs to be emphasized that in real crystals electronic excitation can be delocalized over several chromophores and that entropy needs to be considered. The correct description of SF then becomes considerably more complicated than in a dimer [50]. The structural guidance obtained from theoretical work on dimers should therefore be viewed as only approximate.

2 Synthesis of 1, its Analogs and Derivatives

The classical synthetic approach to **1** and similar compounds is the reaction of a 3-arylphthalide with an aryllithium (Fig. 4). This reaction has been recently used to synthesize a series of alkylated (PI Dron et al., submitted for publication) and fluorinated (J Kaleta et al., unpublished results) derivatives for SF studies (Fig. 3) and has also been used to prepare numerous covalent dimers [51] (both aspects to be discussed in following text). Another synthetic approach has been described more recently [52], and the parent **1** is available commercially.

3 Electronic States of Parent 1 and its Ions

Much is known about solution spectroscopy, photophysics, and oxidation–reduction behavior of the isolated molecule of the parent chromophore **1**, its radical cation, and its radical anion [53].

3.1 Molecular Geometry

Steric interference between the ortho hydrogens of the phenyl substituents on the one hand and the oxygen lone pair and the peri hydrogens of the isobenzofuran core on the other hand prevents complete coplanarity and results in the existence of two ground state conformations with phenyl twist angles of about 24°, one with C_2 symmetry and the other with C_s symmetry. These ground state conformations are calculated to have nearly identical energies and geometries, and differ primarily only by the sense of rotation of the phenyl groups. Their calculated vibrational and electronic transition spectra are essentially identical. The crystal structure of the C_2 conformer has been solved [53] (Fig. 5; Table 1), and the alternation of CC bond lengths in its isobenzofuran core suggests that in the ground state, **1** can be

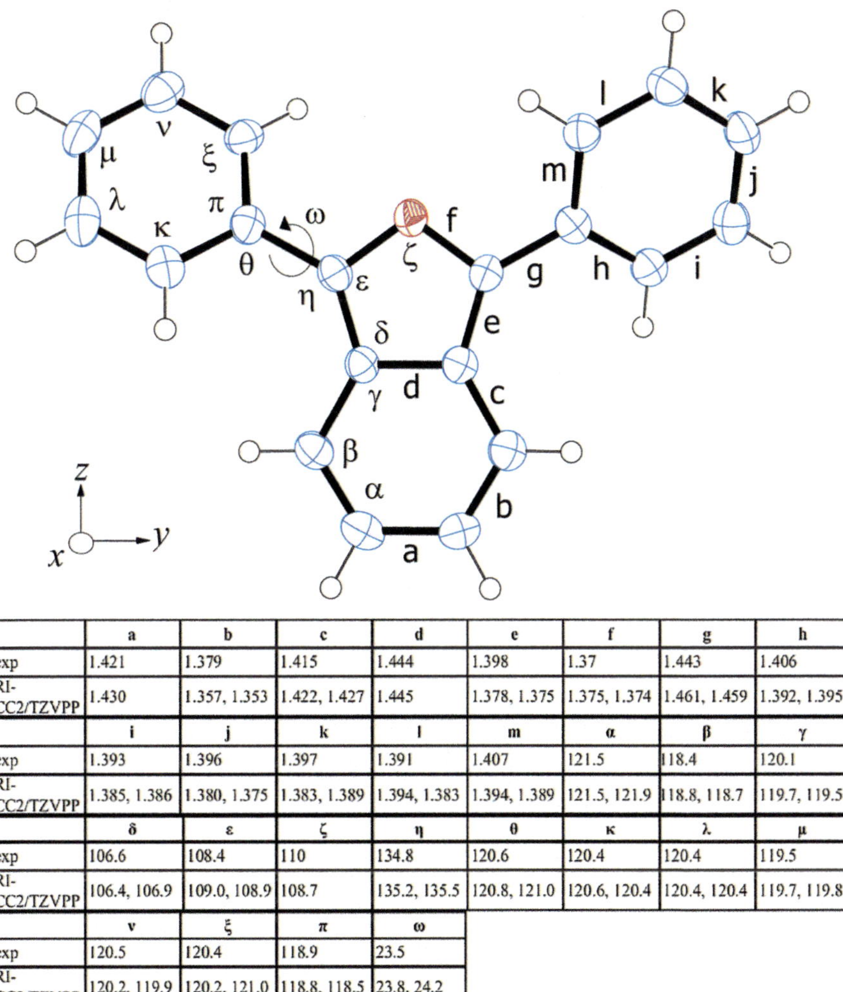

	a	b	c	d	e	f	g	h
exp	1.421	1.379	1.415	1.444	1.398	1.37	1.443	1.406
RI-CC2/TZVPP	1.430	1.357, 1.353	1.422, 1.427	1.445	1.378, 1.375	1.375, 1.374	1.461, 1.459	1.392, 1.395
	i	j	k	l	m	α	β	γ
exp	1.393	1.396	1.397	1.391	1.407	121.5	118.4	120.1
RI-CC2/TZVPP	1.385, 1.386	1.380, 1.375	1.383, 1.389	1.394, 1.383	1.394, 1.389	121.5, 121.9	118.8, 118.7	119.7, 119.5
	δ	ε	ζ	η	θ	κ	λ	μ
exp	106.6	108.4	110	134.8	120.6	120.4	120.4	119.5
RI-CC2/TZVPP	106.4, 106.9	109.0, 108.9	108.7	135.2, 135.5	120.8, 121.0	120.6, 120.4	120.4, 120.4	119.7, 119.8
	ν	ξ	π	ω				
exp	120.5	120.4	118.9	23.5				
RI-CC2/TZVPP	120.2, 119.9	120.2, 121.0	118.8, 118.5	23.8, 24.2				

Fig. 5 Ground-state structure of the C_2 conformer of **1** from single-crystal X-ray diffraction (*ellipsoids drawn at 50% scale*) and from calculations. Distances are given in angstroms, angles are given in degrees Reprinted with permission from [53]. Copyright 2010 American Chemical Society

described approximately as a π-excessive heterocyclic polyene with four double bonds and two phenyl substituents. Calculated geometries of the S_1 and T_1 excited states and D^+ and D^- radical ion states exhibit less bond length alternation in the isobenzofuran ring system, indicating stronger π-electron delocalization. This is compatible with the reduced twist angles of the phenyl substituents, which are only about 7–10° in the S_1, T_1, and D^- states, making the C_2 and C_s conformers nearly identical and effectively having C_{2v} symmetry. Interestingly, in the D^+ cation, the phenyl twist angle remains nearly the same as in the ground state.

Table 1 Optimized RICC2/TZVPP geometries of excited and ionized states of the C_2 conformer of 1,3-diphenylisobenzofuran (**1**), state symmetries in the C_2 group, and experimental geometry of the S_0 ground state

Parameter	S_0 (A)	S_1 (B)	S_2 (A)	T_1 (B)	D^+ (A)	D^- (B)
a	1.421	1.388	1.433	1.380	1.410	1.400
b	1.379	1.414	1.403	1.418	1.388	1.392
c	1.415	1.402	1.391	1.388	1.405	1.412
d	1.444	1.442	1.473	1.432	1.427	1.431
e	1.398	1.421	1.438	1.440	1.418	1.410
f	1.370	1.389	1.378	1.398	1.367	1.385
g	1.443	1.415	1.412	1.410	1.428	1.419
h	1.406	1.418	1.423	1.420	1.400	1.412
i	1.393	1.387	1.391	1.387	1.374	1.376
j	1.396	1.400	1.397	1.400	1.387	1.390
k	1.397	1.400	1.408	1.400	1.388	1.391
l	1.391	1.387	1.385	1.386	1.373	1.374
m	1.407	1.421	1.419	1.422	1.401	1.415
α	121.5	121.4	120.9	120.9	121.4	121.1
β	118.4	118.1	118.8	118.8	117.9	119.0
γ	120.1	120.4	120.2	120.3	120.6	119.9
δ	106.6	106.2	105.9	106.8	106.6	107.0
ε	108.4	110.2	108.7	109.2	108.4	108.5
ς	110.0	107.2	110.5	108.1	109.9	109.0
η	134.8	134.4	134.6	134.7	134.8	134.8
θ	120.6	121.5	120.5	122.1	120.3	123.0
κ	120.4	120.3	120.0	120.5	119.9	121.2
λ	120.4	120.7	120.8	120.7	119.9	121.8
μ	119.5	119.6	119.6	119.4	120.7	117.6
υ	120.5	120.6	120.4	120.7	119.9	121.7
ξ	120.4	120.4	120.5	120.6	119.8	121.2
π	118.9	118.5	118.7	118.1	119.8	116.6
ϖ	23.5	9.9	14.0	9.1	22.7	6.7

See Fig. 5 for the definition of geometrical parameters [53]

3.2 Singlet Excited States [53]

The electronic transition moments in both conformers of the slightly non-planar **1** are calculated to lie within a few degrees of the y and z axes shown in Fig. 5, and it is acceptable to describe the electronic states of **1** in terms of the C_{2v} group.

Panels a and c in Fig. 6 show the absorption, fluorescence, fluorescence excitation, and magnetic circular dichroism (MCD) spectra of **1** in cyclohexane (CH) at room temperature. Panel d shows the absorption, fluorescence, fluorescence excitation, and the anisotropy of fluorescence and fluorescence excitation in 3-methylpentane (3-MP) at low temperature (77 K). It is believed that the two conformers contribute equally to the spectra shown. Panels b and f show the results of semi-empirical and ab initio calculations, both of which account well for the observations.

Fig. 6 Spectra of **1**. **a** Magnetic circular dichroism spectrum in cyclohexane (CH) [at room temperature (rt). **b** Calculation according to the Pariser–Parr–Pople (PPP) method. *Bar height* indicates value of the B term, bar color indicates polarization (*blue*, *y*; *red*, *z*). **c** Absorption (*black*), fluorescence (*green*), and fluorescence excitation (*red*) in CH (rt). **d** Absorption (*black*), fluorescence (*green*), fluorescence excitation (*red*), fluorescence anisotropy (*bright red*), and fluorescence excitation anisotropy (*blue*) in 3-methylpentane (3-MP) (at 77 K). **e** Absorption in 3-MP (at 77 K) decomposed into its *y*-polarized (*blue*) and *z*-polarized (*red*) components. **f** CC2/TZVPP calculation (all transitions have been shifted to lower values by 3000 cm^{-1}). *Bar length* indicates the oscillator strength of *y*-polarized (*blue*) and *z*-polarized (*red*) transitions. Reprinted with permission from [53]. Copyright 2010 American Chemical Society

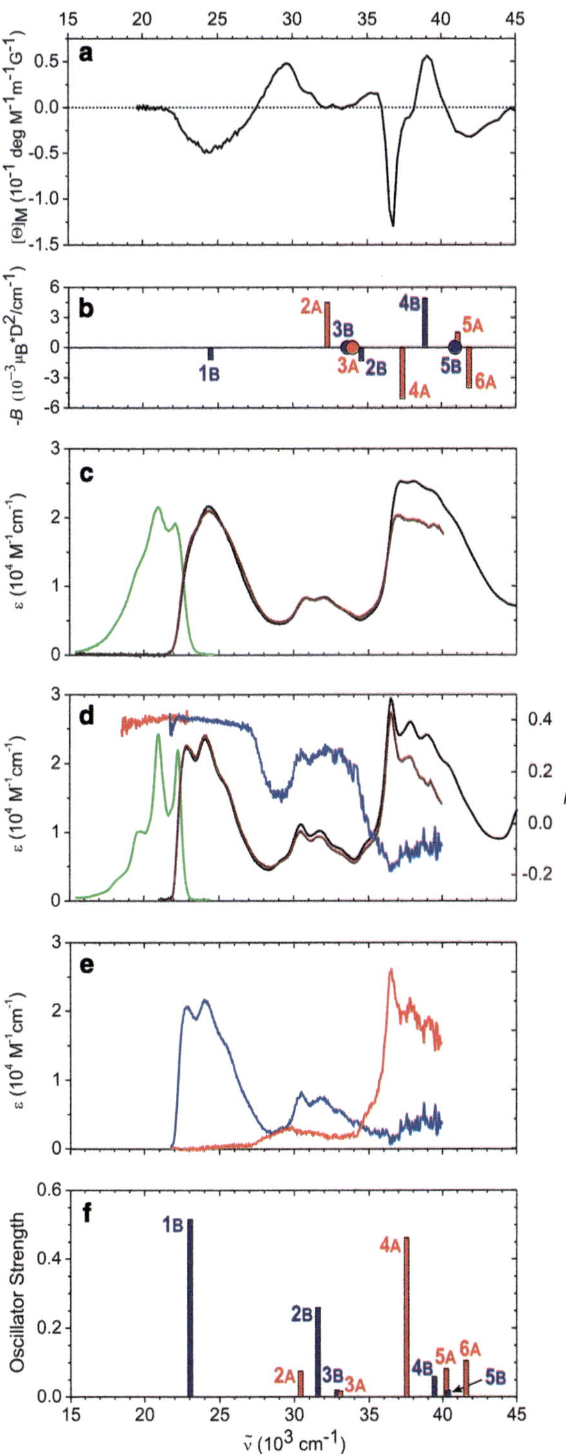

The photophysical data for **1** are nearly solvent independent: the first absorption peak lies at $\sim 24{,}300$ cm^{-1} and is relatively intense ($\varepsilon = 23{,}400$ M^{-1} cm^{-1}), and its vibronic features are spaced at ~ 1400 cm^{-1}. Fluorescence is an approximate mirror image of the first absorption band. Its first peak is found at $\sim 22{,}100$ cm^{-1}, and its quantum yield is 0.96 ± 0.03. The absence of detectable internal conversion or intersystem crossing makes **1** a favorable candidate for SF [36]. The fluorescence lifetime is ~ 6 ns and corresponds to a fluorescence rate constant of ~ 0.15 ns^{-1}. At low temperature, the vibronic structure is more pronounced, making it clear that the true Stokes shift is small (<500 cm^{-1}).

The essentially constant fluorescence anisotropy of 0.38 for excitation anywhere in the first absorption band and for observation anywhere in the emission demonstrates that the first absorption band and the fluorescence are purely polarized along the same axis; all methods of calculation agree that its direction is y in Fig. 5. The pure polarization of the first absorption band permitted the use of linear dichroism in stretched polyethylene to prove that two distinct conformers are actually present and orient to different degrees. Their Franck–Condon envelopes in the first transition are slightly different.

The second electronic transition from the ground state occurs near 30,000 cm^{-1} and is z-polarized (Fig. 6e). It is weak and cannot be seen in an ordinary absorption spectrum. The somewhat stronger broad band around 32,000 cm^{-1} is due to a y-polarized transition, and calculations predict two additional very weak transitions near this energy, one of each polarization. The first intense z-polarized transition is found at $\sim 37{,}000$ cm^{-1}, where a non-radiative decay channel opens and the fluorescence excitation spectrum stops following the absorption spectrum faithfully.

The absorption spectrum of the S_1 state appears immediately following resonant pulsed excitation and decays as a single exponential with the same lifetime as the fluorescence. The most prominent bands are located near 14,500 and 22,000 cm^{-1} (Fig. 7).

The nature of the excited states and of the molecular orbitals involved in the excitations is shown in Fig. 8. The lower energy states can be understood simply in terms of the perimeter model [54, 55], which allows the S_1 state to be given the label L_1 and the S_2 state, the label L_2. The perimeter model also leads to a recognition of **1** as a "positive hard MCD chromophore" [56] and correctly accounts for signs in the MCD spectrum.

3.3 Triplet States

As might be expected from a fluorescence quantum yield of nearly unity, no phosphorescence is observed in a low-temperature glass, and no transient triplet–triplet absorption is seen in a laser flash photolysis experiment without a sensitizer. The excitation energy of T_1 has been reported as $\sim 11{,}900$ cm^{-1} (adiabatic) in benzene solution based on triplet sensitization [37], as $\sim 12{,}900$ cm^{-1} in Dimethylformamide (DMF) based on electrogenerated chemiluminescence [38], and as $\sim 11{,}400$ cm^{-1} in a solid film from electron energy loss spectrum [53]. Triplet transitions T_1–T_n were identified by diffusion-controlled energy transfer from the triplet of photoexcited anthracene to ground state **1** (Fig. 7). A strong and

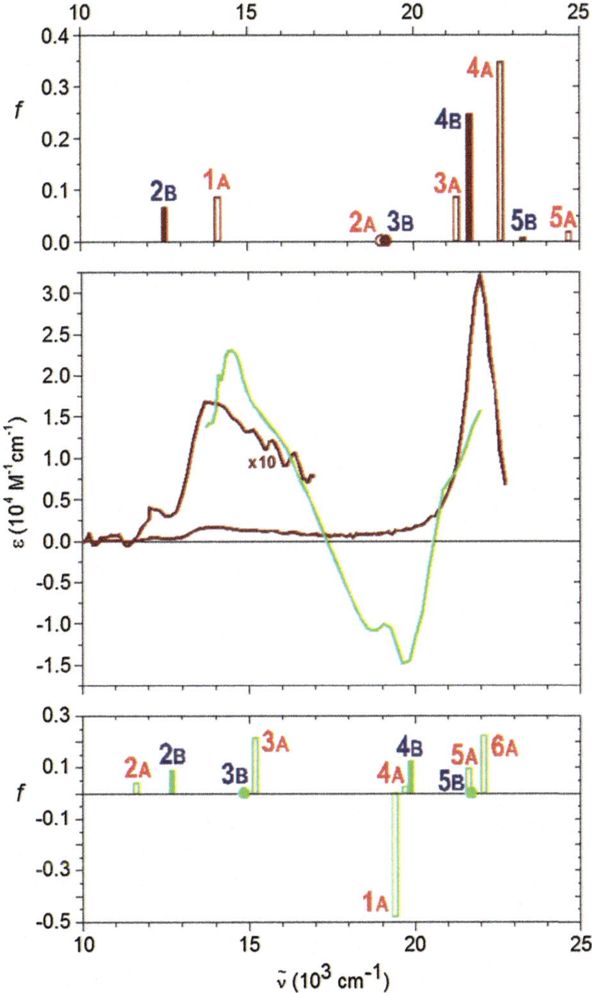

Fig. 7 Transient absorption spectra of the first excited singlet state (S$_1$, *chartreuse*, $\tau = \sim 5$ ns) and the first excited triplet state (T$_1$, *brown*, $\tau = \sim 230$ μs) of **1** in dimethyl sulfoxide (DMSO) Reprinted with permission from [53]. Copyright 2010 American Chemical Society

narrow band at $\sim$21,500 cm^{-1} has been attributed to overlapping transitions from the 1B lowest triplet state to the 3A, 4B, and 4A states, and should therefore be of mixed polarization. Lower lying triplet transitions are calculated and observed near 14,000 cm^{-1} but are considerably weaker than those at higher energy. The triplet lifetime in dilute solution is $\sim$230 μs.

The most important gap in our knowledge is the absence of the electron paramagnetic resonance spectrum of the triplet of **1**, which means that the D and E parameters ordinarily used for the description of the zero-field-splitting tensor are unknown.

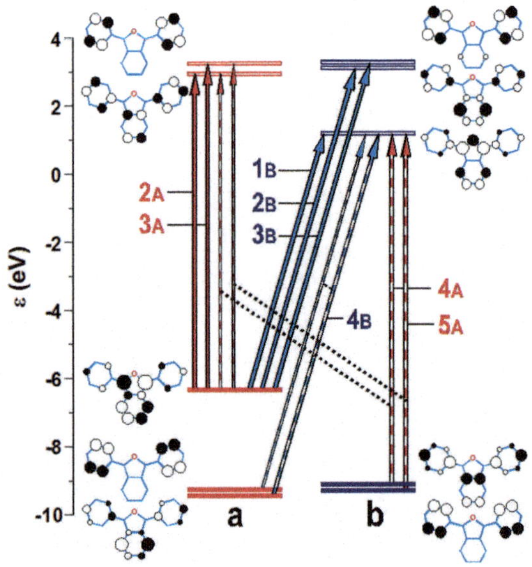

Fig. 8 Orbital energies in **1**. Chief contributions to $S_0 \rightarrow S_x$ excitations calculated at the CC2/TZVPP level are shown. *Striped arrows* indicate that more than one electron promotion has a significant amplitude. Reprinted with permission from [53]. Copyright 2010 American Chemical Society

In summary, the vertical excitation energies S_0–S_1 ($\sim$24,300 cm^{-1}), S_0–T_1 ($\sim$11,400 cm^{-1}), and S_0–T_2 (estimated at 25,500 cm^{-1}) satisfy the conditions $E(S_1), E(T_2) \geq 2E(T_1)$, considered desirable for SF.

3.4 Electrochemistry and Radical Ions

As would be expected for a π-excessive heterocycle, **1** is relatively easy to oxidize and difficult to reduce. In liquid SO_2, reversible one-electron oxidation takes place at +0.35 V against ferrocene/ferricinium ($Fc^{0/1}$) and at +0.85 V against a saturated calomel electrode (SCE). Reversible one-electron reduction in *N,N*-dimethylformamide takes place at −1.85 V against SCE [51].

The absorption spectra of the radical cation **1**$^{+\bullet}$ and radical anion **1**$^{-\bullet}$, produced by pulse radiolysis in 1,2-dichloroethane and tetrahydrofuran solution, respectively, are shown in Fig. 9, revealing bands at 15,000 and 18,500 cm^{-1}, respectively. As might be expected [57], the transitions in the absorption spectra of the singlet and triplet states of **1** and of the radical ions **1**$^{+\bullet}$ and **1**$^{-\bullet}$ are inter-related.

3.5 Substituted Derivatives

Various derivatives of **1** carrying weakly interacting substituents have been synthesized (Fig. 3), mostly with the intention of keeping the chromophore nearly the same but modulating its packing in the solid state dramatically. Measurements of SF rates would then presumably primarily provide information on the effect of crystal packing alone. In solution, the optical properties primarily reflect the degree of twisting of the phenyl groups that the substitution engenders. For the fluorinated derivatives, the twisting increases when fluorine atoms are present at the ortho or

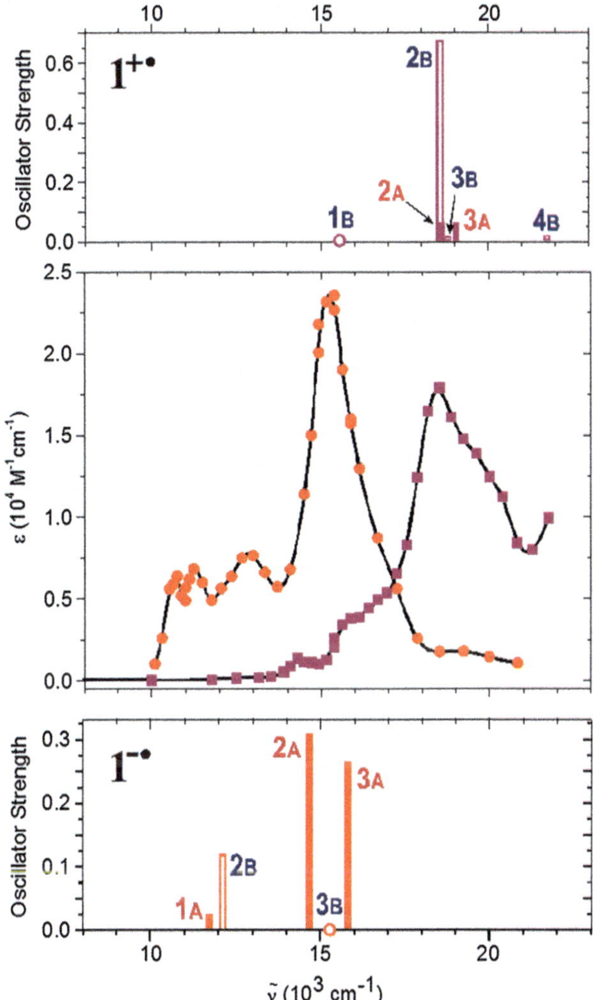

Fig. 9 Absorption of the radical anion $1^{-\bullet}$ (*orange*) and radical cation $1^{+\bullet}$ (*rose*) obtained by pulse radiolysis of 5 mM **1** in tetrahydrofuran and 1,2-dichloroethane, respectively. CC2/TZVPP calculated spectra on top (cation) and bottom (anion). Full bars indicate z polarization, empty bars indicate y polarization. Calculated transitions have been shifted to the lower energies by 3000 (cation) or 2000 (anion) cm^{-1} Reprinted with permission from [53]. Copyright 2010 American Chemical Society

meta positions in the phenyl groups, whereas substitution in the para position has negligible effect. A larger twist reduces conjugation and causes a blue shift in the onset of absorption and fluorescence. As the degree of fluorination increases, the fluorescence lifetimes and quantum yields are slightly reduced. The triplet spectra are largely unaffected. For the alkyl derivatives **1f–1h**, the deviations of absorption spectra from **1** are minor. All derivatives show a slight red shift of <200 cm^{-1} and remain highly fluorescent.

4 Photophysics of Thin Films of 1 and its Derivatives

Thin films of **1** were fabricated using thermal evaporation of powdered material or casting from solution [33]. The films employed for photophysical measurements were between 25 and 150 nm thick and tended to be highly polycrystalline, based on sharp and strong X-ray diffraction (XRD) peaks, but the crystal structure depended on the deposition conditions. Two dominant progressions of XRD peaks can be seen, separated by about 0.1° at the first observable reflection ($2\theta = \sim 9°$) (Fig. 10a). The films with lower 2θ values are labeled α and those with higher 2θ values are labeled β. Films of a mixed type can be made directly or through strategic annealing steps. Rapid deposition of **1** (>5 Å/s) favors the polymorph α, whereas slower deposition and mild thermal annealing favor the β form. Most solution cast films were of the α form, but exposure to ambient conditions converted them to β. Crystal views are shown in Fig. 10b. Crystals grown from solution were isolated and studied with XRD and structural analysis. Two forms were discovered, matching the forms seen in thin films [49]. The crystal habits also roughly match crystallite shapes observed via atomic force microscopy for thin films. The ability to fabricate films of nearly pure polymorph composition and to compare these with known bulk structures was a major milestone for the pursuit of understanding the role of interchromophore coupling on SF [45]. The use of naturally occurring polymorphs in this fashion has become common among other SF systems [58, 59].

Despite the similarity of the unit cell parameters and nearest neighbor interactions, the two film types exhibit very different photophysics. Ground state absorption is best judged by extracting absorbance from integrating sphere measurements of transmission and reflection due to considerable scattering and wavelength-dependent reflectivity of the thin films. The resultant spectra are shown

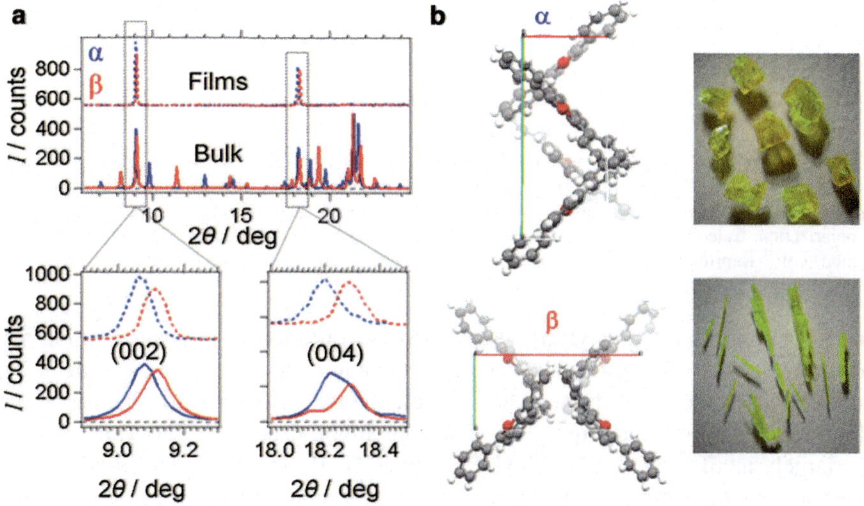

Fig. 10 **a** X-ray diffraction of two thin film types of **1**. **b** Unit cells of α (lower 2θ values) and β (higher 2θ values) thin films looking down the c axis and photographs of their respective single crystals

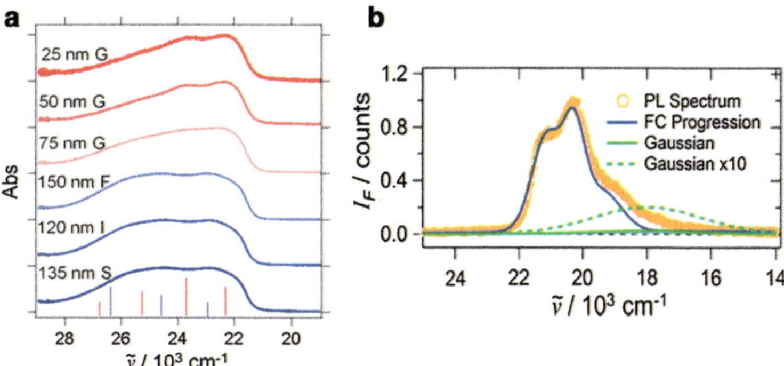

Fig. 11 Absorption and fluorescence spectra of thin films of **1**. **a** Absorption vs. thickness and substrate type. *Blue* indicates type α, *red* indicates type β. Substrates are: glass (G), fused silica (F), indium tin oxide (I), and sapphire (S). Stick spectra associated with fit to a Franck–Condon (FC) progression are shown on the horizontal axis. **b** Fluorescence for a type β thin film, showing a fit with a FC progression and residual Gaussian-shaped emission. *PL* Photoluminescence

in Fig. 11a. They are shifted to the red of those of isolated **1** by 300 and 1000 cm^{-1} for α and β, respectively. The vibronic envelopes are also altered, with the α form possessing a larger amplitude for the 0–1 vibronic transition. The fluorescence retains a similar vibronic progression for both film types, but also contains an underlying red-shifted Gaussian band (Fig. 11b). This feature is stronger for β than for α film types, as is the overall room temperature fluorescence yield (59 vs. 16%, respectively).

The triplet yields of crystalline thin films of **1** were determined from transient absorption data by taking the ratio of signal amplitudes associated with the maximum triplet signal ($\Delta t = 200$ ps) and the initial bleach ($\Delta t = 0.2$ ps). This ratio was corrected by the ratio of ground state bleach to triplet absorption molar extinction coefficients at 22,500 and 21,500 cm^{-1}, respectively, and was assumed to be $\varepsilon_G{:}\varepsilon_T = 0.7$, the same as in solution. The corrected signal ratio was further modified by an orientation factor that reflects the relative alignment of triplet and S_0–S_1 bleach transition dipole moments at the respective wavenumbers. The transition moment of the ground state absorption lies in the molecular y axis, and XRD results imply that in thin films it is therefore oriented parallel to the substrate surface. The triplet absorption has a mixed polarization (y, z) in the spectral region of interest, and only the y-polarized fraction contributes significantly to the observed film absorption at normal incidence. Tilting the sample and detecting the triplet signal E^T with s- and p-polarized light as a function of the tilt angle ω yielded no change in triplet absorption strength. The fraction of absorption that is polarized along the y axis is given by $2E_{sT} \sin^2 \omega \big/ \left[E_p^T - E_s^T \left(1 - 3\sin^2 \omega\right) \right]$. Since measurements suggest $E_p^T \simeq E_s^T$, this quantity reduces to two-thirds. The two corrections cancel each other out, leaving the ratio of net signal amplitudes for triplet and bleach equal to the triplet yield.

The triplet yield Φ_T for the β form is ~ 1.4 at room temperature (Fig. 12a). It depends on temperature, rising to 2.0 at 80 K before falling slightly by 10 K. It is

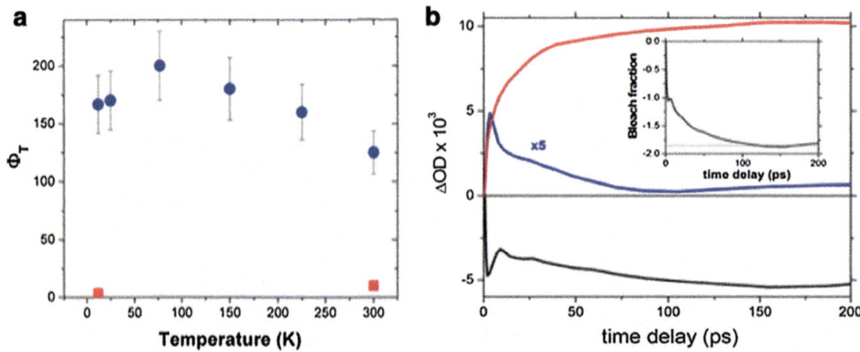

Fig. 12 **a** Triplet quantum yield versus temperature for α (*blue*) and β (*red*) films. **b** Transient absorption profiles for α film, corresponding to singlet (*blue*), triplet (*red*), and ground state bleach (*black*) populations. Copyright 2010 American Chemical Society

much smaller for the ß form, ~ 0.1 at room temperature [33]. The dramatic difference in Φ_T between the two forms is underpinned by the kinetic profiles of singlet and triplet state populations determined from transient absorption spectroscopy (Fig. 12b). Transition assignments are made using information obtained from solution studies of **1** and are justified by the weak electronic coupling displayed in the films.

In α films, the T_1–T_n absorption features increase with an ~ 15 ps time constant, which is identical to the decay time constant of stimulated emission and S_1–S_n absorption. Concomitantly, the bleach strength increases with the same time constant, signaling the consumption of additional ground states during SF. The increase in bleach amplitude can be quantified and provides a second estimate of Φ_T, which matches the yield found from direct counting of triplets produced after excitation. For β films, a large stimulated emission component is present from early delay times and decays multiexponentially. A small residual triplet signature is observed at long delay times. We have identified several factors that limit Φ_T for β-type crystals: (1) an intrinsically slower SF process due to lower values of the SF matrix elements, as calculated from molecules in the unit cells of α and β; (2) the propensity for fast excimer formation; and (3) a larger Davydov splitting that leads to a depressed S_1 energy that introduces a significant unfavorable energy difference between $E(S_1)$ and $2 \times E(T_1)$ [60]. Evidence for excimer formation is found in the fast rise of a broadened stimulated emission feature for β films.

Crystal engineering strategies have been considered for modulating SF rates, but many intrinsically alter the electronic structure of the molecule to an excessive degree [61]. The series of alkylated (PI Dron et al., submitted for publication) and fluorinated (J Kaleta et al., unpublished results) derivatives of **1** shown in Fig. 3 provides an interesting variety of intermolecular packing geometries in polycrystalline thin films (Fig. 13), without significant perturbation of the chromophores and their S_1–$T_1 T_1$ energy alignment.

Whereas films of the parent **1** and its monofluoro derivative **1a** have nearly identical intermolecular packing and both exhibit a SF rate of ~ 15 ps (Fig. 14c

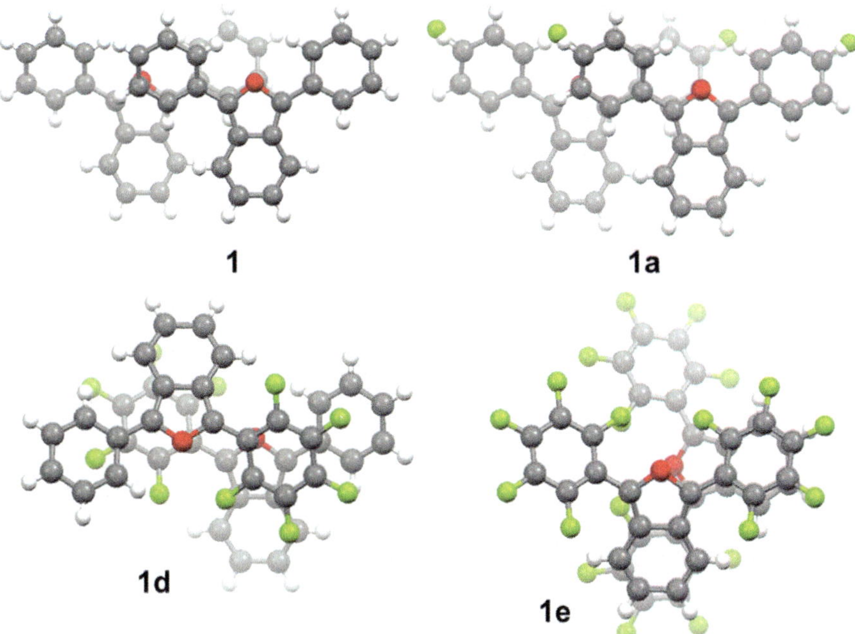

Fig. 13 Views of mutual disposition of closest pairs of molecules in selected crystals of **1** and its derivatives. Red spheres indicate oxygen, yellow spheres indicate fluorine, blue spheres indicate nitrogen, gray spheres indicate carbon, white spheres indicate hydrogen. For **1a** the para substituents are displayed as half hydrogen and half fluorine due to crystal symmetry

shows the associated spectrum), the pentafluoro derivative **1d**, with a slightly modified slip-stack structure, undergoes SF in ~ 7 ps (J Kaleta et al., unpublished results). In these instances, S_0 bleach and S_1 and T_1 absorption are the only features observed in the transient absorption spectrum, and these times reflect SF rates. The crystal structure for the decafluoro derivative **1e** reveals a different packing motif, in which the molecules are not slip-stacked but arranged with their y axes close to mutually perpendicular. An excimer is not formed, and transient absorption measurements of films of **1e** show that SF still occurs with a time constant of ~ 12 ps, which is slightly faster than for **1**.

Derivatives of **1** carrying one or two alkyl (methyl or tert-butyl) groups on one of the phenyl rings (**1f–1h**) have also been investigated (PI Dron et al., submitted for publication). Their absorption and fluorescence spectra in solution are nearly identical with that of **1**, and the fluorescence quantum yields are essentially unity. The films cast from solution are at best partially crystalline, with only **1f** exhibiting clear peaks in XRD. The absorption spectra are slightly broadened compared with the solution, most notably for **1g** and **1h**. In all films, two stages of excited state evolution are detected (listed in the order **1f, 1g, 1h**): one with ~ 25, 35, and 75 ps time constants and the other with ~ 300, 500, and 800 ps (Fig. 15a). In each case, triplet formation is most clearly seen in the slower step, while excimer formation occurs with the faster time constant. The relative yields of triplet versus excimer can be qualitatively discerned by their strengths of absorption (at 21,500 and

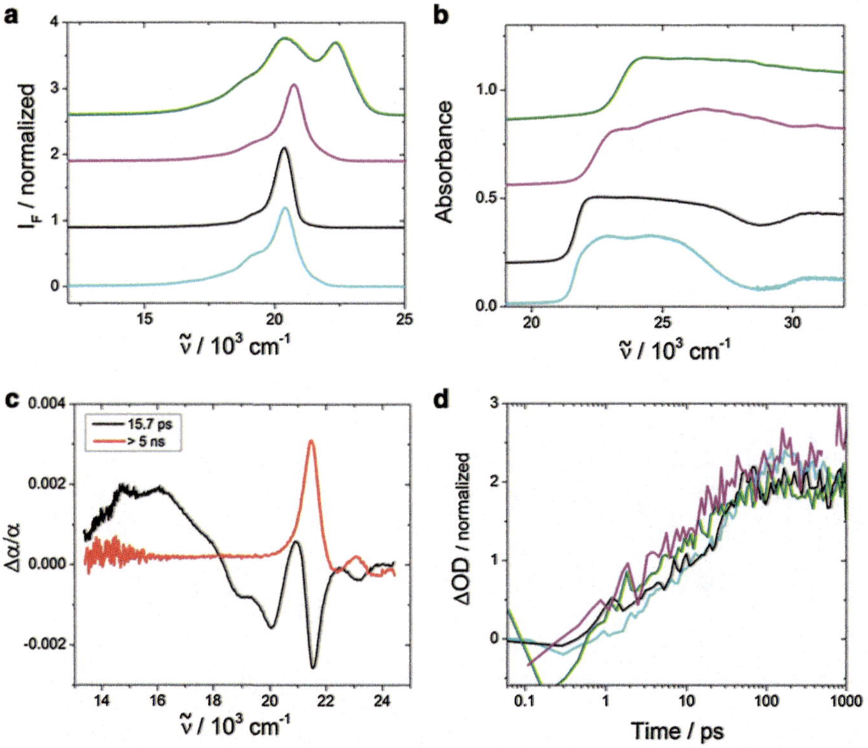

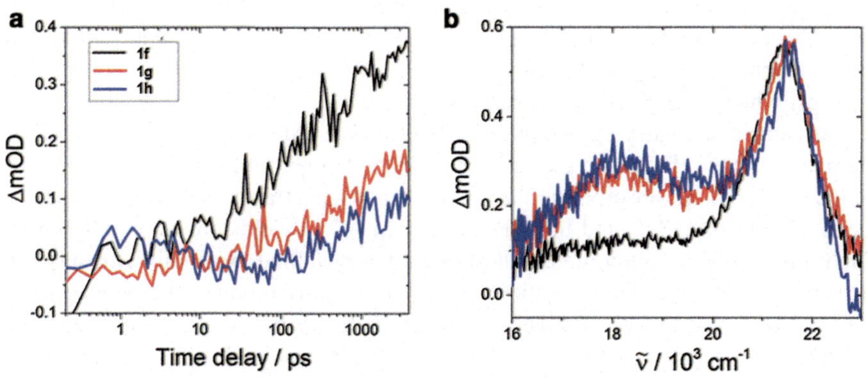

Fig. 14 Fluorescence (**a**) and UV–VIS (**b**) absorption for films of fluorinated derivatives of **1** (offset for clarity, **1**, **1a**, **1d**, **1e**, bottom to top). **c** Decay associated spectra for a film of **1a**, showing exclusively triplet bands in the long-lived component and mixed singlet/triplet bands corresponding to the S_1–T_1T_1 conversion in the 15-ps component. **d** Kinetic profiles of triplet population probed at $\sim 21{,}500$ cm^{-1} for **1**, **1a**, **1d**, and **1e**. For color scheme, see **a** and **b**

Fig. 15 a Triplet rise profiles of alkyl-substituted derivatives obtained from transient absorption. **b** Transient absorption spectra at delay time of 5 ns

18,000 cm^{-1}) in the long delay time spectra (Fig. 15b). The values of Φ_T are 0.75 ± 0.2, 0.55 ± 0.18, and 0.34 ± 0.15, respectively. The parallel formation of excimers and triplets suggests a "two-site" model for excited state dynamics in these quasi-amorphous films, similar to the behavior found in quasi-amorphous diphenyltetracene films [62], in which slow exciton diffusion to preferred sites leads to a delayed rise in triplets. The trend in triplet formation time and Φ_T for alkyl-substituted **1** is correlated with the substitution of increasingly bulkier groups. The disruption of close packing potentially affects both the concentration of SF active nearest neighbor sites (most prevalent in regions where type α phases are present), as well as the opportunity for fast diffusion to such sites. Excitations residing and remaining at the majority of locations in a disordered film are likely to quickly form excimers.

5 Covalent Dimers of 1

5.1 Weakly Coupled Dimers

With two molecules of **1** covalently bound to each other, a range of interchromophore dispositions and electronic couplings has been explored (Fig. 16). Weak electronic coupling, needed to avoid excessive triplet–triplet interaction, is assured either by the inclusion of an aliphatic spacer between chromophores, by enforcing a large twist between dimer halves via steric effects, or by connecting the monomers through their meta positions (**2**, **3**, **4**) [61]. In these dimers, excitonic effects slightly modify the solution vibronic envelope of **1**. Additional strength is observed near the lowest 0–0 transition in those dimers that have the y axes of **1** (Fig. 3) aligned head-

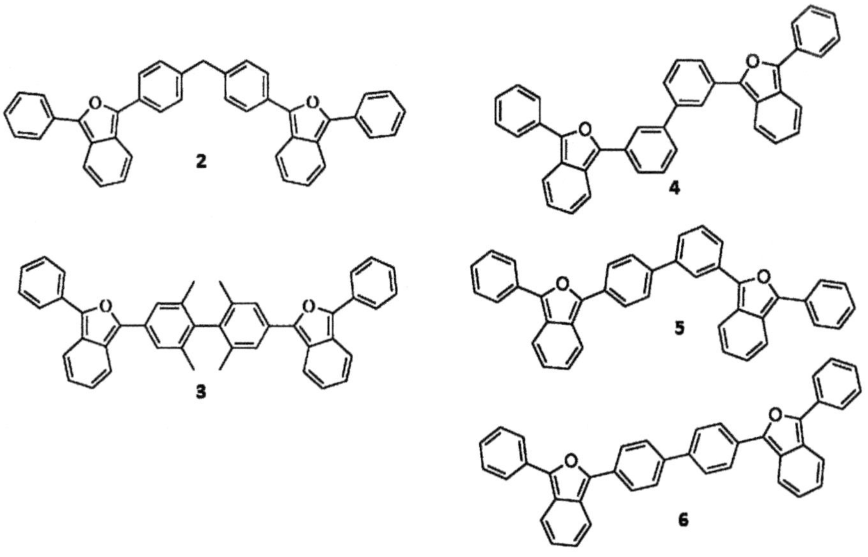

Fig. 16 Covalent dimers of **1**

to-tail. Molar absorption is approximately doubled in the dimers relative to **1**, as expected for two independent chromophores. Fluorescence remains the dominant excited state pathway for the weakly coupled dimers in nonpolar solvents, with Φ_F in excess of 0.90. Triplets are undetectable after direct excitation ($\Phi_T < 1\%$). The fluorescence lifetimes are slightly reduced compared to **1**, but the spectral features S_1–S_n (obtained by transient absorption after direct excitation) and T_1–T_n (obtained by transient absorption following triplet sensitization) remain unchanged.

In polar solution the situation changes dramatically. Φ_F is strongly quenched to a value between 0.1 and 0.3. The fluorescence lifetimes become multiexponential with fast (0.2 ns), intermediate (1–2 ns), and slow (4–6 ns) components (Fig. 17a). In pump-probe spectroscopy, absorption bands at 15,000 and 18,000 cm^{-1} are observed. Through comparison with the spectra of radical ions of **1** described above, the bands are assigned to **1**$^{\bullet+}$ and **1**$^{\bullet-}$, and it becomes clear that the charge-separated, or so-called "charge-transfer (CT) state" of the dimer is being observed. The rise time of the CT features characterizes the charge transfer event as relatively fast (50–200 ps) and dependent on the interchromophore geometry and solvent polarity. The CT and S_1 species are in equilibrium, and the fluorescence lifetimes thus reflect all of the relevant rates. A small yield of triplet states Φ_T is detected at delay times of >1 ns, and the triplet rise time is correlated with CT population decay (Fig. 17b). Φ_T varies from 0.01 to 0.10 depending on temperature and solvent. Due to the low yields, a definitive assignment of the triplets as being born from SF is not possible, and radical pair intersystem crossing also is a plausible pathway (Fig. 17c). In any event, it is apparent that the SF time is much longer than the natural decay time of the CT state (1–2 ns). The authors of a more recent study on a terrylenediimide dimer system come to similar conclusions about the deleterious effect of populating the real CT state in polar solvents; however, in that case the beneficial involvement of the virtual CT state in nonpolar solvents was also shown [63].

Measurement of the temperature-dependent rate of charge transfer between halves of the dimer yielded an energy difference between the localized S_1 and CT states of several kilocalories per mole. Shifting of the S_1–CT equilibrium toward the CT states at lower temperatures forces a shift of additional population to T_1, which

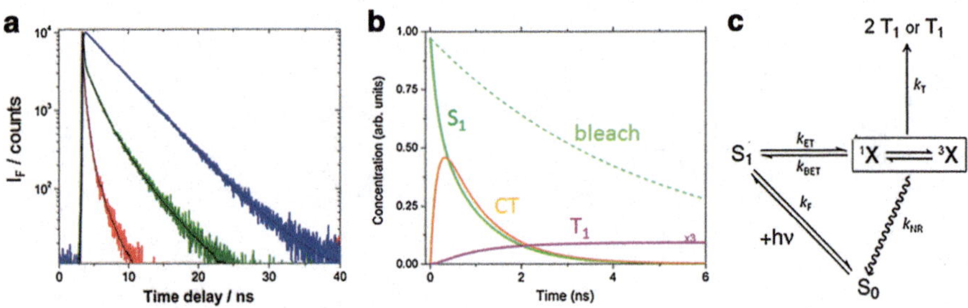

Fig. 17 **a** Fluorescence decay kinetics for **1** (*blue*), **2** (*red*), and **3** (*green*) in DMSO. **b** Concentration profiles from transient absorption showing the rise and decay of the charge-transfer (CT) state population correlated with rise time of triplets. **c** Scheme for triplet formation in **2** and **3** involving intermediate X = CT state. Copyright 2013 American Chemical Society

increases the triplet quantum yield at lower temperatures. This trend continues as the temperature is lowered until the CT is arrested as the solvent freezes and the solvent stabilization of the CT state upon photoexcitation is too slow to compete with radiative relaxation from S_1. This type of behavior mimics observations of so-called twisted intramolecular charge-transfer (TICT) states in classical systems [64].

5.2 Strongly Coupled Dimers

Covalent dimers with stronger electronic coupling have no spacer between chromophores, no substitution for increased steric hindrance, and a para–para or meta–para attachment regiochemistry (**5** and **6**). These directly conjugated dimers behave differently than **1**, even in nonpolar solvents. A red-shifted and broadened ground state absorption is observed, particularly for **6** [65]. Fluorescence is also red-shifted and strongly solvatochromic (Fig. 18a), with widely varying yields of 20–85%. Fluorescence decay kinetics detected at various wavenumbers are generally biexponential, and the amplitude spectra associated with the faster and slower components are unique. The spectrum associated with the slower decay is blue-shifted and more strongly structured, similar to that of isolated **1**. The spectrum associated with the faster decay is red-shifted, which suggests a species with larger oscillator strength and more extended delocalization than isolated **1**. The corresponding excitation spectra reveal similar trends: a monomer-like excited state species dominates at higher energies whereas evidence for a state with extended conjugation is found at lower energies.

Triplets form in both polar and nonpolar solvents. For **5**, the triplet yield is 0.08 in toluene and 0.12 in DMSO. In DMSO the triplets clearly form via a CT intermediate state, as in **2** and **3**, whereas in toluene the CT state formation is less clear. For **6**, the initially formed triplet spectrum resembles the T_1–T_n of **1** with a slight red shift (maximum at $\sim 20,000$ cm^{-1}). However, over the course of 1–50 μs, the spectrum changes and contains only a strong band at 14,000 cm^{-1} (Fig. 18b). This strong band is also observed upon triplet energy transfer from anthracene. The action spectrum of triplet formation in toluene (Fig. 18c) reveals a shift in the onset of triplet production from that of the ground state absorption, implying that a higher

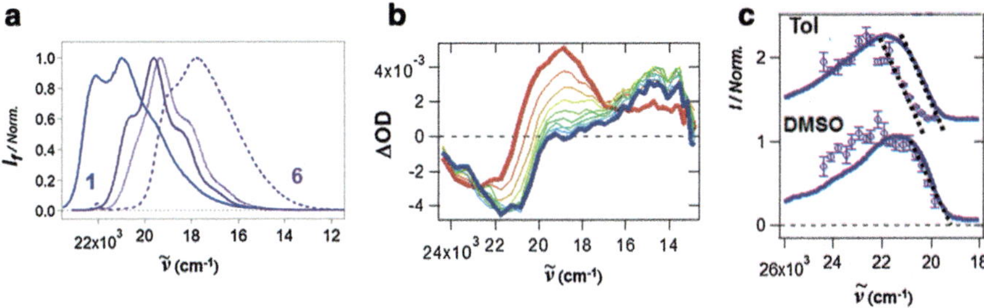

Fig. 18 **a** Fluorescence of **6** in cyclohexane (*solid line*), toluene (*dotted line*), and DMSO (*dashed line*). Fluorescence of **1** in toluene is shown for comparison. **b** Transient absorption spectra of **4** in toluene from delay times of 100 ns (*red*) through to 50 μs (*blue*). **c** Action spectrum for triplet production for **6** in toluene (Tol) and DMSO

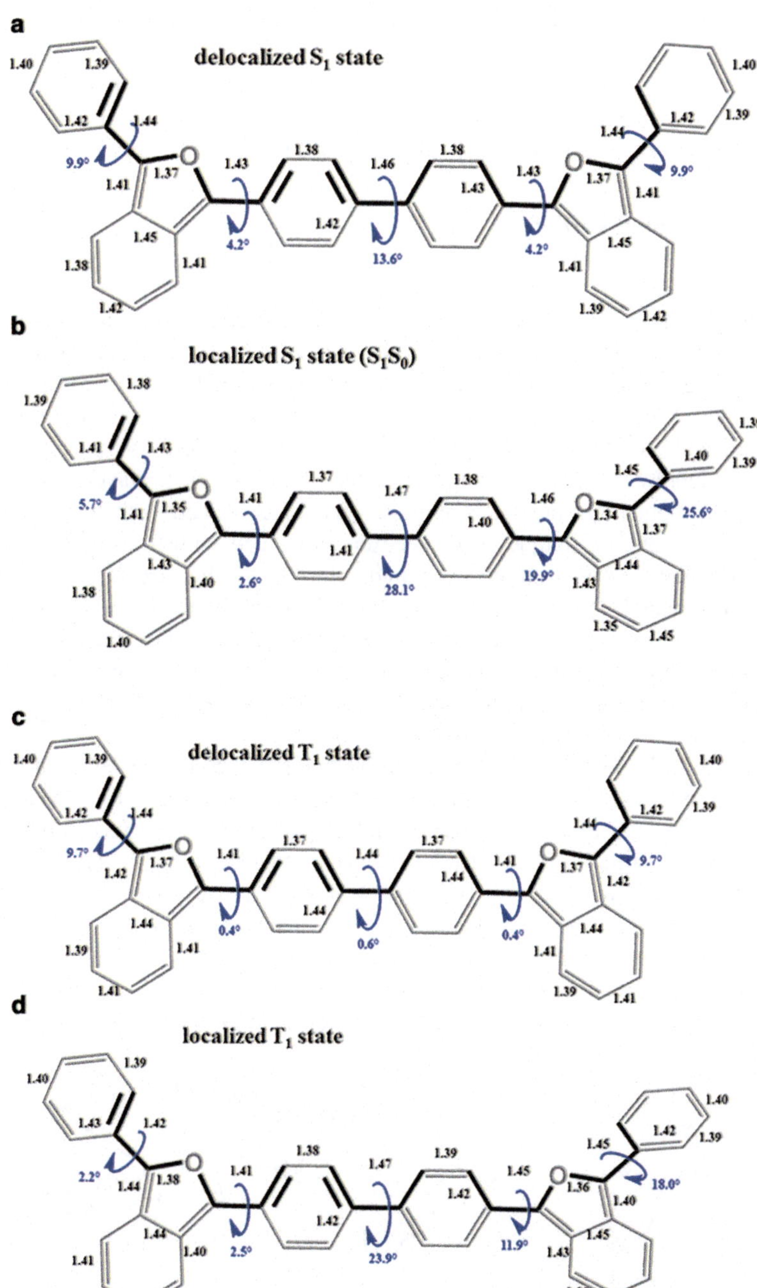

Fig. 19 **a, b** A comparison of the calculated geometries of the S_1^* delocalized S_1 and localized S_0S_1 states for the "syn" conformer of **6** (B3LYP/SVP and TD-HF/SVP, respectively). **c, d** A comparison of the calculated geometries of the delocalized T_1^* and localized S_0T_1 states for the "anti" conformer of **6** (B3LYP/SVP and HF/SVP, respectively)

energy singlet forms triplets more efficiently than a lower energy singlet. Calculations of minimized singlet and triplet geometries (Fig. 19) further point toward a dual triplet model in which a localized triplet resembling that of monomeric **1** forms relatively efficiently (Φ_T 0.30–0.40) upon excitation above 22,000 cm^{-1}. A delocalized triplet forms with low efficiency (Φ_T <0.05) after lower energy excitation. The latter triplet is also formed via diffusion-controlled energy transfer between the localized triplet and the ground state of **6**.

The excitation localization/delocalization isomerism, present in both the triplet state and the first singlet excited but not the ground state, is quite unusual. It appears to result from the excitation of ground state species with particular geometries favored to produce an excited singlet that is predominantly localized (higher energy excitation) or delocalized (lower energy excitation) (compare Fig. 19). These singlets can then produce triplets that are correspondingly localized or delocalized. The excited singlet with twisted geometry is found to be less prevalent when **6** is dissolved in polar solvents, and concomitantly the dual behavior is more difficult to detect (Fig. 18c) than in less polar solvents like toluene. SF appears to be possible from the localized singlet, but it would probably be highly endothermic from the delocalized singlet, which from the temperature and excitation energy dependence of fluorescence is found to lie 2300 cm^{-1} below the localized singlet [66]. This type of isomerism may be an important feature of other SF dimer systems which have attracted increasing attention recently [63, 67, 68].

6 Photovoltaic Devices with the Chromophore 1

In order to observe evidence of SF in a situation where charges are collected, one must couple the SF chromophores to a charge (or energy) acceptor [69]. Dye-sensitized solar cells (DSSCs) are common platforms for extracting charge from an organic dye following photoexcitation. We employed **1** and **1n** as "sensitizers" of nanocrystalline TiO_2 for the collection of photocurrent [70]. Without functionalization **1** should not have strong affinity for binding to TiO_2; however, we discovered that a small amount of a colorless photo-oxidation product that formed in oxygenated solutions was able to facilitate adsorption of **1** on the surface of TiO_2. The sensitized photoelectrodes were brightly colored and stable in an electrolyte (Fig. 20a). The devices produced a high yield of photocurrent (roughly 80% internal quantum efficiency) and an overall power conversion efficiency of 1.1%. Electron transport through the nanocrystalline TiO_2 films was found to be on par with that seen in high-performing dye/TiO_2 systems, resulting in diffusion lengths (60–100 μm) that were at least eight times longer than the typical film thickness ($\sim$7 μm). Thus, the photocurrent yield was not limited primarily by recombination but rather by the yield of injected charges, which is potentially influenced by SF. However, the electron injection time was measured to be <200 fs, which is far faster than SF.

In order to demonstrate that triplets born from SF could be dissociated and lead to photocurrent, a dielectric barrier of varying thickness was deposited on the TiO_2 surface (Fig. 20b). ZrO_2 was deposited via chemical bath deposition in cycles that added $\sim$5 Å of thickness each. The dielectric barrier reduced the electron injection

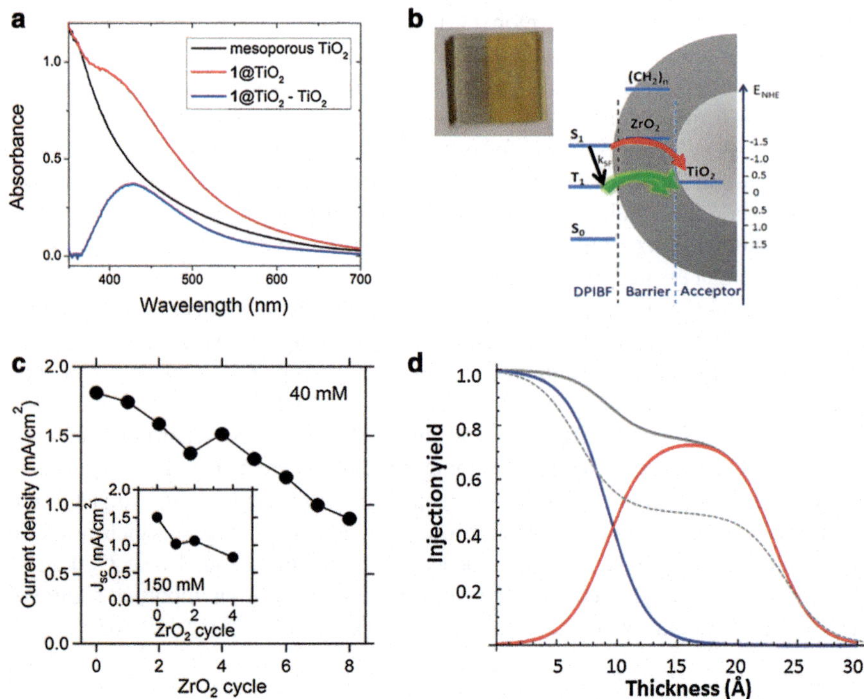

Fig. 20 **a** Net UV–VIS absorption of **1** on nanocrystalline TiO$_2$ photoelectrodes. *Inset* Photograph of stained TiO$_2$. **b** SF and electron injection scheme with anticipated energy level alignments. **c** Photocurrent density versus ZrO$_2$ cycle (roughly 5 Å per cycle). **d** Calculated photocurrent curve with singlet (*blue*) and triplet (*red*) contributions. Solid and dashed gray curves are derived assuming that the percentage of SF-active surface-bound **1** is 50% and 25%, respectively

rate from S_1, allowing SF to occur in the chromophore assembly near the surface. Subsequent injection from long-lived triplets, created at a yield of roughly 1.5 times that of singlets, led to a "kink" in the photocurrent versus barrier thickness profile (Fig. 20c). The rise in photocurrent corresponds with the thickness of ZrO$_2$ needed to slow electron injection from S_1 beyond $\sim$30 ps, the estimated SF time for the molecules near the TiO$_2$ surface (Fig. 20d).

A five-carbon saturated chain terminated with a carboxylic acid group was attached to **1** to yield **1n** and covalently bind the dye to the surface. Now, binding was strong even in the absence of the photo-oxidation products, and the carbon chain further slowed the S_1 injection process. The "kink" in the photocurrent spectrum occurred for a much smaller thickness of ZrO$_2$ than with the unsubstituted **1**. The photocurrent quantum efficiency of the device never exceeded 100% due to loss channels that advance with increasing ZrO$_2$ thickness. However, the clear demonstration of the influence of triplets born from SF on the photocurrent represents an important step toward the practicality of SF-based solar photoconversion devices.

7 Outlook

1,3-Diphenylisobenzofuran (**1**) represents the first successful demonstration of a molecule specifically designed to exhibit efficient SF. The observation of high triplet quantum yield and the associated fast triplet rise times in thin films has left little doubt about the validity of the design principles used to arrive at small molecule candidate structures based on biradicaloids for SF. The potential for utilizing SF in **1** has also been demonstrated, although true practicality would require further functionalization to endow the chromophore with greater photostability and optical absorption that covers a larger portion of the solar spectrum. In particular, schemes in which triplets born from SF in biradicaloids like **1** undergo energy transfer to high-performing narrow band gap bulk semiconductors should be explored [71, 72]. The S_0–T_1 energy gap of **1**, roughly 1.42 eV, is significantly larger than the band gap of Si (1.1 eV), but it may be an excellent match for gallium arsenide (GaAs; $\sim$1.4 eV), from which world record single-junction solar cells have been made. This type of very simple tandem solar cell shows promise for practical utilization.

Acknowledgements This material is based on work supported by the U. S. Department of Energy, Office of Basic Energy Sciences, Division of Chemical Sciences, Biosciences, and Geosciences. J.C.J. acknowledges Contract No. DE-AC36-08GO28308 with NREL and J.M. acknowledges Award Number DOE DE-SC0007004. J.M. also acknowledges support from the Grant Agency of the Czech Republic, Grant No. GA15-19143S.

References

1. Shockley W, Queisser HJ (1961) J Appl Phys 32:510–519
2. Hanna MC, Nozik AJ (2006) J Appl Phys 100:074510–074518
3. Smith MB, Michl J (2010) Chem Rev 110:6891–6936
4. Smith MB, Michl J (2013) Annu Rev Phys Chem 64:361–386
5. Buchanan EA, Havlas Z, Michl J (2017) In: Sabin JR, Brändas EJ (eds) Advances in quantum chemistry: Ratner volume, vol 17. Academic Press, UK
6. Johnson JC, Michl J (2014) In: Nozik AJ, Conibeer G, Beard MC (eds) Advanced concepts in photovoltaics. Royal Society of Chemistry, Oxfordshire
7. Swenberg CE, Geacintov NE (1973, 1975) In: Birks JB (ed) Organic molecular photophysics, vol 1 (pp 489–564) and vol 2 (pp 395–408). Wiley, London
8. Pope M, Swenberg CE (1999) Electronic processes in organic crystals and polymers, 2nd edn. Oxford University Press, Oxford
9. Singh S, Joned WJ, Siebrand W, Stoichef BP, Schneider WG (1965) J Chem Phys 42:330–342
10. Swenberg CE, Tracy WT (1968) Chem Phys Lett 2:327–328
11. Geacintov NE, Pope M, Vogel F (1969) Phys Rev Lett 22:593–596
12. Pope M, Geacintov NE, Vogel F (1969) Mol Cryst Liquid Cryst 6:83–104
13. Merrifield RE, Aviakan P, Groff RP (1969) Chem Phys Lett 3:386–388
14. Groff RP, Avakian P, Merrifield RE (1970) Phys Rev B 1:815
15. Tomkiewicz Y, Groff RP, Avakian P (1971) J Chem Phys 54:4504–4507
16. Geacintov NE, Burgos J, Pope M, Strom C (1971) Chem Phys Lett 11:504–508
17. López-Delgado R, Miehé JA, Sipp B (1976) Opt Commun 19:79–82
18. Yarmus L, Rosenthal J, Chopp M (1972) Chem Phys Lett 16:477–481
19. Swenberg CE, Van Metter R, Ratner M (1972) Chem Phys Lett 16:482–485
20. Swenberg CE, Geacintov NE (1973) In: Birks JB (ed) Organic molecular photophysics. Wiley, New York
21. Klein G, Voltz R, Schott M (1972) Chem Phys Lett 16:340–344

22. Klein G, Voltz R, Schott M (1973) Chem Phys Lett 19:391–394
23. Arnold S, Alfano RR, Pope M, Yu W, Ho P, Selsby R, Tharrats J, Swenberg CE (1976) J Chem Phys 64:5104–5114
24. Fleming GR, Millar DP, Morris GC, Morris JM, Robinson GW (1977) Aust J Chem 30:2353–2359
25. Frankevich EL, Lesin VI, Pistupa AI (1978) Chem Phys Lett 58:127–131
26. Von Burg K, Zchokke-Gränacher I (1979) J Chem Phys 70:3807–3811
27. Arnold S, Whitten WB (1981) J Chem Phys 75:1166–1169
28. Katoh R, Kotani M (1992) Chem Phys Lett 196:108–112
29. Jundt C, Klein G, Sipp B, Le Moigne J, Joucla M, Villaeys AA (1995) Chem Phys Lett 241:84–88
30. Müller AM, Avlasevich YS, Müllen K, Bardeen CJ (2006) Chem Phys Lett 421:518–522
31. Müller AM, Avlasevich YA, Schuller WW, Müllen K, Bardeen CJ (2007) J Am Chem Soc 129:14240–14250
32. Lee J, Jadhav P, Baldo MA (2009) Appl Phys Lett 95:192
33. Johnson JC, Nozik AJ, Michl J (2010) J Am Chem Soc 132:16302–16303
34. Wang C, Tauber MJ (2010) J Am Chem Soc 132:13988–13991
35. Burdett JJ, Müller AM, Gosztola D, Bardeen C (2010) J Chem Phys 133:144506
36. Paci I, Johnson JC, Chen X, Rana G, Popoviæ D, David DE, Nozik AJ, Ratner MA, Michl J (2006) J Am Chem Soc 128:16546–16553
37. Herkstroeter WG, Merkel PB (1985) J Photochem 16:331–341
38. Ziebig R, Pragst FZ (1979) Z Phys Chem 260:795–803
39. Howard JA, Mendenhall GD (1975) Can J Chem 53:2199–2201
40. Krieg M (1993) J Biochem Biophys Methods 27:143–149
41. Minami T, Nakano M (2012) J Phys Chem Lett 3:145–150
42. Akdag A, Havlas Z, Michl J (2012) J Am Chem Soc 134:14624–14631
43. Zeng T, Ananth N, Hoffmann R (2014) J Am Chem Soc 136:12638–12647
44. Wen J, Havlas Z, Michl J (2015) J Am Chem Soc 137:165–172
45. Johnson JC, Nozik AJ, Michl J (2013) Acc Chem Res 46:1290–1299
46. Feng XT, Luzanov AV, Krylov AI (2013) J Phys Chem Lett 4:3845–3852
47. Tamura H, Huix-Rotllant M, Burghardt I, Olivier Y, Beljonne D (2015) Phys Rev Lett 115:107401
48. Havlas Z, Michl J (2016) Isr J Chem 56:96–106
49. Ryerson J, Schrauben JN, Ferguson AJ, Sahoo SC, Naumov P, Havlas Z, Michl J, Nozik AJ, Johnson JC (2014) J Phys Chem C 118:12121–12132
50. Teichen PE, Eaves JD (2015) J Chem Phys 143:044118
51. Akdag A, Wahab A, Beran P, Rulíšek L, Dron PI, Ludvík J, Michl J (2015) J Org Chem 80:80–89
52. Jacq J, Einhorn C, Einhorn J (2008) Org Lett 10:3757–3760
53. Schwerin AF, Johnson JC, Smith MB, Sreearunothai P, Popović D, Černý J, Havlas Z, Paci I, Akdag A, MacLeod MK, Chen X, David DE, Ratner Ma, Miller JR, Nozik AJ, Michl J (2010) J Phys Chem A 114:1457–1473
54. Platt JR (1949) J Chem Phys 17:484–495
55. Moffitt W (1954) J Chem Phys 22:320–333, 1820–1829
56. Michl J (1984) Tetrahedron 40:3845–3934
57. Nelsen SF, Weaver MN, Yamazaki D, Komatsu K, Rathore K, Rathore R, Bally T (2007) J Phys Chem 111:1667–1676
58. Dillon RJ, Piland GB, Bardeen CJ (2013) J Am Chem Soc 135:17278–17281
59. Margulies EA, Logsdon JL, Miller CE, Ma L, Simonoff E, Young RM, Schatz GC, Wasielewski MR (2017) J Am Chem Soc 139:663–671
60. Schrauben J, Ryerson J, Michl J, Johnson J (2014) J Am Chem Soc 136:7363–7373
61. Johnson JC, Akdag A, Zamadar M, Chen X, Schwerin AF, Paci I, Smith MB, Havlas Z, Miller JR, Ratner MA, Nozik AJ, Michl J (2013) J Phys Chem B 117:4680–4695
62. Roberts ST, McAnally RE, Mastron JN, Webber DH, Whited MT, Brutchey RL, Thompson ME, Bradforth SE (2012) J Am Chem Soc 134:6388–6400
63. Margulies EA, Miller CE, Wu Y, Ma L, Schatz GC, Young RM, Wasielewski MR (2016) Nat Chem 8:1120–1125
64. Grabowski ZR, Rotkiewicz K, Rettig W (2003) Chem Rev 103:3899–4032
65. Schrauben J, Akdag A, Wen J, Ryerson J, Havlas Z, Michl J, Johnson JC (2016) J Phys Chem A 120:3473–3483
66. Greyson EC, Stepp BR, Chen X, Schwerin AF, Paci I, Smith MB, Akdag A, Johnson JC, Nozik AJ, Michl J, Ratner MA (2010) J Phys Chem B 114:14223–14232

67. Sanders SN, Kumarasamy E, Pun AB, Trinh MT, Choi B, Xia J, Taffet EJ, Low JZ, Miller JR, Roy X, Zhu XY (2015) J Am Chem Soc 137:8965–8972
68. Lukman S, Chen K, Hodgkiss JM, Turban DHP, Hine NDM, Dong S, Wu J, Greenham NC, Musser AJ (2016) Nat Commun 7:13622
69. Congreve DN, Lee JY, Thompson NJ, Hontz E, Yost SR, Reusswig PD, Bahlke ME, Reineke S, Van Voorhis T, Baldo MA (2013) Science 340:334–337
70. Schrauben JN, Zhao Y, Mercado C, Dron PI, Ryerson JM, Michl J, Zhu K, Johnson JC (2015) ACS Appl Mater Interfaces 7:2286–2293
71. Tabachnyk M, Ehrler B, Gelinas S, Bohm ML, Walker BJ, Musselman KP, Greenham NC, Friend RH, Rao A (2014) Nat Mater 13:1033–1038
72. Thompson NJ, Wilson MW, Congreve DN, Brown PR, Scherer JM, Bischof TS, Wu M, Geva N, Welborn M, Van Voorhis T, Buloviæ V (2014) Nat Mater 13:1039–1043

Printed by Printforce, the Netherlands